COMPLEX PHYSICAL, BIOPHYSICAL AND ECONOPHYSICAL SYSTEMS

WORLD SCIENTIFIC LECTURE NOTES IN COMPLEX SYSTEMS

AIMS AND SCOPE

The aim of this new interdisciplinary series is to promote the exchange of information between scientists working in different fields, who are involved in the study of complex systems, and to foster education and training of young scientists entering this rapidly developing research area.

The scope of the series is broad and will include: Statistical physics of large nonequilibrium systems; problems of nonlinear pattern formation in chemistry; complex organization of intracellular processes and biochemical networks of a living cell; various aspects of cell-to-cell communication; behaviour of bacterial colonies; neural networks; functioning and organization of animal populations and large ecological systems; modeling complex social phenomena; applications of statistical mechanics to studies of economics and financial markets; multi-agent robotics and collective intelligence; the emergence and evolution of large-scale communication networks; general mathematical studies of complex cooperative behaviour in large systems.

Published

Vol. 1 Nonlinear Dynamics: From Lasers to Butterflies
Vol. 2 Emergence of Dynamical Order: Synchronization Phenomena in Complex Systems
Vol. 3 Networks of Interacting Machines
Vol. 4 Lecture Notes on Turbulence and Coherent Structures in Fluids, Plasmas and Nonlinear Media
Vol. 5 Analysis and Control of Complex Nonlinear Processes in Physics, Chemistry and Biology
Vol. 6 Frontiers in Turbulence and Coherent Structures
Vol. 7 Complex Population Dynamics: Nonlinear Modeling in Ecology, Epidemiology and Genetics
Vol. 8 Granular and Complex Materials
Vol. 9 Complex Physical, Biophysical and Econophysical Systems

World Scientific Lecture Notes
in Complex Systems – Vol. 9

editors

Robert L Dewar

Frank Detering

The Australian National University, Australia

COMPLEX PHYSICAL, BIOPHYSICAL AND ECONOPHYSICAL SYSTEMS

Proceedings of the 22nd Canberra International Physics Summer School

The Australian National University, Canberra   8 – 19 December 2008

World Scientific

NEW JERSEY • LONDON • SINGAPORE • BEIJING • SHANGHAI • HONG KONG • TAIPEI • CHENNAI

Published by

World Scientific Publishing Co. Pte. Ltd.
5 Toh Tuck Link, Singapore 596224
USA office: 27 Warren Street, Suite 401-402, Hackensack, NJ 07601
UK office: 57 Shelton Street, Covent Garden, London WC2H 9HE

British Library Cataloguing-in-Publication Data
A catalogue record for this book is available from the British Library.

World Scientific Lecture Notes in Complex Systems — Vol. 9
COMPLEX PHYSICAL, BIOPHYSICAL AND ECONOPHYSICAL SYSTEMS
Proceedings of the 22nd Canberra International Physics Summer School

ISBN-13 978-981-4277-31-0
ISBN-10 981-4277-31-2

Printed in Singapore by World Scientific Printers

Preface

This book arose from a summer school held in Canberra in December 2008, primarily funded by the Australian Research Council's Complex Open Systems Research Network (COSNet), which is hosted by The Australian National University (ANU) Centre for Complex Systems. The summer school was the 22nd Physics Summer School organised under the auspices of the Centre for Complex Systems and its predecessors,[1] and was also generously supported by the Asia-Pacific Center for Theoretical Physics.

COSNet started in 2004 with the vision statement:

> Complexity is the common frontier in the physical, biological and social sciences. This Network will link specialists in all three sciences through five generic conceptual and mathematical theme activities. It will promote research into how subsystems self-organise into new emergent structures when assembled into an open, non-equilibrium system. Outcomes will include new technologies and software tools and deeper understanding of fundamental questions in science. An essential function of the network will be introducing researchers and end users to new tools and broadening the horizons of graduate students.

[1]The summer schools were started by the ANU Department of Theoretical Physics in 1988, and were continued from 1994 by the Centre for Theoretical Physics at ANU (initially with the aid of seed funding from the ARC as a pilot for a National Institute for Theoretical Physics) until it became the ANU Centre for Complex Systems in 2001, which has continued this tradition as part of the COSNet vision.

The chapters of this book have been written by the Summer School lecturers, including those in computer laboratory sessions introducing software tools. The book presents up-to-date discussions of new theoretical approaches and tools (in particular from *statistical and nonlinear physics*)—fractional diffusion, dynamical systems analysis, entropy approaches, network analysis, agent-based modelling—and is unique in the scope of its coverage of applications of complex systems science:

- Extraterrestrial complex systems science—astrophysical, solar and space plasmas;
- Earth system science—from the technicalities of global warming to the Gaia big picture;
- Living and man-made systems–financial systems, genomics, brain dynamics, social networks, use of chaos theory in technologies.

All chapters were peer reviewed, and indexed using a common lexicon of keywords, in particular ones bringing out the generic features of complex systems, such as the following properties: Complex systems

- exhibit *emergence*: some properties present at system level are not present at a lower level, e.g. a cell is alive, but is made of inanimate elements;
- are *open*: energy and information are constantly being imported and exported across system boundaries;
- have a *history*: the history cannot be ignored, even a small change in circumstances can lead to large deviations in the future;
- can *adapt*: in response to external or internal changes, systems can self-organise without breaking;
- are *not completely predictable*: when a system is adaptive, unexpected behaviours can emerge—prediction becomes statistical expectation;
- are *multi-scale and hierarchical*: system size and structure scale are over several orders of magnitude and distinct properties and functions are associated with different scales—dynamics can propagate through scales and exhibit avalanches and cascade effects;
- are *not simply ordered*: there is no compact and concise way to encode the whole information contained in the system;
- have *multiple (meta) (stable) states*: small perturbations lead to recovery, larger ones can lead to radical changes of properties—dynamics do not average simply.

The editors wholeheartedly thank the chapter authors and peer reviewers for making this book possible. We also thank all the summer school participants, the sponsors and the support people, in particular Mr Daniel Chan, COSNet Manager for assistance with all aspects of the organisation, and Ms Fanny Lemaitre-Detering for help with graphics design.

Robert L. Dewar, organiser, editor, and Convenor of COSNet
Frank Detering, organiser and editor

Canberra, October 2009

Participants in the 22nd Canberra International Physics Summer School

Standing, from left: Eriita Jones, Gordon Briggs, Robert Dewar, Roderick Dewar, Felix Fung, Frank Detering, Yacov Salomon, Robert Niven, Steven Lade, Rowena Ball, Waiho Wong, Richard Dendy, Rowan MacQueen, Myall Hingee, Carl Kneipp, David Liley, Federico Frascoli, Anne-Marie Grisogono, Ian Wood, Ian Enting, Vanja Radenovic, Tomaso Aste, Roy Williams, Mathew McGann.

Sitting/kneeling, from left: Francesco Pozzi, Pan Yu, Sacha van Albada, Hawoong Jeong, Tim Baynes, Navin Doloswala, Kass Hingee, Matthew Berryman, Guy Metcalfe, Kyuyong Lee, Hyungtae Kook.

Contents

Chapter 1

Introduction to Complex and Econophysics Systems: A navigation map

T. Aste[1,2,3] and T. Di Matteo[1,2]

[1] *Department of Applied Mathematics, School of Physical Sciences, The Australian National University, Canberra, ACT 0200, AUSTRALIA;*
[2] *Department of Mathematics, King's College, The Strand, London, WC2R 2LS, UK;*
[3] *School of Physical Sciences, University of Kent, Canterbury, Kent, CT2 7NH, UK.*

This Chapter is an introduction to the basic concepts used in complex systems studies. Our aim is to illustrate some fundamental ideas and provide a *navigation map* through some of the cutting edge topics in this emerging science. In particular, we will focus our attention to *econophysics* which mainly concerns the application of tools from statistical physics to the study of financial systems.

Contents

1.1. Introduction

We all experience complexity in everyday life, where simple answers are hard to find and the consequences of our actions are difficult to predict. Modern science recognizes that problems involving the collective behavior of many interacting elements are often "complex." These systems typically display collective, organized behaviors that cannot be predicted from traditional atomistic studies of their components in isolation. As a result, phenomena in complex systems are often counterintuitive. Their explanations require the use of new tools and new paradigms which range from network theory to multi-scale analysis.

There is no uniquely agreed definition of complex system, but if we look up the word "complex" in a dictionary we might find *"a whole made up of complicated or interrelated parts"* (from Merriam-Webster Online Dictionary). Indeed, complex systems are in general made up of several parts, which are interrelated and often complex themselves. Financial systems provide great examples of complex systems, being systems with a very large number of agents that interact in complicated ways, the agents themselves being complex individuals who act following rules and feelings, applying both knowledge and guesswork.

To properly introduce complexity from a common, well established, ground it is probably better to start from a time when the universe was perceived to be "harmonious" and it was believed that the laws of nature should tend to produce order. In the introduction of the *Principia* (1687) Newton writes: *"I wish we could derive the rest of the phenomena of nature by the same reasoning from mechanical principles for I am induced by many reasons to suspect that they may all depend on certain forces."* Laplace (1749-1827) was indeed persuaded that *"An infinitely intelligent mathematician would be able to predict the future by observing the present state of the universe and using the laws of motion."* This idea has been central in the foundation of modern science and we can still find a very similar idea expressed by Einstein and Infield in 1938:[1] *"(...) all these things contributed to the belief that it is possible to describe all natural phenomena in terms of simple forces between unalterable objects."* Within this framework, the final aim of a scientific endeavor is to *"(...) reduce the apparent complexity of natural phenomena to some simple fundamental ideas and relations."*[1]

However, together with the evolution of our scientific knowledge it has become increasingly clear that sometimes the reduction of the system into smaller and simpler parts with known relations might not lead to any valuable knowledge about the overall system's properties. Probably, in this respect, the most revealing examples are living biological systems. There are some simple organisms of which we know essentially everything from the molecular level, through the cellular organization, up to the animal behavior. However, we still miss one fundamental emerging point: the animal is *alive* and, although we can understand each single part of it, we cannot describe the property of being alive from the simple assemblage of the inanimate parts. Even Newton, well after the glorious years of the Principia, had to admit that *"I can calculate the motions of the heavenly bodies, but not the madness of people."* This was in 1720, when after making a very good profit from the stocks of the South Sea Company, he reinvested right at the top of what is now known the "South Seas Bubble," which crashed and made him loose 20,000 pounds.

The study of complex systems is a very challenging endeavor which requires a paradigmatic shift. Indeed, in these systems the global behavior is an "emergent property" which is not simply related to the local properties of its components. Energy and information are constantly imported and exported across system boundaries. History cannot be ignored, even a small change in the current circumstances can lead to large deviations in the future. In complex systems the "equations of motion" themselves can evolve and change: in response to external (or internal) changes, the system can reorganize itself without breaking and it can adapt to the new environment. Complex systems have multiple (meta) (stable) states where small perturbations lead to recovery and larger ones can lead to radical changes of properties. In these systems, dynamics does not average simply. Complex systems are multi-scale and hierarchical. Complex systems are characterized by, and often dominated by, unexpected, unpredictable, adaptive, emerging behaviors, which can span over several orders of magnitude. Dynamics can propagate through scales, both upwards, when the system is hierarchically organizing, and downwards, when large fluctuations may melt down such hierarchy. The presence of large, scale-free, power-law fluctuations makes often impossible from a finite set of measures to calculate the parameters characterizing these statistical fluctuations. Complex systems are disordered in the sense that there is no compact and concise way to encode the whole information contained in the system. Even the measure of complexity is a complex task per se. We can say that the complexity of a system scales by its number of elements, its number of interactions, by the complexity of the element, by the complexity of the interaction. This is a "recursive" measure which reflects the multi-scale nature of these systems.

We are all aware that accurate predictions of real world complex phenomena are very challenging. Forecasting tomorrow's weather or the house market trends for the next five years, seems to be more a form of art than an exact science. Indeed, as Neils Bohr emphatically warned, *"Prediction is very difficult, especially about the future."* As the physicist Giorgio Parisi pointed out, the meaning of the word *prediction* has evolved with the evolution of science and it has assumed a new meaning in the context of complex systems.[2] At the times of Newton or Laplace, when *classical mechanics* was founded, prediction meant the accurate forecast of the position and velocity of any object for the infinite future, given a precise knowledge of the present position and velocity. The first change in the meaning of prediction happened already in the 19th century with the beginning of *statistical mechanics.* In that context, prediction becomes a probabilistic concept where the theory no longer aims to describe the behavior of each single molecule, but the statistical properties of the ensemble of the molecules. With the advent of *quantum mechanics* the probabilistic description of natural phenomena became the predictive framework for atomic and subatomic phenomena, and the uncertainty principle introduced the idea that some variables might not be measurable simultaneously with arbitrary precision. More recently, the word prediction assumed another new significance in the context of the theory of nonlinear dynamics. In deterministic chaos, despite the fact that the system behavior is fully determined by the initial conditions, the high sensitivity of trajectories over large time intervals to infinitesimal changes of the initial conditions makes such a deterministic prediction meaningless. In these systems, prediction concerns the identification of classes of long-time behaviors associated with given regions of the space of initial conditions, and statistical statements about these behaviors.

The science of complex systems has introduced another paradigmatic change to the concept of prediction and consequently it has changed the meaning of physical investigation. For these systems the main question is the dependence of the emerging behaviors at system level on the set of rules and constraints imposed at local level. Hierarchy and emergence imply that we must describe the system at different levels of abstraction. In these systems, models and theories often apply only to a given abstraction level and different theories might apply to different levels. This is a different and somehow lesser power of predictability compared with traditional physical theories. On the other hand, this opens the way to apply physical investigation to new kinds of systems: the modeling of brain functions, the study of financial markets, or the study of the influence of a given gene on some biological functions, are among the topics currently investigated by physicists.[3,4]

A few years ago, Stephen Hawking declared that *"the next century will be the century of complexity."* Indeed, science is changing in response to the new problems arising from the study of complex systems. The scientific community now faces new expectations and challenges, the nature of problems has forced modern scientists to move beyond the conventional reductionist approaches. Complex systems studies have opened a new scientific frontier for the description of the social, biological, physical, and engineered systems on which human society has come to depend.

1.2. An Example of Complex Systems: Financial Markets

Financial markets are open systems in which many subunits interact nonlinearly in the presence of feedback. Financial systems are archetypal of complexity. In markets, several different individuals, groups, humans and machines, generically called "agents," operate at different frequencies with different strategies. The agents interact both individually and collectively at different levels within an intricate network of complicated relations. The emerging dynamics continuously evolves through bubbles and crashes with unpredictable trends. Although intrinsically "complex," financial systems are very well suited for statistical studies. Indeed, the governing rules are rather stable and the time evolution of the system is continuously monitored, providing therefore a very large amount of data for scientists to investigate.

Since the mid '90s a growing number of physicists have undertaken the challenge of understanding financial and economic systems. A new research area related to complex systems research has emerged and it has been named "Econophysics." This is a relatively recent discipline, but it has already a rich history, with a variety of approaches, and even controversial trends.[5–11] Econophysics is an interdisciplinary field which applies the methods of statistical physics, nonlinear dynamics, and network theory to macro-micro/economic modeling, to financial market analysis and social problems.

There are several open questions that econophysicists are actively investigating. Some of the main topics concern:

- development of theoretical models able to encompass empirical evidence;
- statistical characterization of the stochastic process describing price changes of a financial asset;
- search for scaling and universality in economic systems;
- implementation of models for wealth and income distributions;
- use of network theory and statistical physics tools to describe collective fluctuations in financial assets prices;
- development of statistical mechanics approaches in socio-economic systems;
- exploration of novel theoretical approaches for interacting agents;
- investigation of new tools to evaluate risk and understand complex system behaviors under partial information and uncertainty.

In this Chapter we will focus on a few examples of financial system studies from an Econophysics perspective.

1.3. Probabilities and Improbabilities

Let us start our "navigation" from some fundamental concepts and theorems from probability theory that are of relevance to the study of complex systems. In particular we focus our attention on large fluctuations and the probability of their occurrence, that, as we shall see shortly, is an important characterizing aspect of complex systems phenomena.

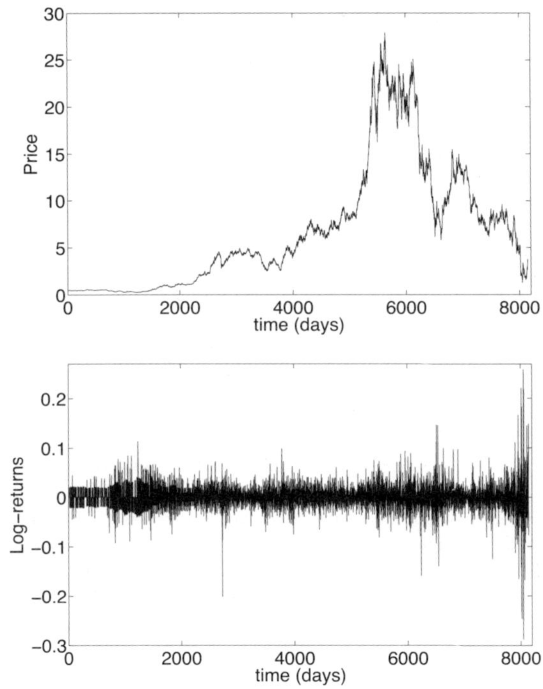

Fig. 1.1. (top) Daily closing adjusted prices for the stock Ford Motor in the New York Stock Exchange market. The time-period ranges from 3 Jan 1977 to 7 April 2009 (there are 8142 points in total). (bottom) Log-returns over the same time-period. The data are from http://finance.yahoo.com.

1.3.1. *Log-returns*

In order to let the reader focus on a practical example, let us start with the study of the statistical properties of daily closing prices of an equity over a given period of time (let us, for instance, analyze the prices of Ford Motor Company in the New York Stock Exchange as shown in Fig. 1.1). From a statistical point of view we want to characterize the daily variations of these prices, and to this end it is convenient to look at the so-called log-returns, defined as:[5,12]

$$r(t,\tau) = \log\left(\text{price}(t+\tau)\right) - \log\left(\text{price}(t)\right) \quad ; \tag{1.1}$$

where, in this case, we take $\tau = 1$ day. We can calculate from the data plotted in Fig. 1.1 that these log-returns fluctuate around zero with a sample

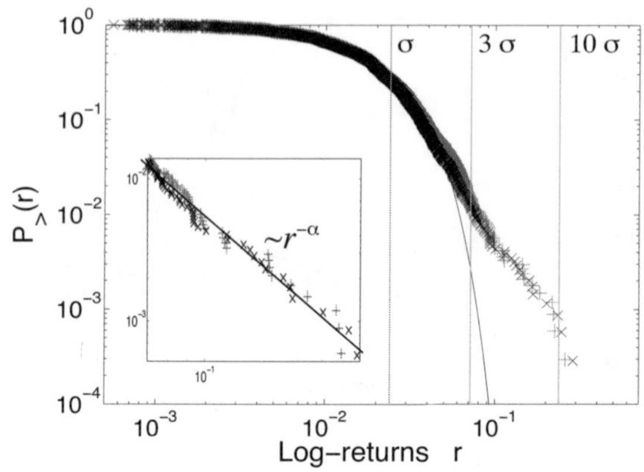

Fig. 1.2. Complementary distribution of the log-returns for the Ford Motor stock prices reported in Fig. 1.1. The "+" is the distribution of the positive returns $P(R \geq r)$ and the "×" the distribution of the negative ones $P(R \leq -r)$. The line is the comparison with a normal distribution with the same mean and variance. The inset is the tail region and the linear behavior in log-log scale highlights that there is a characteristic power-law decreasing trend $P_>(r) \sim r^{-\alpha}$. The best fit reveals an exponent $\alpha \sim 2.4$. The vertical lines correspond to deviation from the mean of respectively one, three and ten standard deviations (σ).

mean[a] $\mu = \langle r(t,\tau) \rangle \sim 2.4 \times 10^{-4}$ which is very small if compared with the estimated standard deviation $\sigma \sim 0.0237$. On the other hand, the distribution has a rather large fourth central moment[13] $\mu_4 = \left\langle (r(t,\tau) - \mu)^4 \right\rangle \sim 16.9$.

1.3.2. *Leptokurtic distributions*

The fact that higher moments have increasingly large relative values is a good indication that the distribution might deviate from a normal distribution. Such a deviation is often measured by using the excess kurtosis:

$$\gamma_2 = \frac{\mu_4}{\sigma^4} - 3 \ ; \tag{1.2}$$

In the case of Ford Motor we obtain $\gamma_2 \sim 5 \times 10^7$, which is a very sizable deviation. In fact, let us stress that the excess kurtosis of a *normal* distribution is zero. Distributions with large kurtosis are called *leptokurtic*. They are characterized by larger-than-normal probability of very small

[a]The "sample mean" $\langle x \rangle$ of a given set of values $\{x_1, ... x_n\}$ is calculated as the sum of all entries divided by their number $\langle x \rangle = (1/n) \sum_{i=1}^{n} x_i$. More generally, the sample mean of any function $f(x)$ is $\langle f(x) \rangle = (1/n) \sum_{i=1}^{n} f(x_i)$.

fluctuations, but also by larger-than-normal probabilities of very large fluctuations.

1.3.3. *Distribution tails*

The deviation of the fourth central moment μ_4 from that expected for a normal distribution is a very good indication that these systems should have special non-normal statistical properties. However, we must stress that it is very important in complex systems studies to look at the whole distribution and not only at the moments. In particular, for financial analysis and risk assessment the most important part of a distribution, which must be studied with attention, is is that describing large fluctuations, the so-called "tail of the distribution."

An idea of the deviation from the normal distribution in the tail region is given in Fig. 1.2 where we plot the complementary cumulative distribution, which is defined as:

$$P(R \geq r) = P_{>}(r) = \int_r^{\infty} p(s)ds \tag{1.3}$$

with $p(s)$ the probability density function. One can see from Fig. 1.2 that both the positive and negative tails deviate from a normal probability function with the same average and standard deviation. We see that the probability to find a deviation from the mean of the order of 1 standard deviation is about 30% (once every few days) and it is comparable for both the measured distribution and for the normal one. A sizable deviation is instead observed if we look at the probability to observe fluctuations larger or equal than 3 standard deviations. The normal distribution predicts 0.3% which is essentially once every year or two, in average. On the other hand, the observed frequency is in average once every 4 months. The deviation between the normal statistics and the observed one becomes huge if we move further away from the average. For instance, fluctuations larger than 10 standard deviations are observed several times during the investigated period 03/01/77 - 07/04/09 ($\sim$ 8000 days). Conversely, the normal statistics predicts an extremely small probability for such event (probability $\sim 10^{-23}$). In practice, it predicts that it would be very unlikely to observe such a fluctuation even if one waits for a time longer than the age of the universe.

1.4. Central Limit Theorem(s)

In the previous section we have compared the observed probability distributions with the normal one. Indeed, normal distributions are commonly observed in a very wide range of natural and artificial phenomena throughout statistics, natural science, medicine and social science. One of the reasons for this wide occurrence of normal distributions is that in several phenomena the observed quantities are sums of other (hidden) variables that contribute with different weights to the observed value. The Central Limit Theorem guarantees us that, *under some conditions*, the aggregate distribution in this kind of additive process tends towards the normal distribution. Therefore, a deviation from normal distribution is a good indication that we are dealing with a particular class of phenomena.

Let us first discuss the case where the distribution converges towards the normal one and then let us understand the origin of the observed deviations.

1.4.1. *Tendency towards normal distribution*

The Central Limit Theorem (CLT) states that if the various contributing variables are distributed independently and if they all follow the same identical distribution [i.e. they are *independent and identically-distributed* (i.i.d.) random variables] with *finite variance*, then the sum of a large number of such variables will tend to be normally distributed.[13] Which is, given n i.i.d. variables $\{x_i\}$ with mean[b] $E(x_i) = \mu$ and finite variance $E((x_i - \mu)^2) = \sigma^2$, the probability distribution of the sum

$$y = \sum_{i=1}^{n} x_i \tag{1.4}$$

is approximated by the probability density function:

$$p_n(y) \sim \Phi(y) = \frac{1}{\sqrt{2\pi n\sigma^2}} \exp\left(-\frac{(y - n\mu)}{2n\sigma^2}\right) , \tag{1.5}$$

for large n. The Berry–Esseen theorem guarantees that, if the third absolute moment $\mu_3 = E(|x_i - \mu|^3)$ is finite, then such a convergence is such that the difference goes as $1/\sqrt{n}$. More precisely: $|p_n(y - E(y)) - \Phi(y - E(y))| \leq 3c\mu_3/(\sigma^3\sqrt{n})$ with c a constant smaller than 1 and larger than 0.409.

The conditions dictated by the Central Limit Theorem to obtain normal distributions are quite broad and normal distributions are indeed

[b] The symbol $E(X)$ is the expectation value (or mean, or first moment) of the random variable X.

widespread. However, they are not commonly observed in complex systems, where strong deviations from the normal behavior are routinely found especially for large fluctuations.

1.4.2. *Violation of central limit theorem*

The central limit theorem (CLT) applies to a sum of random variables and it relies on three assumptions. These conditions are often violated in complex systems. Let us discuss each one of these conditions schematically.

- The CLT applies to a *sum* of variables. However, there are several processes which are not purely additive. One extreme case is a purely multiplicative process where the observable is a product of random variables. Incidentally, this extreme case is also a particular one because the product can be transformed into a sum by applying the logarithm and the CLT can be applied on the distribution of the log of the variable resulting in a log-normal distribution. However, in general, the process can be a mix of multiplicative and additive terms. Moreover, several different variables can contribute in an interrelated way through a network of "interactions."
- A condition is that the variables should be *independent*. On the other hand, often the variables are correlated and therefore not independent. Sometimes these correlations are associated with cascading events (one event triggers the other, which causes another, etc.) that can produce "avalanches" characterized by very large fluctuations in sizes with distributions having power law behaviors.
- A second condition requires the variables to be *identically distributed*. Again, often this is not the case and a broad range of distributions can sometimes shape the resulting aggregate distribution in an almost arbitrary way. However, we will see in the following Sections that the statistical behavior simplifies if one limits the study to extreme fluctuations only.
- The last condition requires *finite variance*. On the other hand, it is now widely recognized that in many complex systems the probability of large fluctuations often decreases slower than an exponential, usually with a power law trend $p(x) \sim x^{-\alpha-1}$, and the variance becomes undefined when $\alpha \leq 2$. To this class of distributions with non-defined variance an extension of the CLT applies.

1.4.3. *Extension of central limit theorem*

The Central Limit Theorem can be extended to a more general class of additive processes by dropping the condition of finite variance. Given n i.i.d. variables the sum $y = \sum_{i=1}^{n} x_i$ tends to a stable probability density function $f(y)$ which has characteristic function[14,15]

$$\hat{f}(q) = \exp\left(iq\mu - |cq|^{\alpha} \left(1 - i\beta \frac{q}{|q|} \Phi(\alpha, q) \right) \right) , \tag{1.6}$$

for $0 \leq \alpha \leq 2$ and with $\Phi(\alpha, q) = \tan(\pi\alpha/2)$ when $\alpha \neq 1$ or $\Phi(\alpha, q) = 2/\pi \log(|q|)$ when $\alpha = 1$. Let us recall that the characteristic function is defined by the transformation: $\hat{f}(q) = \frac{1}{2\pi} \int_{-\infty}^{+\infty} e^{-iqy} f(y) dy$ and, vice versa, $f(y) = \int_{-\infty}^{+\infty} e^{iqy} \hat{f}(q) dq$. The parameter $c > 0$ is a scale factor which is a measure of the width of the distribution. The parameter $-1 \leq \beta \leq 1$ is called skewness and is associated with asymmetry. In the symmetric case, when $\beta = 0$, the distribution becomes a stretched exponential function. In the case $\alpha = 2$, Eq. (1.6) gives the normal distribution. When $\alpha > 2$ the variance is finite and the central limit theorem applies predicting therefore the convergence towards the normal distribution. In general, the distribution defined by the characteristic function in Eq. (1.6) has no compact analytic form for $f(y)$ in the direct space. However, it is rather simple to show that, in the asymptotic limit of large fluctuations, the probability density function decreases as a power law, $f(y) \sim y^{-\alpha-1}$, where the exponent α is the same exponent as the one from the tails of the distributions of the variables x_i.

1.4.4. *Stable distributions*

The normal distribution and the distribution in Eq. (1.6) are "stable distributions." As a general property, stable distributions must satisfy the following condition: a distribution is stable if and only if, for any $n > 1$, the distribution of $y = x_1 + x_2 + ... + x_n$ is equal to the distribution of $n^{1/\alpha} x + d$ with $d \in \mathbb{R}$.[14] This implies

$$p_n(y) = \frac{1}{n^{1/\alpha}} p\left(\frac{y - d}{n^{1/\alpha}} \right) , \tag{1.7}$$

where $p_n(y)$ is the aggregate distribution of the sum of the n i.i.d. variables and $p(x)$ is the distribution of each of the variables x_i, with $i = 1, ...n$. The distribution is called strictly stable if $d = 0$. It is rather simple to prove that the distribution in Eq. (1.6) satisfies the scaling in Eq. (1.7) and it is indeed a stable distribution.

1.5. Looking for the Tails

A key question that we all might wish to answer is: *what is the maximum loss that we might possibly experience from our financial investments?* When dealing with risk we must be aware that in the market extremely large fluctuations can happen with finite probability. We often underestimate risk because extreme fluctuations are rather unusual in natural phenomena described by normal statistics. Large fluctuations described by non-normal statistics are instead rather common in financial systems, and in complex systems in general, representing one of the distinctive features of these systems. Indeed, a crucial key to many risk management problems is the understanding of the occurrence of extreme losses. It is for instance important in the evaluation of insurance losses from natural disasters such as hurricanes or earthquakes. Extreme losses happen rarely but they can be catastrophically deadly. However, the very fact that they are rare means that there is little statistics to rely on, which makes very difficult to predict the probability of their occurrence with precision. It is therefore very important to implement methods which can help to precisely estimate the behavior of the probability distribution in the region of large and rare variations, the "tail" of the distribution.

1.5.1. *Extreme fluctuations*

Let us consider a sequence of events $x_1, .., x_n$ characterized by a probability distribution function $p(x)$. We are interested in estimating the probability of the *maximum value* of such events $\max\{x_1, .., x_n\}$ for a given number n of occurrences. (For instance the largest size of the loss in the most catastrophic hurricane over a series of n hurricanes.)

A priori the probability of the events x_i can follow any kind of distribution. However, we are asking for the probability of the *maximum* and, in this case, we have an important general result, which is valid for asymptotic distributions of extreme order statistics.

The *Fisher–Tippet–Gnedenko, extreme value theorem*[6,13] states that the maximum of a sample of independent and identically distributed random variables, after proper renormalization, converges in distribution to one of three possible distributions, the *Gumbel* distribution, the *Fréchet* distribution, or the *Weibull* distribution.

1.5.2. *Extreme value distribution*

These distributions are particular cases of the generalized extreme value distribution (GEV), whose complementary cumulative distribution is:

$$G(x) = \exp\left(-\left(1+\frac{1}{\alpha}\left(\frac{x-\mu}{\sigma}\right)\right)^{-\alpha}\right) \tag{1.8}$$

for $1+(1/\alpha)(x-\mu)/\sigma > 0$. This is the general limit distribution of properly normalized maxima of a sequence of i.i.d. random variables. The subfamilies defined by $\alpha > 0$ and $\alpha < 0$ correspond, respectively, to the *Fréchet* and *Weibull* distributions whereas the *Gumbel* distribution is associated with the limit $\alpha \to \infty$.

1.5.3. *Fat-tailed distributions*

For the study of price fluctuations in financial markets, and specifically for risk analysis, we are interested in "fat-tailed" distributions where the complementary cumulative distribution $P_>(x)$ tends to $1-G(x)$ (Fréchet) in the tail region of large x. It is easy to show that for large x this distribution behaves as a power law with

$$1 - G(x) \sim a x^{-\alpha} \ . \tag{1.9}$$

Therefore, the tail of the distribution is associated with one parameter only: the exponent α which fully characterizes the kind of extreme value statistics.

From a general perspective, as far as extreme event statistics is concerned, we can classify probability distributions in three broad categories with respect to the value of the tail index α.

1) Thin-tailed distributions, for which all moments are finite and whose cumulative distributions decrease at least exponentially fast in the tails, they have $\alpha \to \infty$.
2) Fat-tailed distributions whose cumulative distribution function declines as a power law in the tails. For these distributions only the first k moments with $k < \alpha$ are bounded, and in particular:
   - for $\alpha > 2$ the standard deviation is finite and the distribution of a sum of these variables will converge towards the normal form in Eq. (1.5) (Central Limit Theorem[13]);
   - for $0 < \alpha \leq 2$ the standard deviation is not defined and the distribution of a sum of these variables will converge towards the Levy

stable distribution in Eq. (1.6) (extension of the Central Limit Theorem[13]).

3) Bounded distributions, which have no tails. They can be associated with $\alpha < 0$.

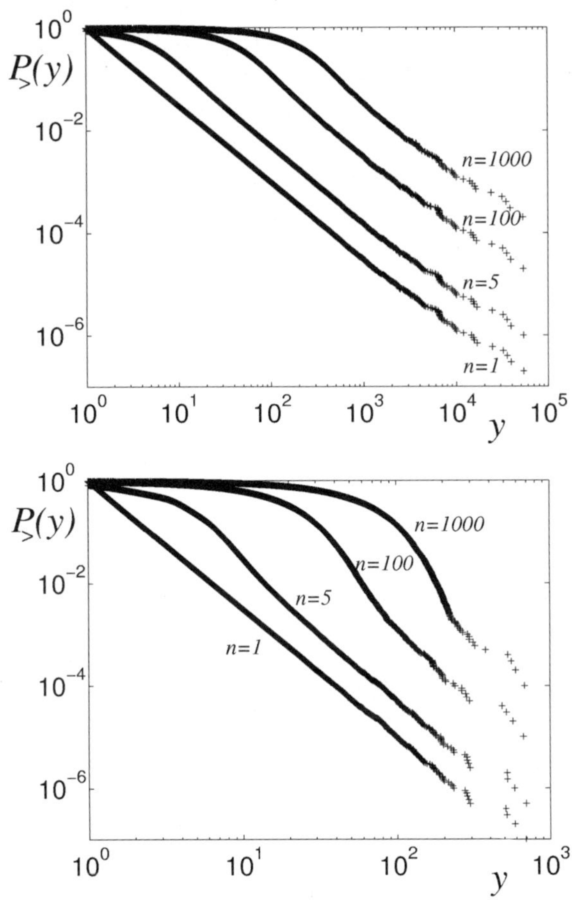

Fig. 1.3. Complementary cumulative distribution of the aggregate statistics resulting from a sum of n i.i.d. power law distributed variables. Specifically, we have, $y = \sum_{i=1}^{n} x_i$ with x_i independent random variables with probability distribution $p(x) = ax^{-\alpha-1}$. The top figure refers to the case $\alpha = 1.5$ whereas the bottom to the case $\alpha = 2.5$. Different aggregation sizes ($n = 1, 5, 100, 1000$) are shown.

1.5.4. *Sum of power-law-tailed distributions*

A very important consequence of the extreme value theorem is that the tails of a fat-tailed distribution (for i.i.d. processes) are invariant under addition even if the distribution as a whole is varying with aggregation. For instance, if we observe that daily returns are well fitted with a Student-t distribution,[13] then the Central Limit Theorem tells us that the monthly returns should be well approximated by a normal distribution and *not* a Student-t distribution. Yet the tails of the monthly returns are like the tails of the daily returns with the same exponent α. However, we must be aware that estimation of the tail exponent is not an easy task and a precise measurement of α requires a large sample size. This is why the use of data recorded every few seconds, or even tick by tick data (high frequency data), is highly recommended in this kind of analysis.[6]

Evidences of heavy tails in financial assets return distributions is plentiful ever since the seminal work of Mandelbrot on cotton prices.[16] However, the debate is still highly active and controversial, in particular on whether the second moment of the distribution of returns converges. Which requires to establish whether the exponent α is larger than 2 (σ defined) or smaller than 2 (σ not defined). It is clear that this question is central to many models in finance that specifically rely on the finiteness of the variance of returns. Indeed, as discussed in Section 1.4, there is a fundamental difference between additive i.i.d. processes with finite or infinite variance.

Let us here investigate further these differences with a simple example. We take the sum of n i.i.d. random variables x_i characterized by the following power-law probability density function:

$$p(x) = \frac{\alpha x_{\min}^{\alpha}}{x^{\alpha+1}} \ , \tag{1.10}$$

with $x \geq x_{\min}$ for some arbitrary $x_{\min} > 0$. Let us study the two cases $\alpha < 2$ and $\alpha > 2$. Figure 1.3 (top) shows that in the first case, for $\alpha = 1.5$, the distribution of the sum of the variables rests persistently a power law in most of the tail region (the complementary cumulative distribution decreases linearly in log-log scale) with the same exponent α of the individual random variables in Eq. (1.10) (which is the case $n = 1$ in Fig. 1.3). Indeed, in this case the distribution tends to the stable distribution [Eq. (1.6)] which behaves as a power law in the tail region $p_n(y) \to f(y) \sim y^{-\alpha-1}$.

We can see from Fig. 1.3 (bottom) that the second case, for $\alpha = 2.5$, is remarkably different. The shape of the distribution changes rapidly with the gathering of variables displaying a steeper decrease with x than for a power law distribution. Indeed, in this case, the Central Limit Theorem predicts a convergence of the aggregate distribution towards the normal one. However, in the tail region, below the Berry–Esseen convergence threshold, the extreme value theorem predicts a Fréchet distribution for the extreme variations and therefore we observe persistence of the power law trend in this extreme region. This is indeed evident from Fig. 1.3.

1.6. Capturing the tail

Have we already seen the worst or are we going to experience even larger losses? The answer of this question is essential for any good management of risk. We now have the instruments to answer this question. Indeed, we can apply the extreme value theory *outside* our sample to consider extreme events that have not yet been observed. To this purpose it is essential to be able to properly measure the tail index α.

1.6.1. *Power law in the tails*

We can see from the inset in Fig. 1.2, that the tails of the distributions of both the positive and negative fluctuations for the daily log-returns of the Ford Motor prices are decreasing linearly in log-log scale. This is an indication of a power law kind of behavior [i.e. $P_>(r) \sim ar^{-\alpha}$]. The main issue is how to quantify precisely the tail exponent α. There are several established methods to estimate the exponent α.[6,12] Let us here mention that a good practical rule is first to look qualitatively for the signatures of a linear trend in the log-log plot of $P_>(r)$ and afterwards, check the goodness of the quantitatively estimated α by comparing the data in the plot with the straight line from the power law function $ar^{-\alpha}$.

1.6.2. *Rank frequency plot*

Let us first point out that the plot of $P_>(r)$ in Fig. 1.2 is a so-called "rank-frequency" plot. This is a very convenient, and simple, method to analyze the tail region of the distribution without any loss of information which would instead derive from gathering together data points with an artificial binning. In order to make this plot one first sorts the n observed values in ascending order, and then plot them against the vector $[1, (n-1)/n, (n-2)/n, ..., 1/n]$. Indeed, for a given set of observations $\{x_1, x_2, ..., x_n\}$, we have that $\text{Rank}(x_i)/n = 1 - P_>(x_i)$.

A best fit of the exponent for the data in Fig. 1.2 reveals a value $\alpha \sim 2.4$. Values of exponents between 2 and 4 are very typical for these kinds of systems. These distributions typically have finite second moment but diverging fourth moment, and this is the reason why they reveal very high excess kurtosis.

1.7. Random Walks

So far we have discussed some relevant statistical features typically associated with complex system phenomena and in particular with stock price fluctuations. In this section we introduce a technique to model such fluctuations.

There are many factors that contribute to the "formation" of the price of a given equity and to its variation during time. This is, per se, the subject of several books and huge efforts have been dedicated to better understand this issue. From a very general and simple perspective we can say with some confidence that the price of a given equity is changing every time it is traded. At each successful transaction the price is fixed for that given time. A future transition will be processed at a marginally different price depending on the market expectations (rational or irrational) regarding that specific asset. This information is in part reflected in the bid and ask and their volumes on the order book and in part it lies in the mind and in the hearts of the human traders, and in the algorithms of the automatic traders as well.

1.7.1. *Price fluctuations as random walks*

An asset price model that has been widely used assumes that the logarithm of the asset price $x(t) = \log[\text{price}(t)]$ at a given time t results with some probability $p(\eta)$ at some value η above or below the logarithm of the price at the previous trading time. Formally we can write:

$$x(t + \Delta) = x(t) + \eta(t) \;; \tag{1.11}$$

where $\Delta > 0$ is the time-step. Equation 1.11 defines a *random walk* which is a particular case of a stochastic process. Sometime the random variable η is called "noise."

Random walk kinds of processes have been widely used in modeling complex systems. The term "random walk" was first used by Karl Pearson in 1905. He proposed a simple model for mosquito infestation in a forest:

at each time step, a mosquito moves a fixed length at a randomly chosen angle. Pearson wanted to know the mosquitos distribution after many steps. The paper (a letter to Nature[17]) was answered by Lord Rayleigh, who had already tackled the problem for sound waves in heterogeneous materials. As a matter of fact, the theory of random walks was developed a few years before (1900) in the PhD thesis of a young economist: Louis Bachelier. He proposed the random walk as the fundamental model for financial time series. Bachelier was also the first to draw a connection between discrete random walks and the continuous diffusion equation. Curiously, in the same year as the paper of Pearson (1905), Albert Einstein published his paper on Brownian motion, which he modeled as a random walk, driven by collisions with gas molecules. Smoluchowski in 1906 also published very similar ideas.

Note that Eq. (1.11) assumes discrete time and uses equally spaced time-steps Δ. In reality, the market time is indeed not continuous since transactions are registered at discrete times. However, these transaction times are not equally spaced, having periods with high activity and others with a relatively small number of transactions. Furthermore, the price variations at two consecutive times might be related. For instance, in periods of large volatility (large price fluctuations) the size of $|\eta(t)|$ is likely to be consistently larger than average for extended periods of time (a phenomenon called volatility clustering). Generally speaking, Eq. (1.11) must be considered as a basic model, which has however the advantage of being relatively easy to treat both analytically and numerically. The model can then be extended to consider continuous time and/or non-uniform spacing between time-steps and/or time correlations.

One more specific question about the random walk model in Eq. (1.11) concerns the size of the discrete time step Δ. In the market a stock can be traded several times in a second, however there can be intervals of several seconds where the stock is not traded. This "granularity" of the trading time is difficult to handle; as a general rule we must consider Δ of the order of a few seconds. The exact value is not particularly relevant in the present context, but the order of magnitude is very important, as we shall see hereafter.

1.7.2. *Log-return as sum of random variables*

Given that $x(t)$ in Eq. (1.11) is the log-price, the log-returns are $r(t,\tau) = x(t+\tau) - x(t)$ [Eq. (1.1)] and they can be written as

$$r(t,\tau) = \sum_{s=0}^{\tau/\Delta-1} \eta(s\Delta + t) \ . \tag{1.12}$$

They are therefore sums of $n = \tau/\Delta$ random variables and, if the $\eta(t)$ are i.i.d., the Central Limit Theorem must apply to $r(t,\tau)$. We have seen in Section 1.4 that we have two broad cases: (1) the probability distribution function of $\eta(t)$ has finite variance and therefore the distribution of $r(t,\tau)$ should approximate a normal distribution for large τ; (2) the variance is not defined and therefore the distribution of $r(t,\tau)$ should approximate a Levy Stable distribution for large τ. If we have fat-tailed distributions, as the ones described in Sections 1.5 and 1.6, then the parameter that distinguishes between these two classes is the tail index α. The case $\alpha \geq 2$ leads to normal distributions, whereas $\alpha < 2$ yields to Levy Stable distributions.

We have seen in our example with the Ford Motor data (Section 1.6.2), that in this specific example the tail index is best fitted with $\alpha \sim 2.4$ which is therefore larger than 2. In this case, the Central Limit Theorem tells us that a sum of n of these variables (where $\tau = n\Delta$) will converge towards a normal form and the Berry–Esseen theorem guarantees that this convergence is in $1/\sqrt{n}$. This implies that if we look for deviations from the normal statistics, we should explore the tail region where, roughly speaking, $P_>(x) < 1/\sqrt{n}$. Since in Fig. 1.2 we are reporting the statistics of daily returns, we have $\tau = 1$ day which corresponds to about 6 market hours and therefore ~ 22000 seconds, we expect to still observe power law behaviors in the tail region where $P_>(x) < 1/\sqrt{22000} \sim 10^{-2}$, which is indeed where the distribution starts to differ substantially from the normal statistics as one can clearly see in Fig. 1.2.

1.7.3. *High frequency data*

It is clear that a better estimate of the tail exponent can be obtained by reducing the interval τ, and this requires the use of infra-day data. Nowadays, there is a great availability of "high" frequency financial data, up to the whole order book, where every single bid, ask and transaction price is registered together with the volumes (amount of capital traded). However, to work with infra-day data poses some new technical challenges.[6]

For instance, the opening prices are affected by the events occurred during the closure and by the night electronic trading. The closure prices are also affected by the same reasons, in expectations. There are periods in the day that are very active and others that are instead pretty gloomy. For instance, large activities and sudden price variations are likely at times when other markets open or close. It is beyond the purposes of this Chapter to give any account of the so-called "seasonality" effects in the infra-day data. However, it is important that readers bear in mind that infra-day data should be handled with care. A highly recommended "minimal-trick" is to eliminate from the analysis data the first 20 min after opening and the last 20 min before closure.[6]

1.8. Scaling

We have seen that the statistical analysis of price fluctuations can be performed over different time scales. In general, the overall statistical properties of the log-returns are changing with the time interval τ. Indeed, there are different factors governing the variation at short- or long- time scales. However, it is also clear that the two must be related. Such a relation between the different probability distributions of the fluctuations at different time intervals is called *scaling* of the distribution.

The presence of large fluctuations and in particular power law noise with exponent $\alpha < 2$ affects dramatically the overall dynamics of the process and it is reflected in the *scaling* properties of the aggregate statistics. Let us first note that the log-returns $r(t,\tau)$ from the random walk process in Eq. (1.11) can be written as a sum of $n = \tau/\Delta$ noise terms [as explicitly shown in Eq. (1.12)]. Therefore, the changes of the statistical properties of $r(t,\tau)$ with τ (the so-called scaling of the distribution) correspond to the changes of the aggregate statistics of the sum of $n = \tau/\Delta$ i.i.d. variables. We know already from the previous discussion in Sections 1.4 and 1.6 that there is a difference in the aggregate statistics of random variables with finite or undefined variance. Such a difference is reflected in the diffusion dynamics of the random walker.

1.8.1. *Super-diffusive processes*

In additive stochastic processes [such as Eq. (1.12)] with fat-tailed noise (i.e. $p(\eta) \sim |\eta|^{-\alpha-1}$, with $0 < \alpha < 2$), all the motion is dominated by the large fluctuations and it results in a super-diffusive behavior where the mean square displacement increases faster than τ. Let us here give an heuristic simple derivation of this fact which shows clearly the origin of anomalous diffusion behavior in presence of power law noise.

Given a power law probability $p(\eta) \sim |\eta|^{-\alpha-1}$, the probability of a jump of size L which is larger or equal than a given $L_{\max}$ is given by the complementary cumulative distribution:

$$P(L \geq L_{\max}) = P_{>}(L_{\max}) = \int_{L_{\max}}^{\infty} p(\eta) d\eta \sim \frac{1}{L_{\max}^{\alpha}} \ . \tag{1.13}$$

We can infer an idea of the time-dependence of $L_{\max}$ by noticing that if we monitor the process for an interval of time $\tau = n\Delta$ we will have a *finite* probability to observe a jump of size $L_{\max}$ if $nP(L \geq L_{\max}) \sim 1$ and therefore, from Eq. (1.13), $n/L_{\max}^{\alpha} \sim 1$, yielding

$$L_{\max} \sim \left(\frac{\tau}{\Delta}\right)^{1/\alpha} \ . \tag{1.14}$$

We can now use the same argument to calculate the mean square displacement after $n = \tau/\Delta$ time steps: $E(r^2) - E(r)^2 = E(r^2) = nE(\eta^2)$ [having $E(r) = 0$]. When $0 < \alpha < 2$, we have $E(\eta^2) = \int_{L_{\min}}^{L_{\max}} \eta^2 p(\eta) d\eta \sim L_{\max}^{2-\alpha}$ and therefore $E(r^2) \sim nL_{\max}^{2-\alpha} = L_{\max}^2$. This indicates that the whole average movement in the process is the size of the largest jump. In other words, the evolution is entirely dominated by the largest jumps. By using Eq. (1.14) we have

$$E(r^2) \sim \left(\frac{\tau}{\Delta}\right)^{2/\alpha} \ . \tag{1.15}$$

We see that for $0 < \alpha < 2$ the mean square displacement increases faster than τ and the system is "super-diffusive." For $\alpha \geq 2$ the arguments above do not hold any longer and the mean square displacement grows linearly with τ as for any diffusive process.

1.8.2. *Sub-diffusive processes*

Let us here also mention that an opposite kind of scaling is observed when the mean square displacement increases slower than τ. This case is referred to as "sub-diffusive" behavior and it can be obtained from additive kinds of models when the time-step intervals between subsequent variations are unequally distributed following a power-law kind of distribution. It is also the result of time-correlated processes.

1.8.3. *Uni-scaling*

The random walk is a very simple and useful model to introduce and study stochastic processes. However, it must be stressed that most of the real stochastic processes are *correlated*, and therefore they are not random walks.[18] We have seen that random walk processes are associated with scaling laws that describe the way the distribution changes when the variables are aggregated. For instance for a stable process the probability distribution of the log-returns should scale with τ accordingly with:

$$p_\tau(r) = \left(\frac{\Delta}{\tau}\right)^{1/\alpha} p\left(\left(\frac{\Delta}{\tau}\right)^{1/\alpha} r\right) . \tag{1.16}$$

This is a direct consequence of Eq. (1.7) and the fact that $r(t,\tau)$ is the sum of τ/Δ random variables [Eq. (1.12)]. Accordingly, the q-moments scale as $E(|r(t,\tau)|^q) = (\tau/\Delta)^{q/\alpha} E(|r(t,1)|^q)$. This is one particular form of scaling, which applies to stable distributions and is analogous to that discussed in the previous section.

More generally, in analogy with the previous scaling law, one can define a stochastic process where the probability distribution of $\{x(ct)\}$ is equal to the probability of $\{c^H x(t)\}$. Such a process is called self-affine.[19] In self-affine processes, with stationary increments, the q moments must scale as

$$E(|r(t,\tau)|^q) = E(|x(t+\tau) - x(t)|^q) = c(q)\left(\frac{\tau}{\Delta}\right)^{qH} . \tag{1.17}$$

The parameter H is called the self-affine index or scaling exponent (or Hurst exponent—see chapter by Henry *et al.* in this volume). It is related to the fractal dimension by $D_f = D+1-H$, where D is the process dimensionality ($D = 1$ in our case). It is clear from Eq. (1.16) that in the case of stable distributions $E(|r(t,\tau)|^q) = E(|x(t+\tau) - x(t)|^q) = (\tau/\Delta)^{q/\alpha} E(|r(t,1)|^q)$ and we can identify $H = 1/\alpha$.

A process satisfying the scaling in Eq. (1.17) is called *uniscaling*.[19]

1.8.4. *Multi-scaling*

The kinds of processes encountered in finance, and in complex systems in general, often scale in an even more complicated way indicating that the kind of observed scaling is not simply a fractal. In contrast to the more conventional fractals, in these processes we need more than one fractal dimension depending on the aspect we are considering. In order to properly model real world processes we must use *multiscaling* processes where, for stationary increments, the q moments scale as[19–21]

$$E(|x(t+\tau) - x(t)|^q) = c(q) \left(\frac{\tau}{\Delta}\right)^{qH(q)} \quad . \tag{1.18}$$

with $H(q)$ a function of q (for $q > -1$). The function $qH(q) - 1$ is called the scaling function of the multi-scaling process and it is in general concave.

Most of the processes in financial systems are multiscaling. This implies that the moments of the distribution scale differently according to the time horizons (i.e. the distribution of the returns changes its shape with τ) revealing that the system properties at a given scale might not be preserved at another scale.

1.9. Complex Networked Systems

Let us now continue our "navigation" by introducing a new fundamental factor which concerns the collective dynamics of the system. As we already mentioned in the introduction, studies of complex systems have ranged from the human genome to financial markets. Despite this breadth of systems—from food chains to power grids, or voting patterns to avalanches—a unifying and characterizing aspect has emerged: all these systems are comprised of many interacting elements. In recent years, it has become increasingly clear that in complex system studies it is of paramount importance to analyze the dynamics of all the elements highlighting their emerging collective properties.

The study of collective dynamics requires the simultaneous investigation of a large number of different variables. So far, in this Chapter, we have focused on the complex behavior of a *single* element (i.e. the price of a given stock), however it is clear that any information from such an individual property is meaningless if not properly compared with the information from the rest of the elements constituting the system (i.e. all the other stocks in the market). Typically, in these systems each element is not evolving in isolation and therefore the collective dynamics is reflected in the individual behavior as much as the individual changes affect the global variations. The understanding of the properties of such a network of interactions and co-variations is one of the key elements to understand complex systems. Indeed, one of the most significant breakthroughs in complex systems studies has been the discovery that all these systems share similar structures in the network of interactions between their constitutive elements.[22–24]

1.9.1. *Scale-free networks*

It results that in a large number of complex systems the probability distribution of the number of contacts per vertex (called *degree distribution* $p(k)$) is "fat tailed" with $p(k) \sim k^{-\alpha-1}$, with exponents α typically ranging between 1 and 2.[25] Such networks are widespread including internet, world wide web, protein networks, citation networks, world trade network etc.

It is important to stress that these distributions are different from what it would result by randomly connecting pair of vertices in a *random graph* which will instead yield a Poissonian distribution with exponentially-fast-decreasing tails. The power law degree distribution implies that there is no "typical" scale and all scales of connectivity are represented. These "scale-free" networks are characterized by a large number of poorly connected vertices but also by a few very highly connected "hubs."[25]

1.9.2. *Small and ultra-small worlds*

The properties of such networks are also different from the properties of regular or random networks. For instance, the average distance ($\langle d \rangle$) between two vertices scales with the total number of vertices in the network (V) as $\langle d \rangle \sim \log \log V$ ($1 < \alpha < 2$). This means that the network is very compact and only very few steps are necessary to pass from an individual to another. This property is called "ultra small world" in contrast with the "small world" property where $\langle d \rangle \sim \log V$, which holds for $\alpha \geq 2$ or for random networks ($\alpha \to \infty$). In contrast, regular lattices in D dimensions are "large worlds" with $\langle d \rangle \sim V^{1/D}$. Without entering into any detail, it should be quite clear to anyone just from an intuitive perspective that the difference between a large world and an ultra small world is huge, and can have very dramatic implications. For instance, the dynamical properties, such as the rate of spreading of a disease through the network, are strongly affected by the network structure. A world pandemic on a "large world" contact network will require a very long chain made of hundreds of thousands of contacts to infect all the individuals from a single source. By contrast, on an "ultra small world" it would take a chain of only a few steps to infect all the individuals in the world. Alarmingly, it has been shown[26] that the airport-network system has a small world structure.

1.9.3. *Extracting the network*

The extraction of the network of interrelations associated with a given complex system can be a challenging task. Indeed, except for a few cases where the network has an unambiguously given property, there are a large number of other cases where the network is not as clearly defined. The network of relations between elements of the systems can have *weights*, can be *asymmetric*, there might be *missed* unknown links, there might be asynchronous interactions, feedback and fluctuations. In this section we investigate a specific case which is rather general and widespread in the study of these systems.

We consider a system with a large number of elements where, a priori, every element can be differently affected by any other and we aim to infer the network of most relevant links by studying the mutual dependencies between elements from the analysis of the collective dynamics. In other words, we search for variables that behave similarly and we want to link them with edges in the network. Conversely, we do not want to directly connect the variables that behave independently. To this purpose we must first look at methods to define and quantify dependency among variables.

1.9.4. *Dependency*

Generally speaking, the mutual dependence between two variables x and y should be measurable from the "difference" between the probability to observe them simultaneously and the probability to observe them separately. Let us call $P(X \leq x, Y \leq y)$ the joint cumulative distribution to observe both the values of the variables X and Y to be less than or equal to two given values x and y. We must compare this joint probability with the marginal cumulative probabilities $P_X(X \leq x)$ and $P_Y(Y \leq y)$ of observing the variables independently from each other. A theorem[12] guarantees us that two random variables X and Y are *independent* if and only if:

$$P(X \leq x, Y \leq y) = P_X(X \leq x)P_Y(Y \leq y) \ . \tag{1.19}$$

This identity reflects the intuitive fact that when the variables are independent the occurrence of one event must make it neither more nor less probable that the other occurs. Therefore, given two variables X and Y, the difference, or the distance between the joint cumulative probability $P(X \leq x, Y \leq y)$ and the product of the two marginal cumulative probabilities $P_X(X \leq x)$ and $P_Y(Y \leq y)$, should be a measure of dependence between the two variables. Indeed, when $P(X \leq x, Y \leq y) > P_X(X \leq$

$x)P_Y(Y \leq y)$ we have the so called *positive quadrant dependency*, which expresses the fact that "when X is large, Y is also likely to be large." Conversely, $P(X \leq x, Y \leq y) < P_X(X \leq x)P_Y(Y \leq y)$ we have the so called *negative quadrant dependency*, which expresses the fact that "when X is large, Y is likely to be small." One simple quantification of such a measure of dependency is the *covariance*:

$$\operatorname{Cov}(X,Y) = \iint [P(X \leq x, Y \leq y) - P_X(X \leq x)P_Y(Y \leq y)]\, dxdy\,. \tag{1.20}$$

A positive covariance indicates that the two variables are likely to behave similarly whereas a negative covariance indicates that the two variables tend to have opposite trends. However, it should be stressed that this is a measure of linear dependency and there are nonlinear cases where dependent variables have zero covariance [e.g. $y = x^2 - 1$, with $E(x) = 0$ and $E(x^2) = 1$].

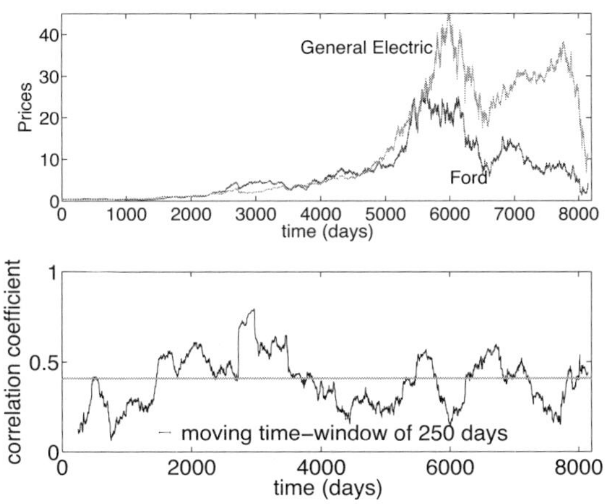

Fig. 1.4. (Top) Daily closing adjusted prices for the two stocks Ford Motor and General Electric in the New York Stock Exchange. The time-period ranges from Jan 3 1977 to April 7 2009. (Bottom) Correlation coefficient calculated over a moving window of 250 days ($\sim$ 1 year). The horizontal line is the correlation coefficient calculated over the whole period.

1.9.5. *Correlation coefficient*

A measure of dependency directly proportional to the covariance is the Pearson product-moment correlation coefficient $\rho_{i,j}$. Given two random variables x_i and x_j with expectation values μ_i and μ_j, and standard deviations σ_i and σ_j, the correlation coefficient is defined as

$$\rho_{i,j} = \frac{\mathrm{Cov}(x_j, x_j)}{\sigma_i \sigma_j} = \frac{E\left((x_i - \mu_i)(x_j - \mu_j)\right)}{\sigma_i \sigma_j} \quad . \tag{1.21}$$

Analogously to the covariance, positive values for $\rho_{i,j}$ indicate that the two variables are likely to behave similarly, whereas negative $\rho_{i,j}$ indicate that the two variables tend to have opposite trends. The correlation coefficient has however the advantage of being bounded between $[-1, 1]$, with the two limits corresponding to perfectly anti-correlated and perfectly correlated variables. For example, one can verify that the two variables $x_j = a + bx_i$ have $\rho_{i,j} = b/|b|$ giving $\rho_{i,j} = +1$ when $b > 0$ and $\rho_{i,j} = -1$ when $b < 0$.

1.9.6. *Significance*

In practice, the correlation coefficient is estimated over a finite set of data points: the time series $x_i(t)$ with $t = t_0 + s\Delta$ with $s = 1, 2, ...T$. The Pearson estimator $\rho_{i,j}$ is calculated from Eq. (1.21) by substituting the expectation values $E(...)$ with the sample averages $\langle(...)\rangle$ and by using the sample means and standard deviations. Clearly, the smaller the observation time T, the larger will be the inaccuracy on the estimated coefficient.

The use of the correlation coefficient to measure dependence between variables is very common and widespread, and it turns out to be a very efficient measure in a large number of domains. However, this measure can be very problematic and it might sometimes lead to serious faults. We have already mentioned that, in nonlinear cases, completely dependent variables can have zero covariance and consequently zero correlation coefficient. Other problems might arise with non-normally distributed variables. Indeed, we already noticed that the standard deviation is not defined for random variables with fat-tailed power law distributions and tail exponent smaller than or equal to 2. This implies that for these variables the correlation coefficient is not defined as well. Moreover, when the tail index belongs to the interval $\alpha \in (2, 4]$, the correlation coefficient exists but its Pearson estimator is highly unreliable because its distribution is fat tailed, with undefined second moments, and therefore it can have unbounded large variations.

Moreover, in complex systems studies we are often observing systems that are not stationary and the interrelations between the elements are themselves changing during the observation time. As a general rule, one must assume that these changes happen on a longer time scale than the one within which the correlations are measured. Therefore, T should not be too small, in order to improve the statistics, but it should not be too long either, in order to avoid being influenced by the long-term changes.

A practical example is given in Fig. 1.4 (top) where the historical data for Ford Motor (same as in Fig. 1.1) are plotted together with the data for General Electric. One can see that there are similarities and differences. The cross correlation coefficient for the log-returns over the entire period is $\rho \sim 0.4$. On the other hand, Fig. 1.4 (bottom) shows that the values over sub-periods calculated on a moving window of 1 year ($\sim$ 250 days) fluctuate around this value, showing significant variations depending on the market evolution.

1.9.7. *Building the network*

In practice, we often have more than two variables and the dependency problem is in general a high dimensional challenge. However, the extension from two to n variables is straightforward, with the exception that the names change and the joint distribution takes the name of *multivariate distribution* when $n > 2$ (bivariate for $n = 2$). As far as we are interested in the dependencies between couples of variables x_i and x_j we can apply straightforwardly Eq. (1.21) to each couple of variables, obtaining an $n \times n$ correlation matrix which is symmetric and has all ones on the diagonal. We have therefore $n(n-1)/2$ distinct entries.

Let us here concentrate on one precise example: the system of companies quoted on the same equity market as Ford Motor (the NYSE) and, for simplicity, let us take only a set of 100 among the most capitalized ones. Even with such a reduction, the system of correlations between the various stock prices has 4,950 entries, most of which are redundant. In terms of the network of interactions we are now looking at the *complete graph* where every node is connected by one *weighted* edge to every other node in the network. We must simplify such a system extracting only a subset of the most relevant interrelations: the "backbone" network.

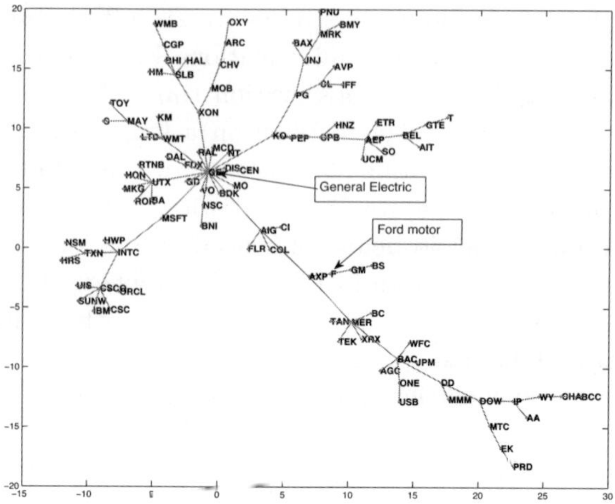

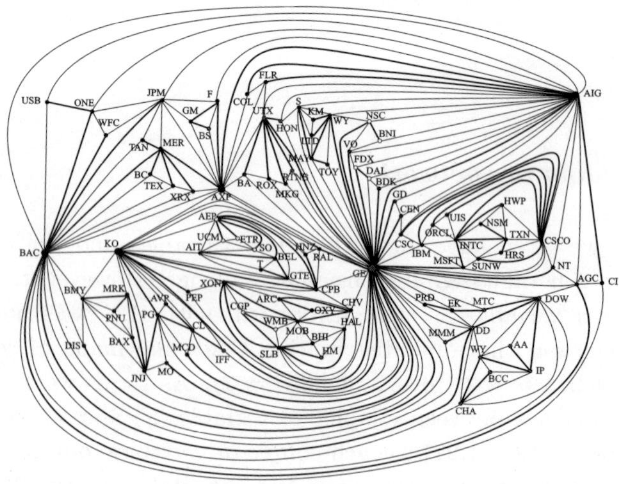

Fig. 1.5. (top) Minimum Spanning Tree network built from the cross-correlation matrix for 100 stocks in the US equity market. (bottom) Planar Maximally Filtered Graph built for the same cross-correlation data.

1.9.8. *Disentangling the network: minimum spanning tree*

We want to build a network whose topological structure represents the correlation among the different elements. All the important relations must be represented, but the network should be as "simple" as possible. The simplest connected graph is a spanning tree (a graph with no cycles that connects all vertices). It is therefore natural to choose as representative network a spanning tree which retains the maximum possible correlations. Such a network is called the Minimum Spanning Tree (MST).

There are several algorithms to build a MST, the two most common being Prim's algorithm[27] and Kruskal's algorithm[28] both from the '50s, but there are also older ones. Remarkably, there are also very recently discovered ones such as the one proposed by Chazelle[29] in the year 2000, which is, so far, the algorithmically most efficient, running in almost linear time with the number of edges.

The general approach for the construction of the MST is to connect the most correlated pairs while constraining the network to be a tree. Let us here describe a very simple algorithm (similar to Kruskal's) which is very intuitive and will help to clarify the concept.

(1) Make an ordered list of pairs i, j, ranking them by decreasing correlation $\rho_{i,j}$ (the largest first and the smallest last).
(2) Take the first element in the list and add the edge to the graph.
(3) Take the next element and add the edge if the resulting graph is still a forest or a tree, otherwise discard it.
(4) Iterate the process from step 3 until all pairs have been exhausted.

The resulting MST has $n - 1$ edges and it is the spanning tree that maximizes the sum of the correlations over the connected edges.

The resulting network for the case of the 100 stocks quoted in the NYSE studied during the period 3 January '95 to 31 December '98 is shown in Fig. 1.5.[30,31] We can see that in the MST the stock Ford Motor (F) is linked to the stock General Motor (GM). They form a separate branch together with Bethlehem Steel (BS), and the branch is attached to the main "trunk" through the financial services provider American Express (AXP). This structure of links that we have here extracted with the MST is economically very meaningful because we know that cars need steel to be built and consumers need credit from financial companies to buy the cars. What is remarkable is that these links have been extracted from the cross-correlation matrix without any a priori information on the system.

It is clear that the same method can potentially be applied to a very broad class of systems, specifically in all cases where a correlation (or even, more simply, a similarity measure) between a large number of interacting elements can be assigned.

1.9.9. *Disentangling the network: planar maximally filtered graph*

Although we have just shown that the MST method is extremely powerful, there are some aspects that might be unsatisfactory. In particular the condition that the extracted network should be a tree is a strong constraint. Ideally, one would like to be able to maintain the same powerful filtering properties of the MST but also allowing the presence of cycles and extra links in a controlled manner.

A recently proposed solution consists in building graphs embedded on surfaces with given genus.[32] (Roughly speaking the genus of a surface is the number of holes in the surface: $g = 0$ corresponds to the embedding on a topological sphere; $g = 1$ on a torus; $g = 2$ on a double torus; etc.) The algorithm to build such a network is identical to the one for the MST discussed previously except that at step 3 the condition to accept the link now requires that the resulting graph must be embeddable on a surface of genus g. The resulting graph has $3n - 6 + 6g$ edges and it is a triangulation of the surface. It has been proved that the MST is always a subgraph of such a graph.[31]

It is known that for large enough genus any network can be embedded on a surface. From a general perspective, the larger the genus, the larger is the complexity of the embedded triangulation. The simplest network is the one associated with $g = 0$ which is a triangulation of a topological sphere. Such planar graphs are called Planar Maximally Filtered Graphs (PMFG).[31] PMFG have the algorithmic advantage that planarity tests are relatively simple to perform.

The PMFG network for the case of the 100 stocks studied previously is reported in Fig. 1.5.[31] We can observe that in this network Ford Motor (F) acquires a direct link with Bethlehem Steel (BS), it acquires a new link with the bank JPMorgan Chase (JPM) and it also acquires a link with the very influential insurance services American International Group (AIG). We note that F, AXP and BS form a 3-clique (a triangle), which becomes a 4-clique (a tetrahedron)[c] by adding GM. As one can see from Fig. 1.5, the PMFG is a network richer in links and with a more complex structure than the MST, of which it preserves and expands some hierarchical properties.

[c] An r-clique is a complete graph with r vertices where each vertex is connected with all the others.

1.10. Conclusion

In conclusion, in this Chapter we have introduced a few fundamental notions useful for the study of complex systems, with the aim of providing a sort of referential "navigation map." We have presented and discussed a large number of concepts, theorems and topics. However, the most important aspects that we have treated under different perspectives can be summarized in the following two points:

1) the ubiquitous presence in these systems of fat-tail probability distributions and their effects on the statistics of extreme events and on the scaling properties of additive processes;
2) the importance in these systems of the collective cross-correlated dynamics and the need of novel investigation techniques which combine statistical methods with network theory.

Let us here conclude by summarizing all the different aspects discussed in this Chapter by including them within a single, compact formula that is a sort of *constitutive equation for complex systems*:

$$x_i(t_{s+1}) = x_i(t_s) + \eta_i(t_s) + \sum_{u=0}^{s}\sum_{k=1}^{n} J_{i,k}(t_s, t_u) x_k(t_u) \quad , \tag{1.22}$$

with $J_{i,k}$ an exchange matrix associated with a weighted, directed network of interactions between the variables. This equation describes a process where n variables $x_1(t), ..., x_n(t)$ start at t_0 with values $x_i(t_0) = x_i^0$ and evolve in time through the time points $t_1, t_2, ...$, which are not necessarily equally spaced. The term $\eta_i(t_s)$ is an additive noise equivalent to the one described in Eq. (1.11). On the other hand, the last term describes the interaction between variables and can be used to introduce multiplicative noise, feedback and autocorrelation effects. All the characterizing features of complex systems that we have been discussing in this Chapter can be modeled and accounted by means of Eq. (1.22). This equation is compact but not simple and cannot be solved in general. On the other hand, it is rather straightforward to implement numerically and can be applied to model a very broad range of systems. Indeed, equations of the form of Eq. (1.22) have been proposed in the literature to model very different kinds of complex systems, from the spread of epidemics to productivity distribution. However, to our knowledge, the great generality and the wide applicability of such an equation has not so far been pointed out.

There are of course several other important topics, methods and techniques very relevant in the study of complex systems but which we have been unable to squeeze inside this Chapter. In this respect, the rest of this book provides a great example of the variety, diversity and breadth of this fast-expanding field.

Acknowledgments

We wish to acknowledge the strong support from the ARC network RN0460006. This work was partially supported by the ARC Discovery Projects DP0344004 (2003), DP0558183 (2005) and COST MP0801 project.

References

1. A. Einstein and L. Infield, *The Evolution of Physics: The Growth of Ideas from Early Concepts to Relativity and Quanta.* (Simon and Schuster, New York, 1938).
2. G. Parisi, Complex systems: a physicist's viewpoint, *Physica A.* **263**, 557–564, (1999).
3. N. Boccara, *Modeling Complex Systems.* (Springer-Verlag, Berlin, 2004).
4. fURL `www.santafe.edu/research/topics-physics-complex-systems.php`.
5. R. N. Mantegna and H. E. Stanley, *An Introduction to Econophysics: Correlations and Complexity in Finance.* (Cambridge University Press, Cambridge, UK, 2000).
6. M. M. Dacorogna, R. Gençay, U. A. Müller, R. B. Olsen, and O. V. Pictet, *An Introduction to High Frequency Finance.* (Academic Press, San Diego CA, 2001).
7. T. Di Matteo, T. Aste and S. T. Hyde. Exchanges in complex networks: income and wealth distributions. In *The Physics of Complex Systems (New advances and Perspectives)*, Proceedings of the International School of Physics "Enrico Fermi." Eds. F. Mallamace and H. E. Stanley, 435–442 (IOS Press, Amsterdam 2004).
8. P. Ball, Culture crash, *Nature.* **441**, 686–688, (2006).
9. M. Gallegati, S. Keen, T. Lux, and P. Ormerod, Worrying trends in econophysics, *Physica A.* **370**, 1–6, (2006).
10. T. Di Matteo and T. Aste, "No Worries": Trends in Econophysics, *Eur. Phys. J. B.* **55**, 121–122, (2007).
11. T. Di Matteo and T. Aste, Econophysics Colloquium, *Physica A.* **370**, xi–xiv., (2006).
12. E. Platen and D. Heath, *A Benchmark Approach to Quantitative Finance.* (Springer-Verlag Berlin, 2006).
13. W. Feller, *An Introduction to Probability Theory and Its Applications.* (Wiley, New York, 1968).

14. J. P. Nolan. URL academic2.american.edu/~jpnolan/stable/stable.html.
15. J.-P. Bouchaud and M. Potters, *Theory of Financial Risk and Derivative Pricing: From Statistical Physics to Risk Management (2nd ed.)*. (Cambridge University Press, Cambridge, UK, 2003).
16. T. V. of Certain Speculative Prices, B. B. Mandelbrot, *Journal of Business.* **36**, 394–419, (1963).
17. K. Pearson, The problem of the random walk, *Nature.* **72**, 294–342, (1905).
18. A. W. Lo and A. C. MacKinlay, *A Non-Random Walk Down Wall Street.* (Princeton University Press, 2001).
19. T. Di Matteo, Multi-scaling in finance, *Quantitative Finance.* **7**, 21–36, (2007).
20. T. Di Matteo, T. Aste, and M. M. Dacorogna, Scaling behaviors in differently developed markets, *Physica A.* **324**, 183–188., (2003).
21. T. Di Matteo, T. Aste, and M. M. Dacorogna, Long term memories of developed and emerging markets: using the scaling analysis to characterize their stage of development, *Journal of Banking & Finance.* **29/4**, 827–851, (2005).
22. M. E. J. Newman, The structure and function of complex networks, *SIAM REVIEW.* **45**, 167–256, (2003).
23. G. Caldarelli, *Scale-Free Networks Complex Webs in Nature and Technology.* (Oxford Univesity Press, 2007).
24. L. Amaral and J. Ottino, Complex networks, *Eur. Phys. J. B.* **38**, (2004).
25. R. Albert and A.-L. Barabási, Statistical mechanics of complex networks, *Rev. Mod. Phys.* **74**, 47–97, (2002).
26. R. Guimerà, S. Mossa, A. Turtschi, and L. A. N. Amaral, The worldwide air transportation network: Anomalous centrality, community structure, and cities' global roles, *Proceedings of the National Academy of Sciences of the United States of America.* **102**(22), 7794–7799, (2005). doi: 10.1073/pnas.0407994102. URL http://www.pnas.org/content/102/22/7794.abstract.
27. R. C. Prim, Shortest connection networks and some generalizations, *Bell System Technical Journal.* **36**, 1389–1401, (1957).
28. J. B. Kruskal, On the shortest spanning subtree of a graph and the traveling salesman problem, *Proceedings of the American Mathematical Society.* **7**, 48–50, (1956).
29. B. Chazelle, A minimum spanning tree algorithm with inverse-ackermann type complexity, *Journal of the ACM (JACM).* **47**, 1028–1047, (2000).
30. R. N. Mantegna, Hierarchical structure in financial markets, *Eur. Phys. J. B.* **11**, 193–197, (1999).
31. M. Tumminello, T. Aste, T. Di Matteo, and R. N. Mantegna, A tool for filtering information in complex systems, *PNAS.* **102**, 10421–10426, (2005).
32. T. Aste, T. Di Matteo, and S. T. Hyde, Complex networks on hyperbolic surfaces, *Physica A.* **346**, 20–26, (2005).

Chapter 2

An Introduction to Fractional Diffusion

B. I. Henry, T. A. M. Langlands and P. Straka

School of Mathematics and Statistics,
University of New South Wales,
Sydney, NSW 2052
Australia
b.henry@unsw.edu.au

The mathematical description of diffusion has a long history with many different formulations including phenomenological models based on conservation of mass and constitutive laws; probabilistic models based on random walks and central limit theorems; microscopic stochastic models based on Brownian motion and Langevin equations; and mesoscopic stochastic models based on master equations and Fokker–Planck equations. A fundamental result common to the different approaches is that the mean square displacement of a diffusing particle scales linearly with time. However there have been numerous experimental measurements in which the mean square displacement of diffusing particles scales as a fractional order power law in time. In recent years a great deal of progress has been made in extending the different models for diffusion to incorporate this fractional diffusion. The tools of fractional calculus have proven very useful in these developments, linking together fractional constitutive laws, continuous time random walks, fractional Langevin equations and fractional Brownian motions. These notes provide a tutorial style overview of standard and fractional diffusion processes.

Contents

2.1. Mathematical Models for Diffusion

2.1.1. *Brownian motion and the Langevin equation*

> *Having found motion in the particles of the pollen of all the living plants which I had examined, I was led next to inquire whether this property continued after the death of the plant, and for what length of time it was retained.*
> *Robert Brown (1828)*[1]

When microscopic particles are suspended in a fluid they appear to vibrate around randomly. This phenomenon was investigated systematically by Robert Brown in 1827[1] after he observed the behaviour in pollen grains suspended in water and viewed under a microscope. Brown's interest at the time was concerned with the mechanisms of fertilization in flowering plants. Brown noticed that the pollen grains were in a continual motion that could not be accounted for by currents in the fluid. One possibility favoured by other scientists at the time was that this motion was evidence of life itself, but Brown observed similar motion in pollen grains that had been denatured in alcohol and in other non-living material (including *"molecules in the sand tubes, formed by lightning"*[1]) .

The explanation for Brownian motion that is generally accepted among scientists today was first put forward by Einstein in 1905.[2] The motion of the suspended particle (which, for simplicity, was considered in one spatial direction) arises as a consequence of random buffeting from the thermal motions of the enormous numbers of molecules that comprise the fluid. This buffeting provides both the driving forces and the damping forces (the effective viscosity of the fluid) that are experienced by the suspended particle. The central result of Einstein's theory is that in a given time t, the mean square displacement $r(t)$ of a suspended particle in a fluid is given by

$$\langle r^2(t)\rangle = 2Dt \tag{2.1}$$

where the angular brackets denote an ensemble average obtained by repeating the experiment many times and the constant

$$D = \left(\frac{RT}{6N\pi a\eta}\right) = \left(\frac{k_B T}{\gamma}\right). \tag{2.2}$$

Here T is the temperature of the fluid, $R = Nk_B$ is the universal gas constant, a is the radius of the suspended particle, η is the fluid viscosity, N is Avogadro's number (the number of molecules in an amount of mass equal to the atomic weight in grams) and

$$\gamma = 6\pi\eta a \tag{2.3}$$

is Stokes' relation for the viscous drag coefficient. The results in Eqs. (2.1), (2.2) are known as the Einstein relations. In an interesting footnote to this literature, the Einstein relation in Eq. (2.2), was also derived independently by Sutherland.[3]

A very simple derivation (*infinitely more simple*[4]) of the Einstein relations for motion in one spatial dimension was provided by Langevin a few years later[4] based on Newton's second law applied to a spherical particle in a fluid. The mass times the acceleration is the sum of the random driving force and the frictional viscous force, both arising from the thermal motions of the molecules of the fluid:

$$m\frac{d^2x}{dt^2} = F(t) - \gamma\frac{dx}{dt}. \tag{2.4}$$

The random driving force is assumed to have zero mean, $\langle F(t)\rangle = 0$, and to be uncorrelated with position, $\langle xF(t)\rangle = \langle x\rangle\langle F(t)\rangle = 0$. Equation (2.4) can be simplified by multiplying by $x(t)$, re-writing the left hand side as

$$mx\frac{d^2x}{dt^2} = m\frac{d}{dt}\left(x\frac{dx}{dt}\right) - m\left(\frac{dx}{dt}\right)^2 \tag{2.5}$$

and then taking the ensemble average. This results in

$$m\left\langle\frac{d}{dt}\left(x\frac{dx}{dt}\right)\right\rangle - m\left\langle\left(\frac{dx}{dt}\right)^2\right\rangle = -\gamma\left\langle x\frac{dx}{dt}\right\rangle. \tag{2.6}$$

A further simplification can be made using Boltzmann's Principle of Equipartition of Energy[5] which asserts that the average kinetic energy of each particle in the fluid is proportional to the temperature of the fluid; independent of the mass of the particle. The suspended particle being much larger in mass than the molecules of the fluid will have much smaller velocity according to this result. Applying this principle to the suspended particle we have

$$\frac{1}{2}m\left\langle\left(\frac{dx}{dt}\right)^2\right\rangle = \frac{1}{2}k_BT \tag{2.7}$$

and Eq. (2.6) can be rearranged as

$$\frac{dy}{dt} + \frac{\gamma}{m}y = \frac{k_BT}{m} \tag{2.8}$$

where

$$y = \left\langle x\frac{dx}{dt}\right\rangle. \tag{2.9}$$

Equation (2.8) is straightforward to integrate yielding

$$y = \frac{k_BT}{\gamma}\left(1 - \exp(-\frac{\gamma}{m}t)\right). \tag{2.10}$$

To proceed further we note that

$$y \equiv \left\langle x\frac{dx}{dt}\right\rangle = \frac{1}{2}\frac{d}{dt}\langle x^2\rangle,$$

so that

$$\frac{d}{dt}\langle x^2\rangle = \frac{2k_BT}{\gamma}\left(1 - \exp(-\frac{\gamma}{m}t)\right).$$

For a Brownian particle that is large relative to the average separation between particles in the fluid, Langevin notes that $(\gamma/m) \approx 10^8$ and so after observational times $t \gg 10^{-8}$ we have

$$\frac{d}{dt}\langle x^2\rangle \approx \frac{2k_BT}{\gamma}$$

and hence

$$\langle x^2\rangle \sim \left(\frac{2k_BT}{\gamma}\right)t = 2Dt \tag{2.11}$$

in agreement with the Einstein relations in Eqs. (2.1), (2.2). In three spatial dimensions there are three kinetic degrees of freedom and the mean square displacement is $\langle r^2\rangle \sim 6Dt$.

2.1.2. *Random walks and the central limit theorem*

The lesson of Lord Rayleigh's solution is that in open country the most probable place to find a drunken man who is at all capable of keeping on his feet is somewhere near his starting point!
Karl Pearson (1905)[6]

The Brownian motion of the suspended particle in a fluid can also be modelled as a random walk, a term first introduced by Pearson in 1905[6] who sought the probability that a random walker would be at a certain distance from their starting point after a given number of random steps. The problem was solved shortly after by Lord Rayleigh.[7] The idea of a random walk had been introduced earlier though by Bachelier in 1900 in his doctoral thesis (under the guidance of Poincaré) entitled *La Theorie de la Speculation*.[8] In this thesis Bachelier developed a mathematical theory for stock price movements as random walks, noting that ... *the consideration of true prices permits the statement of the fundamental principle – The mathematical expectation of the speculator is zero.*[8]

2.1.2.1. *Random walks and the binomial distribution*

In the simplest problem of a random walk along a line in one-dimension the particle starts from an origin and at each time step Δt the particle has an equal probability of jumping an equal distance Δx to the left or the right. The probability $P_{m,n}$ that the particle will be at position $x = m\Delta x$ at time $t = n\Delta t$ is governed by the recurrence equation:

$$P_{m,n} = \frac{1}{2}P_{m-1,n-1} + \frac{1}{2}P_{m+1,n-1}, \tag{2.12}$$

with $P_{0,0} = 1$. Note too that after any time k the sum of the probabilities must add up to unity and the largest possible excursion of the random walker after n time steps is to position $\pm n\Delta x$ so that

$$\sum_{j=-k}^{k} P_{j,k} = 1 \qquad \text{where} \qquad k = 0, 1, 2, \ldots n. \tag{2.13}$$

The recurrence equation, Eq. (2.12) is a partial difference equation and although a solution could be sought using the method of separation of variables this generally results in complicated algebraic expressions.

An alternate method is to enumerate the number of possible paths in an n step walk from 0 to m. Without loss of generality this occurs through

k steps to the right and $n-k=k-m$ steps to the left. The k steps to the right can occur anywhere among the n steps. There are

$$C(n,k)=\binom{n}{k}=\frac{n!}{k!(n-k)!}$$

ways of distributing these k steps among the n steps. There are 2^n possible paths in an n step walk so that the probability of an n step walk that starts at 0 and ends at m with k steps to the right is given by

$$p(m,n)=\frac{C(n,k)}{2^n} \quad \text{where} \quad k=\frac{n+m}{2}.$$

This simplifies to

$$p(m,n)=\frac{n!}{2^n\left(\frac{n+m}{2}\right)!\left(\frac{n-m}{2}\right)!}. \tag{2.14}$$

Note that we require $n+m$ and $n-m$ to be even which is consistent with the recognition that it is not possible to get from the origin to an even (odd) lattice site m in an odd (even) number of steps n.

The above result assumes an equal probability of steps to the left and right but it is easy to generalize with a probability r to step to the right and a probability $1-r$ to step to the left. The probability of k steps to the right in an n step walk in this case is

$$P(k)=\binom{n}{k}r^k(1-r)^{n-k}. \tag{2.15}$$

This is the probability mass function for the binomial distribution, i.e., if X is a random variable that follows the binomial distribution $B(n,k)$ then $P(k)=\text{Prob}(X=k)$. Note that in this biased random walk generalization we have

$$p(m,n)=\frac{n!}{\left(\frac{n+m}{2}\right)!\left(\frac{n-m}{2}\right)!}r^{\frac{n+m}{2}}(1-r)^{\frac{n-m}{2}}. \tag{2.16}$$

2.1.2.2. *Random walks and the normal distribution*

Most of the results in this article are concerned with long time behaviours. In the case of $p(m,n)$ we consider n large and $n>m$ but m^2/n nonvanishing. It is worthwhile considering the behaviour of $p(m,n)$ in this limit. Here we consider the simple case of the unbiased random walk, Eq. (2.14), but the analysis can readily be generalized.[10] The mean number of steps to the right is $\langle k \rangle = n/2$ and we consider the distribution for the fluctuations $X = k - \langle k \rangle = m/2$. We now have

$$p(m(X), n) = \frac{n!}{(\frac{n}{2} - X)!(\frac{n}{2} + X)!2^n} \tag{2.17}$$

which can be expanded using the De Moivre-Stirling approximation[9]

$$n! \approx \sqrt{2\pi n} n^n e^{-n} \tag{2.18}$$

to give

$$\begin{aligned} P(X,n) &= \frac{\sqrt{\frac{2}{n\pi}}}{\left(1 - \frac{2X}{n}\right)^{(\frac{n}{2}-X+\frac{1}{2})}\left(1 + \frac{2X}{n}\right)^{(\frac{n}{2}+X+\frac{1}{2})}} \\ &= \frac{\sqrt{\frac{2}{n\pi}}}{\exp\left[(\frac{n}{2} - X + \frac{1}{2})\ln(1 - \frac{2X}{n}) + (\frac{n}{2} + X + \frac{1}{2})\ln(1 + \frac{2X}{n})\right]}. \end{aligned}$$

The long time behaviour is now found after carrying out a series expansion of the log terms in powers of $2X/n$. The result is

$$P(X,n) \sim \sqrt{\frac{2}{n\pi}} e^{-\frac{2X^2}{n}} = \sqrt{\frac{2}{n\pi}} e^{-\frac{m^2}{2n}}. \tag{2.19}$$

Thus the probability density function for unbiased random walks in the long time limit is the Gaussian or normal distribution.

2.1.2.3. *Random walks in the continuum approximation*

Further insights into the random walk description can be found by employing a continuum approximation in the limit $\Delta x \to 0$ and $\Delta t \to 0$. In this approximation we first write $P(m,n) = P(x,t)$ and then re-write Eq. (2.12) as follows:

$$P(x,t) = \frac{1}{2}P(x - \Delta x, t - \Delta t) + \frac{1}{2}P(x + \Delta x, t - \Delta t). \tag{2.20}$$

Now expand the terms on the right hand side as Taylor series in x, t:

$$\begin{aligned} P(x \pm \Delta x, t - \Delta t) \approx P(x,t) \pm \Delta x \frac{\partial P}{\partial x} - \Delta t \frac{\partial P}{\partial t} + \frac{(\Delta x)^2}{2}\frac{\partial^2 P}{\partial x^2} + \frac{(\Delta t)^2}{2}\frac{\partial^2 P}{\partial t^2} \\ \pm \Delta t \Delta x \frac{\partial^2 P}{\partial x \partial t} + O((\Delta t)^3) \mp O((\Delta x)^3). \end{aligned}$$

If we substitute these expansions into Eq. (2.20) and retain only leading order terms in Δt and Δx, then after rearranging

$$\frac{\partial P}{\partial t} = D \frac{\partial^2 P}{\partial x^2} \tag{2.21}$$

where

$$D = \lim_{\Delta t \to 0, \Delta x \to 0} \frac{(\Delta x)^2}{2\Delta t} \tag{2.22}$$

is a constant with dimensions of $m^2 s^{-1}$. The above partial differential equation is known as the diffusion equation (see below).

The continuum approximation for the probability conservation law in Eq. (2.13) is

$$\int_{-\infty}^{+\infty} P(x,t)dx = 1 \tag{2.23}$$

where the limits to infinity are consistent with taking the spacing between steps $\Delta x \to 0$.

The fundamental Green's solution $G(x,t)$ of the diffusion equation with initial condition $G(x,0) = \delta(x)$ can readily be obtained using classical methods. The Fourier transform of the diffusion equation yields

$$\frac{d\hat{G}(q,t)}{dt} = -Dq^2\hat{G}(q,t)$$

with solution

$$\hat{G}(q,t) = e^{-Dq^2 t}. \tag{2.24}$$

where we have used the result that $\hat{G}(q,0) = \hat{\delta}(q) = 1$. The inverse Fourier transform now results in

$$G(x,t) = \frac{1}{2\pi}\int_{-\infty}^{+\infty} e^{-Dq^2 t + iqx}\, dq = \frac{1}{2\pi} e^{-\frac{x^2}{4Dt}} \int_{-\infty}^{+\infty} e^{-Dt(q - \frac{ix}{2Dt})^2}\, dq$$

which simplifies to

$$G(x,t) = \frac{1}{\sqrt{4\pi Dt}} e^{-\frac{x^2}{4Dt}}. \tag{2.25}$$

The mean square displacement can be evaluated directly from

$$\langle x^2 \rangle = \int_{-\infty}^{+\infty} x^2 G(x,t)\, dx \tag{2.26}$$

or indirectly from

$$\langle x^2 \rangle = \lim_{q \to 0} -\frac{d^2}{dq^2}\hat{G}(q,t) \tag{2.27}$$

yielding the familiar result $\langle x^2 \rangle = 2Dt$.

2.1.2.4. *Central limit theorem*

The fundamental solution, Eq. (2.25), is an example of the Gaussian normal distribution

$$P(X \in dz) = \frac{1}{\sqrt{2\pi\sigma^2}} \exp\left(-\frac{(z-\mu)^2}{2\sigma^2}\right) \tag{2.28}$$

for random variables X with mean

$$\mu = \langle X \rangle \tag{2.29}$$

and variance

$$\sigma^2 = \langle X^2 \rangle - \langle X \rangle^2 = \langle X^2 \rangle - \mu^2. \tag{2.30}$$

The Gaussian probability distribution, Eq. (2.25), can be derived independently for random walks by appealing to the Central Limit Theorem (CLT): The sum of N independent and identically distributed random variables with mean μ and variance σ^2 is approximately a Gaussian probability density function with mean $N\mu$ and variance $N\sigma^2$. In the case of random walks, each step Δx is a random variable with mean $\mu = \langle \Delta x \rangle = 0$ and variance $\sigma^2 = \langle \Delta x^2 \rangle - \langle \Delta x \rangle^2 = \langle \Delta x^2 \rangle$. The sum of N such random variables is x so that from the CLT we have

$$P(x \in dz) = \frac{1}{\sqrt{2\pi N \langle \Delta x^2 \rangle}} \exp\left(-\frac{z^2}{2N\langle \Delta x^2 \rangle}\right). \tag{2.31}$$

But $\langle \Delta x^2 \rangle = 2D\langle \Delta t \rangle$ and $N = t/\Delta t$ so that we recover Eq. (2.25) for $X = x$.

This treatment of random walks using the CLT can be applied even if the step length Δx varies between jumps, provided that the step lengths $\Delta x(t)$ are independent identically distributed random variables, i.e.,

$$\langle \Delta x_i \Delta x_j \rangle = \delta_{i,j} \langle \Delta x_i \rangle^2.$$

In an N step walk with jumps at discrete times $t_i = (i-1)\Delta t$ we can define the average drift over time $t = N\Delta t$ as

$$\langle x(t) \rangle = \sum_{i=1}^{N} \langle \Delta x_i \rangle$$

and an average drift velocity as

$$v = \frac{\langle x(t) \rangle}{\Delta t}.$$

The variance of the random walk is

$$\langle x(t)^2\rangle - \langle x(t)\rangle^2 = \sum_{i=1}^{N}\sum_{j=1}^{N}\left(\langle \Delta x_i \Delta x_j\rangle - \langle \Delta x_i\rangle\langle \Delta x_j\rangle\right).$$

Since the walks are uncorrelated this simplifies to

$$\langle x(t)^2\rangle - \langle x(t)\rangle^2 = N\left(\langle \Delta x^2\rangle - \langle \Delta x\rangle^2\right) = N\sigma^2 = 2Dt. \tag{2.32}$$

Note that if the walk is biased then the drift velocity is non-zero and the probability density function is the solution of an advective-diffusion equation (see below).

2.1.3. *Fick's law and the diffusion equation*

Equation (2.25) governing the probability of a random walker at position x after time t is the probability distribution that should result if we measured the positions of a large number of particles in many separate experiments. However if the particles did not interact then we could perform measurements of their positions in the one experiment. The number of particles per unit volume at position x and time t is the concentration $c(x, t)$. If there are N non-interacting walkers in total then they all have the same probability of being at x at time t and hence the concentration $c(x, t) = NP(x, t)$ also satisfies the diffusion equation

$$\frac{\partial c}{\partial t} = D\frac{\partial^2 c}{\partial x^2}. \tag{2.33}$$

We now consider a macroscopic derivation of the diffusion equation based on the conservation of matter and an empirical result known as Fick's Law. The derivation is given in one spatial dimension for simplicity (this may describe diffusion in a three dimensional domain but with one-dimensional flow). In addition to the concentration $c(x, t)$, other macroscopic quantities of interest are the mean velocity $\mathbf{u}(x, t)$ of diffusing particles, and the flux $\mathbf{q}(x, t)$, which in one spatial dimension is the number of particles per unit time that pass through a test area perpendicular to the flow in the positive x direction. The three macroscopic quantities are related through the equation

$$\mathbf{q}(x, t) = c(x, t)\mathbf{u}(x, t). \tag{2.34}$$

Note that while the concentration is a scalar quantity both the mean velocity and the flux are vectors.

If no particles are added or removed from the system then, considering a small test volume V of uniform cross-sectional area A and extension δx we have conservation of matter,

$$\begin{bmatrix}\text{number of particles}\\ \text{in volume } V\\ \text{at time } t+\delta t\end{bmatrix} = \begin{bmatrix}\text{number of particles}\\ \text{in volume } V\\ \text{at time } t\end{bmatrix} + \begin{bmatrix}\text{net number of particles}\\ \text{entering volume } V\\ \text{between } t \text{ and } t+\delta t\end{bmatrix}$$

so that

$$c(x,t+\delta t)A\delta x = c(x,t)A\delta x + q(x,t)A\delta t - q(x+\delta x,t)A\delta t. \tag{2.35}$$

Now divide by $A\delta t\delta x$ and re-arrange terms then

$$\frac{c(x,t+\delta t)-c(x,t)}{\delta t} = -\frac{q(x+\delta x)-q(x,t)}{\delta x} \tag{2.36}$$

and in the limit $\delta t \to 0, \delta x \to 0$,

$$\frac{\partial c}{\partial t} = -\frac{\partial q}{\partial x}. \tag{2.37}$$

The equation of conservation of matter (2.37) for flow in one spatial dimension is also called the *continuity equation.*

Fick's Law[11] asserts that the net flow of diffusing particles is from regions of high concentration to regions of low concentration and the magnitude of this flow is proportional to the concentration gradient. Thus

$$q(x,t) = -D\frac{\partial c}{\partial x}, \tag{2.38}$$

in analogy with Fourier's Law of heat conduction and Ohm's Law for ionic conduction. The minus sign expresses the result that if the concentration is increasing in the x direction (i.e., $\partial c/\partial x > 0$) then the flow of particles is in the negative x direction. The constant of proportionality is the diffusion coefficient. If we combine the equation of conservation of matter, Eq. (2.37), with Fick's Law, Eq. (2.38), then we obtain

$$\frac{\partial c}{\partial t} = \frac{\partial}{\partial x}\left(D\frac{\partial c}{\partial x}\right) = D\frac{\partial^2 c}{\partial x^2} \quad \text{for constant} \quad D. \tag{2.39}$$

One of the most significant aspects of Einstein's results for Brownian motion is that the diffusivity can be related to macroscopic physical properties of the fluid and the particle, as in Eq. (2.2).

2.1.3.1. *Generalized diffusion equations*

The macroscopic diffusion equation is easy to generalize to higher dimensions and other co-ordinate systems. Examples are the diffusion equation in radially symmetric co-ordinates in d dimensional space

$$\frac{\partial c}{\partial t} = \frac{1}{r^{d-1}} \frac{\partial}{\partial r} \left(r^{d-1} D \frac{\partial c}{\partial r} \right). \tag{2.40}$$

Other generalizations of the macroscopic diffusion equation are possible by modifying Fick's law. If the media is spatially heterogeneous then an *ad-hoc* generalization would be to replace the diffusion constant in Fick's law with a space dependent function, ie., $D = D(x)$. Concentration dependent diffusivities and time dependent diffusivities have also been considered.

If the diffusing species are immersed in a fluid that is moving with velocity $\mathbf{v}(x,t)$ then this will produce an *advective flux*

$$\mathbf{q}_A(x,t) = c(x,t)\mathbf{v}(x,t), \tag{2.41}$$

which, when combined with the Fickian flux, Eq. (2.38) and the continuity equation, Eq. (2.37), results in the advective-diffusion equation,

$$\frac{\partial c}{\partial t} = D \frac{\partial^2 c}{\partial x^2} - \frac{\partial}{\partial x} (v\, c). \tag{2.42}$$

A possible generalization of the above equation for spatially inhomogeneous systems is then

$$\frac{\partial c}{\partial t} = \frac{\partial}{\partial x} \left(D(x) \frac{\partial c}{\partial x} \right) - \frac{\partial}{\partial x} (v(x)\, c). \tag{2.43}$$

If there are chemicals that attract or repel the diffusing species there will be a chemotactic flux. The term (chemo) *taxis* means directed motion towards or away from an external (chemical) stimulus. The chemotactic flux is modelled by assuming that species move in the direction of a chemical gradient, thus

$$\mathbf{q}_C(x,t) = \chi c(x,t) \frac{\partial}{\partial x} u(x,t), \tag{2.44}$$

where $u(x,t)$ is the concentration of the chemical species that is driving the chemotactic flux. This flux term is positive if the associated flow is from regions of low concentration to high concentration (chemoattractant, $\chi > 0$) and negative otherwise (chemorepellant, $\chi < 0$). The chemotactic diffusion equation in one dimension is

$$\frac{\partial c}{\partial t} = D \frac{\partial^2 c}{\partial x^2} - \chi \frac{\partial}{\partial x} \left(c \frac{\partial u}{\partial x} \right). \tag{2.45}$$

2.1.4. *Master equations and the Fokker–Planck equation*

In his classic 1905 paper on Brownian motion, Einstein[2] derived the diffusion equation from an integral equation conservation law or master equation. The master equation describes the evolution of the probability density function $P(x,t)$ for a random walker taking jumps at discrete time intervals Δt to be at position x at time t. We let $\lambda(\Delta x)$ denote the probability density function for a jump of length Δx then

$$P(x,t) = \int_{-\infty}^{+\infty} \lambda(\Delta x) P(x - \Delta x, t - \Delta t)\, d\Delta x \tag{2.46}$$

expresses the conservation law that the probability for a walker to be at x at time t is the probability that the walker was at position $x - \Delta x$ at an earlier time $t - \Delta t$ and then the walker jumped with step length Δx. The integral sums over all possible starting points at the earlier time. The correspondence between the master equation and the diffusion equation (or a more general Fokker–Planck equation) can be found by considering continuum approximations in the limit $\Delta t \to 0$ and $\Delta x \to 0$, thus

$$P|_{(x,t-\Delta t)} + \Delta t \frac{\partial P}{\partial t}\bigg|_{(x,t-\Delta t)} \approx \int_{-\infty}^{+\infty} \lambda(\Delta x) \left(P|_{(x,t-\Delta t)} - \Delta x \frac{\partial P}{\partial x}\bigg|_{(x,t-\Delta t)} + \frac{(\Delta x)^2}{2} \frac{\partial^2 P}{\partial x^2}\bigg|_{(x,t-\Delta t)} \right) d\Delta x.$$

The integral over Δx is simplified by noting that

$$\int_{-\infty}^{+\infty} (\Delta x)^n \, \lambda(\Delta x)\, d\Delta x = \langle \Delta x^n \rangle, \quad n \in \mathbb{N}.$$

Thus in the continuum limit the master equation yields the Fokker–Planck equation (also called the Kolmogorov forward equation)

$$\frac{\partial P}{\partial t} \approx \frac{\langle \Delta x^2 \rangle}{2\Delta t} \frac{\partial^2 P}{\partial x^2} - \frac{\langle \Delta x \rangle}{\Delta t} \frac{\partial P}{\partial x}, \tag{2.47}$$

with drift velocity

$$v = \lim_{\Delta x \to 0, \Delta t \to 0} \frac{\langle \Delta x \rangle}{\Delta t}, \tag{2.48}$$

and diffusion coefficient

$$D = \lim_{\Delta x \to 0, \Delta t \to 0} \frac{\langle \Delta x^2 \rangle - \langle \Delta x \rangle^2}{2\Delta t} = \lim_{\Delta x \to 0, \Delta t \to 0} \frac{\langle \Delta x^2 \rangle}{2\Delta t} + O\left(\frac{\langle \Delta x \rangle^2}{\Delta t} \right) \tag{2.49}$$

so that

$$\frac{\partial P}{\partial t} \approx D\frac{\partial^2 P}{\partial x^2} - v\frac{\partial P}{\partial x}. \tag{2.50}$$

2.1.4.1. *Generalized Fokker–Planck equation*

If the step length probability density function is also dependent on position then the master equation generalizes to

$$P(x,t) = \int_{-\infty}^{+\infty} \lambda(\Delta x, x - \Delta x)P(x - \Delta x, t - \Delta t)\, d\Delta x. \tag{2.51}$$

In the continuum limit we proceed as above but with the additional expansion

$$\lambda(\Delta x, x - \Delta x) \approx \lambda|_{(\Delta x,x)} \quad \Delta x \left.\frac{\partial \lambda}{\partial x}\right|_{(\Delta x,x)} + \frac{\Delta x^2}{2}\left.\frac{\partial^2 \lambda}{\partial x^2}\right|_{(\Delta x,x)}, \tag{2.52}$$

and

$$\int_{-\infty}^{+\infty} \Delta x^n\, \lambda(\Delta x, x)\, d\Delta x = \langle \Delta x^n(x)\rangle,$$

which leads to the general Fokker–Planck equation

$$\frac{\partial P}{\partial t} = \frac{\partial^2}{\partial x^2}\left(D(x)P(x,t)\right) - \frac{\partial}{\partial x}\left(v(x)P(x,t)\right), \tag{2.53}$$

with drift

$$v(x) = \lim_{\Delta x\to 0, \Delta t\to 0} \frac{\langle \Delta x(x)\rangle}{\Delta t}, \tag{2.54}$$

and diffusivity

$$D(x) = \lim_{\Delta x\to 0, \Delta t\to 0} \frac{\langle \Delta x^2(x)\rangle - \langle \Delta x(x)\rangle^2}{2\Delta t} = \lim_{\Delta x\to 0, \Delta t\to 0} \frac{\langle \Delta x^2(x)\rangle}{2\Delta t}.$$

The generalized Fokker–Planck equation, Eq. (2.53) is slightly different to the generalized Fickian equation, Eq. (2.43) (see further comments in Vlahos *et al*[12] and references therein).

2.1.5. *The Chapman–Kolmogorov equation and Markov processes*

The sequence of jumps $\{X_t\}$ in a random walk defines a stochastic process. A realization of this stochastic process defines a trajectory $x(t)$. A stochastic process has the Markov property if at any time t the distribution of all X_u, $u > t$ only depends on the value X_t and not on any value

X_s, $s < t$. Let $p(x,t)$ denote the probability density function for X_t and let $q(x,t|x',t')$ denote the conditional probability that X_t lies in the interval $x, x+dx$ given that $X_{t'}$ starts at x'. A first order Markov process has the property that

$$q(x,t|x'',t'') = \int q(x,t|x',t')q(x',t'|x'',t'')\,dx'. \tag{2.55}$$

This equation, which was introduced by Bachelier in his PhD thesis,[8] is commonly referred to as the Chapman–Kolmogorov equation, in recognition of the more general equation derived independently by Chapman and Kolmogorov for probability density functions in stochastic processes. We will refer to the special case as the Bachelier equation. Note that if we multiply the Bachelier equation by $p(x'',t'')$ and integrate over x'' then

$$\begin{aligned}\int p(x'',t'')q(x,t|x'',t'')\,dx'' &= \int p(x'',t'')\left(\int q(x,t|x',t')q(x',t'|x'',t'')\,dx'\right)dx'' \\ &= \int \left(\int p(x'',t'')q(x',t'|x'',t'')\,dx''\right) q(x,t|x',t')\,dx'\end{aligned}$$

so that

$$p(x',t') = \int p(x'',t'')q(x',t'|x'',t'')\,dx''$$

or equivalently

$$p(x,t) = \int p(x',t')q(x,t|x',t')\,dx'. \tag{2.56}$$

Note that in the above we consider the times $t > t' > t''$ to be discrete times. There are many different examples of Markov processes that satisfy Eq. (2.55) and Eq. (2.56).

2.1.5.1. *Wiener process*

It is easy to confirm by substitution that the conditional probability

$$q(x,t|x',t') = \frac{1}{\sqrt{2\pi(t-t')}}\exp\left(-\frac{(x-x')^2}{2(t-t')}\right), \quad t > t' \tag{2.57}$$

is a solution of the Bachelier equation and

$$p(x,t) = \frac{1}{\sqrt{2\pi t}}\exp\left(-\frac{x^2}{2t}\right) \tag{2.58}$$

is a solution of Eq. (2.56) with this conditional probability. The corresponding Markov process is referred to as the Wiener process or Brownian motion. It is the limiting behaviour of a random walk in the limit where the time increment approaches zero. That this limit exists was proven by Norbert Wiener in 1923.[13]

The Brownian motion stochastic process B_t satisfies the following properties:
(i) $B_0 = 0$ and B_t is defined for times $t \geq 0$.
(ii) Realizations $x_B(t)$ of the process are continuous but nowhere differentiable. The graph of $x_B(t)$ versus t is a fractal with fractal dimension $d = 3/2$.
(iii) The increments $B_t - B_{t'}$ are normally distributed random variables with mean 0 and variance $t - t'$ for $t > t'$.
(iv) The increments $B_t - B_{t'}$ and $B_s - B_{s'}$ are independent random variables for $t > t' \geq s \geq s' \geq 0$.

2.1.5.2. *Poisson process*

Another important Markov process is the Poisson process. Here the spatial variable is replaced with a discrete variable labelling the occurrence of events (e.g., the numbers of encounters with injured animals on a road trip). The defining equations are

$$q(n, t|n', t') = \frac{(\alpha(t - t'))^{n-n'}}{(n - n')!} e^{-\alpha(t-t')}, \quad t > t' \tag{2.59}$$

and

$$p(n, t) = \frac{(\alpha t)^n}{n!} e^{-\alpha t}, \tag{2.60}$$

where α is called the intensity of the process. The latter equation is interpreted as the probability that n events have occurred in the interval $[0, t]$ and αt is the expected number of events in this interval. For example in an n step random walk the expected number of steps to the right in time t is $np = (t/\Delta t)p$ where p is the probability to step to the right and Δt is the time interval between steps. From the Poisson distribution the probability of k steps to the right is

$$p(k, t) = \frac{1}{k!} \left(\frac{p}{\Delta t} t\right)^k \exp\left(-\frac{p}{\Delta t} t\right). \tag{2.61}$$

This can be reconciled with the Binomial distribution Eq. (2.15) by considering the limit $n \to \infty$ but np and k finite. Note that this requires that the probability p of a step to the right must be very small, $p \to 0$, and the Poisson distribution is thus the distribution law for rare events.

2.2. Fractional diffusion

In the theory of Brownian motion the first concern has always been the calculation of the mean square displacement of the particle, because this could be immediately observed.
George Uhlenbeck and Leonard Ornstein (1930)[14]

Central results in Einstein's theory of Brownian motion are that the mean square displacement of the Brownian particle scales linearly with time and the probability density function for Brownian motion is the Gaussian normal distribution. These characteristic signatures of standard diffusion are consistent across many different mathematical descriptions; random walks, central limit theorem, Langevin equation, master equation, diffusion equation, Wiener processes. The results have also been verified in numerous experiments including Perrin's measurements of mean square displacements[15] to determine Avogadro's number (the constant number of molecules in any mole of substance) thus consolidating the atomistic description of nature.

Despite the ubiquity of standard diffusion it is not universal. There have been numerous experimental measurements of fractional diffusion in which the mean square displacement scales as a fractional power law in time (see Table 2.1). The fractional diffusion is referred to as subdiffusion if the fractional power is less than unity and superdiffusion if the fractional power is greater than unity. Fractional diffusion has been the subject of several highly cited reviews,[16–18] and pedagogic lecture notes,[12,19] in recent decades. Fractional diffusion has been found to occur as the norm in; spatially disordered systems (such as porous media and fractal media), in turbulent fluids and plasmas, and in biological media with traps, binding sites or macro-molecular crowding.

In the remainder of these notes we describe theoretical frameworks based around the physics of continuous time random walks and the mathematics of fractional calculus to model fractional diffusion. For a more complete description the reader should again consult the review articles and references therein.

Table 2.1. Summary table of scaling laws for fractional diffusion.

scaling	diffusion process	environment
$\langle \Delta X^2 \rangle \sim t\,(\ln t)^\kappa$ $1 < \kappa < 4$	ultraslow diffusion	Sinai diffusion deterministic diffusion
$\langle \Delta X^2 \rangle \sim t^\alpha$ $0 < \alpha < 1$	subdiffusion	disordered solids biological media fractal media porous media
$\langle \Delta X^2 \rangle \sim \begin{cases} t^\alpha & t < \tau \\ t & t > \tau \end{cases}$ $0 < \alpha < 1$	transient subdiffusion	biological media
$\langle \Delta X^2 \rangle \sim t$	standard diffusion	homogeneous media
$\langle \Delta X^2 \rangle \sim t^\beta$ $1 < \beta < 2$	superdiffusion	turbulent plasmas transport in polymers Lévy flights
$\langle \Delta X^2 \rangle \sim t^3$	Richardson diffusion[21]	atmospheric turbulence

2.2.1. *Diffusion on fractals*

Experimental simulations and theoretical results have shown that diffusion on self-similar fractal lattices with fractal dimension d_f is anomalous subdiffusion with scaling[20]

$$\langle r^2 \rangle \sim t^{2/d_w}, \quad d_w > 2. \tag{2.62}$$

The scaling exponent d_w is referred to as the dimension of the walk. For standard random walks on Euclidean lattices in $d = 2$ the dimension of the walk is also $d_w = 2$. Equation (2.62) can be re-written as

$$\langle r^2 \rangle \sim D(r)t \tag{2.63}$$

where the diffusion constant is replaced by the space dependent diffusion coefficent $D(r)$. The results in Eq. (2.62) and Eq. (2.63) are consistent provided that

$$D(r) = r^{2-d_w}. \tag{2.64}$$

If we now reconsider the radially symmetric diffusion equation, Eq. (2.40), but replace the space dimension d with the fractal dimension d_f and replace the diffusion constant D with the spatially varying diffusion coefficient $D(r)$

then we arrive at the O'Shaugnessy–Procaccia fractional diffusion equation[22]

$$\frac{\partial c}{\partial t} = \frac{1}{r^{d_f-1}} \frac{\partial}{\partial r}\left(r^{d_f-1} r^{2-d_w} \frac{\partial c}{\partial r}\right). \tag{2.65}$$

2.2.2. *Fractional Brownian motion*

One of the easiest ways to model anomalous subdiffusion is to replace the constant diffusivity with a time dependent diffusivity $D(t) = \alpha t^{\alpha-1} D$. The evolution equation for the concentration in this case is given by

$$\frac{\partial c}{\partial t} = \alpha t^{\alpha-1} D \frac{\partial^2 c}{\partial x^2}. \tag{2.66}$$

The solution is a Gaussian distribution

$$c(x,t) = \frac{1}{\sqrt{4\pi D t^\alpha}} \exp\left(-\frac{x^2}{4Dt^\alpha}\right). \tag{2.67}$$

The probability density function for this stochastic process is non-Markovian due to the power law diffusivity. The mean square displacement is given by

$$\langle x^2 \rangle = 2Dt^\alpha = 2Dt^{2H} \tag{2.68}$$

where H is the Hurst exponent. The fractional diffusion equation in Eq. (2.66) describes the probability density function for fractional Brownian motion.[23] As an aside it is interesting to note that the power law diffusivity may be expressed as a fractional derivative of a constant,

$$D(t) = {}_0\mathcal{D}_t^{1-\alpha}\left(\Gamma(\alpha) D\right), \tag{2.69}$$

where ${}_0\mathcal{D}_t^{1-\alpha}$ denotes a Riemann–Liouville derivative of order $1-\alpha$ [see further details in the Appendix, Eq. (2.166)].

Starting with Mandelbrot and Van Ness[24] there is a vast literature on fractional Brownian motion as a stochastic process. If we let $B^H(t)$ denote a fractional Brownian motion stochastic process with Hurst exponent $H \in [0,1]$ then three properties of particular note are:

(i) Correlations

$$E(B^H(t)B^H(s)) = \frac{1}{2}\left(|t|^{2H} + |s|^{2H} - |t-s|^{2H}\right),$$

(ii) Self similarity

$$B^H(at) \sim |a|^H B^H(t),$$

(iii) Realizations $x_B(t)$ of the process are continuous but nowhere differentiable. The graph of $x_B(t)$ versus t is a fractal with fractal dimension $d = 2 - H$.

Fractional Brownian motion can also be described by an evolution equation of the form

$$x(t) = x_0 + {}_0\mathcal{D}_t^{-\alpha} F(t). \tag{2.70}$$

In this equation $x(t)$ denotes the position of a random walker at time t given that it started at x_0 and $F(t)$ is Gaussian white noise with autocorrelation $\langle F(t)F(s)\rangle = \delta(t-s)$. The evolution equation, Eq. (2.70), defines fractional Brownian motion $x(t) - x_0$ as a fractional integral [see Appendix Eq. (2.158)], of order α, of white noise; and standard Brownian motion as an ordinary integral of white noise.

Fractional Brownian motion can also be derived from a microscopic fractional Langevin equation[25,26]

$$m\frac{dv}{dt} = F^H(t) - m\int_0^t \gamma(t-t')v(t')\,dt' \tag{2.71}$$

where $F^H(t)$ denotes coloured noise with vanishing mean and correlation related to the dissipative memory kernel $\gamma(t)$ through a fluctuation-dissipation theorem[25]

$$\langle F^H(t)F^H(0)\rangle = mk_bT\gamma(t). \tag{2.72}$$

In the particular case $\gamma(t) = (D_\alpha/mk_BT)t^{-\alpha}$ the fractional Langevin equation

$$m\frac{dv}{dt} = F^H(t) - m\,{}_0\mathcal{D}_t^{\alpha-1}v(t) \tag{2.73}$$

describes subdiffusion for $0 < \alpha < 1$. The probability density function for trajectories that satisfy the fractional Langevin equation has been shown to be[23,27] the fractional Brownian motion diffusion equation, Eq. (2.66).

The fractional integral in Eq.(2.73) is a power law weighted average of the velocity over its entire previous history. This aspect of the dynamics is referred to as temporal memory and it is related to the non-Markovian property. The mathematics of fractional calculus has a long history dating back to Leibniz (1965) but it has only been in recent decades that fractional calculus has permeated mainstream physics literature. The recent interest in fractional calculus in physics is largely due to the relevance of fractional calculus for the physical problem of anomalous diffusion. The keen student would be well advised to acquaint themselves with some of the general mathematical results on fractional calculus in the Appendix before proceeding with the remainder of these notes on fractional diffusion.

2.2.3. *Continuous time random walks and power laws*

It was the man from Ironbark who struck the Sydney town, he wandered over street and park, he wandered up and down. He loitered here, he loitered there ...
A.B "Banjo" Paterson
The Bulletin, 17 December 1892.

2.2.3.1. *CTRW master equations*

In the standard random walk the step length is a fixed distance Δx and the steps occur at discrete times separated by a fixed time interval Δt. A more general random walk can be obtained by choosing a waiting time from a waiting time probability density before each step and then choosing the step length from a step length probability density. These more general walks are Continuous Time Random Walks (CTRWs). They were introduced by Montroll and Weiss in 1965[28] (see also Scher and Lax[29] and Montroll and Shlesinger[30]).

The fundamental quantity to calculate is the conditional probability density $p(x,t|x_0,t_0)$ that a walker starting from position x_0 at time $t=0$, *is at* position x at time t. The conditional probability density $q_n(x,t|x_0,t_0)$ that after n steps a walker starting at x_0 at time $t=0$ *arrives at* position x at time t satisfies the recursion relation

$$q_{n+1}(x,t|x_0,0) = \int_{-\infty}^{+\infty} \left(\int_0^t \Psi(x-x',t-t') q_n(x',t'|x_0,0)\, dt' \right) dx' \tag{2.74}$$

where $\Psi(x-x',t-t')$ is the probability density that in a single step a random walker steps a distance $x-x'$ after waiting a time $t-t'$. This arrival density q satisfies the initial condition

$$q_0(x,t|x_0,0) = \delta_{x,x_0}\delta(t)$$

and the normalization

$$\int_{-\infty}^{+\infty} \int_0^{\infty} q_0(x',t'|x_0,0)\, dt'\, dx' = 1.$$

The conditional probability density that a walker arrives at position x at time t after any number of steps is given by

$$q(x,t|x_0,0) = \sum_{n=0}^{\infty} q_n(x,t|x_0,0). \tag{2.75}$$

After summing over n, and using the initial condition, in the recursion equation, Eq. (2.74), we can write[29]

$$q(x,t|x_0,0) = \int_{-\infty}^{+\infty}\int_0^t \Psi(x',t')q(x-x',t-t'|x_0,0)\,dt'\,dx' + \delta(t)\delta_{x,x_0}. \tag{2.76}$$

In the theory of CTRWs it is assumed that waiting times are independent and identically distributed random variables with density $\psi(t)$, $t>0$ and step lengths are independent and identically distributed random variables with density $\lambda(x)$, $x\in\mathbb{R}$. It is further assumed that the waiting times and step lengths are independent of each other so that

$$\Psi(x-x',t-t') = \lambda(x-x')\psi(t-t'). \tag{2.77}$$

It follows from the normalization of the probability density functions that

$$\psi(t) = \int_{-\infty}^{+\infty} \Psi(x',t)\,dx' \tag{2.78}$$

and

$$\lambda(x) = \int_0^{\infty} \Psi(x,t')\,dt'. \tag{2.79}$$

It is also useful to define the survival probability

$$\Phi(t) = 1 - \int_0^t \psi(t')\,dt' = \int_t^{\infty} \psi(t')\,dt' \tag{2.80}$$

which is the probability that the walker does not step during time interval t (i.e., the waiting time is greater than t).

The conditional probability density that a walker starting from the origin at time zero is at position x at time t is now given by[28]

$$p(x,t|x_0,0) = \int_0^t q(x,t-t'|x_0,0)\Phi(t')\,dt'. \tag{2.81}$$

The right hand side considers all walkers that arrived at x at an earlier time t' and thereafter did not step.

The results for the conditional probability densities in Eq. (2.76) and Eq. (2.81) can be combined to yield the fundamental master equation for CTRWs,

$$p(x,t|x_0,0) = \Phi(t)\delta_{x,x_0} + \int_0^t \psi(t-t') \int_{-\infty}^{+\infty} \lambda(x-x')p(x',t'|x_0,0)\,dx'\,dt'. \tag{2.82}$$

The master equation can be justified using temporal Laplace transforms. The Laplace transform of Eq. (2.76) yields

$$\hat{q}(x,u|x_0,0) = \int_{-\infty}^{+\infty} \hat{\Psi}(x',u)\hat{q}(x-x',u|x_0,0)\,dx' + \delta_{x,x_0}.$$

The Laplace transform of Eq. (2.81) now yields

$$\begin{aligned}\hat{p}(x,u|x_0,0) &= \hat{q}(x,u|x_0,0)\hat{\Phi}(u)\\ &= \int_{-\infty}^{+\infty} \hat{\Psi}(x',u)\hat{\Phi}(u)\hat{q}(x-x',u|x_0,0)\,dx' + \hat{\Phi}(u)\delta_{x,x_0}\\ &= \int_{-\infty}^{+\infty} \hat{\Psi}(x',u)\hat{p}(x-x',u|x_0,0)\,dx' + \hat{\Phi}(u)\delta_{x,x_0}.\end{aligned}$$

The master equation, Eq. (2.82), is the inverse Laplace transform of the above equation. The master equation can also be justified using probability arguments. The first term represents the persistence of the walker at the initial position and the second term considers walkers that were at other positions x' at time t' but then stepped to x at time t after waiting a time $t-t'$.

In the original formulation of the master equation, the steps were assumed to take place on a discrete lattice, so that

$$p(x,t|x_0,0) = \Phi(t)\delta_{x,x_0} + \sum_{x'} \int_0^t \psi(t-t')\lambda(x-x')p(x',t'|x_0,0)\,dt'. \tag{2.83}$$

The CTRW can also be described using a generalized (gain-loss) master equation of the form[30]

$$\frac{dP(x,t)}{dt} = \int_0^t \sum_{x'} [K(x,x';t-t')P(x',t') - K(x',x;t-t')P(x,t')]\,dt'. \tag{2.84}$$

In this equation $P(x,t)$ is the probability for a walker to be at x at time t and $K(x,x';t-t')$ is the probability per unit time for a walker to make a transition from x to x' during time $t-t'$. The CTRW master equation can be shown to be equivalent to the generalized master equation if[30]

$$K(x, x'; t-t') = \lambda(x-x')\phi(t-t') \tag{2.85}$$

and

$$\hat{\phi}(u) = \frac{u\hat{\psi}(u)}{1-\hat{\psi}(u)}. \tag{2.86}$$

It is straightforward to extend the CTRW master equation by considering walkers starting from different starting points. The master equation for the expected concentration of walkers at position x and t is then[31]

$$n(x,t) = \Phi(t)n(x,0) + \int_{-\infty}^{+\infty}\int_0^t n(x',t')\psi(t-t')\lambda(x-x')\,dt'\,dx'. \tag{2.87}$$

We now consider different choices for the densities $\psi(t)$ and $\lambda(x)$ which result in different (possibly fractional) diffusion equations. The approach is as follows; decouple the convolution integrals in the master equation, Eq. (2.87), use a Fourier transform in space and a Laplace transform in time; consider asymptotic expansions of the transformed equation for small values of the Fourier and Laplace variables; carry out inverse Fourier-Laplace transforms using fractional order differential operators (if needed). Some general results on fractional order derivatives are provided in the appendix. Here we introduce the operators as needed.

2.2.3.2. *Exponential waiting times and standard diffusion*

The Fourier-Laplace transform of the CTRW master equation yields

$$\hat{\hat{n}}(q,u) = \hat{\Phi}(u)\hat{n}(q,0) + \hat{\psi}(u)\hat{\lambda}(q)\hat{\hat{n}}(q,u) \tag{2.88}$$

where q is the Fourier variable and u is the Laplace variable.

The Laplace transform of the survival probability can be written as

$$\hat{\Phi}(u) = \frac{1}{u} - \frac{\hat{\psi}(u)}{u}. \tag{2.89}$$

To proceed further we assume asymptotic properties for the step length density and the waiting time density. To begin with we assume that the step length density has the asymptotic expansion

$$\hat{\lambda}(q) \sim 1 - \frac{q^2\sigma^2}{2} + O(q^4), \tag{2.90}$$

where

$$\sigma^2 = \int r^2 \lambda(r)\, dr, \tag{2.91}$$

is finite. This is a general expansion for any even function $\lambda(x) = \lambda(-x)$ with a finite variance σ^2. An example of such a density is the Gaussian density

$$\lambda(x) = \frac{1}{\sqrt{2\pi\sigma^2}} \exp\left(-\frac{x^2}{2\sigma^2}\right). \tag{2.92}$$

The Fourier-Laplace CTRW master equation can now be written approximately as

$$u\hat{\hat{n}}(q,u) = (1-\hat{\psi}(u))\hat{n}(q,0) + u\hat{\psi}(u)\left(1-\frac{q^2\sigma^2}{2}\right)\hat{\hat{n}}(q,u) \tag{2.93}$$

Now consider a waiting time density with a finite mean τ then

$$\hat{\psi}(u) = 1 - \tau u + O(u^2). \tag{2.94}$$

An example of such a density is the exponential density

$$\psi(t) = \frac{1}{\tau} \exp\left(-\frac{t}{\tau}\right). \tag{2.95}$$

It is easy to verify the (memoryless) Markov property that the probability of waiting a time $T > t + s$ conditioned on having waited a time $T > s$ is equivalent to the probability of waiting a time $T > t$ at the outset:

$$P(T > t) = \int_t^\infty \frac{1}{\tau} \exp\left(-\frac{t'}{\tau}\right) dt' = e^{-\frac{t}{\tau}}$$

so that

$$P(T > t + s | T > s) = \frac{P(T > t+s)}{P(T > s)} = e^{-\frac{t}{\tau}} = P(T > t).$$

Using the exponential waiting time density we now have, to leading order,

$$u\hat{\hat{n}}(q,u) = \tau u\hat{n}(q,0) + (u - \tau u^2)\left(1-\frac{q^2\sigma^2}{2}\right)\hat{\hat{n}}(q,u) \tag{2.96}$$

which simplifies to

$$u\hat{\hat{n}}(q,u) - \hat{n}(q,0) = -\frac{\sigma^2 q^2}{2\tau}\hat{\hat{n}}(q,u). \tag{2.97}$$

The inverse Fourier and Laplace transforms now yield the standard diffusion equation

$$\frac{\partial n}{\partial t} = D\frac{\partial^2 n}{\partial x^2} \tag{2.98}$$

where

$$D = \frac{\sigma^2}{2\tau}. \tag{2.99}$$

2.2.3.3. *Power law waiting times and fractional subdiffusion*

The Markovian property of the exponential waiting time density contrasts with that of a Pareto waiting time density

$$\psi(t) = \frac{\alpha\tau^\alpha}{t^{1+\alpha}} \quad t \in [\tau, \infty], \quad 0 < \alpha < 1. \tag{2.100}$$

The cumulative distribution is a power law, $1 - \left(\frac{\tau}{t}\right)^\alpha$. Three properties of note are; (i) the mean waiting time is infinite, (ii) the probability of waiting a time $T > t + s$, conditioned on having waited a time $T > s$, is greater than the probability of waiting a time $T > t$ at the outset (the waiting time density has a temporal memory) and (iii) the waiting time density is scale invariant, $\psi(\gamma t) = \gamma^{-(1+\alpha)}\psi(t)$.

The asymptotic Laplace transform as $u \to 0$ for the Pareto density is given by a Tauberian (Abelian) theorem as (see, e.g., Berkowitz *et al*[32])

$$\hat{\psi}(u) \sim 1 - \Gamma(1-\alpha)\tau^\alpha u^\alpha. \tag{2.101}$$

Again we assume that the step length density is an even function with finite variance and we substitute the above expansion into the Fourier-Laplace master equation, Eq. (2.93), retaining only leading order terms. This results in

$$u\hat{\hat{n}}(q,u) - \hat{n}(q,0) = -\frac{q^2\sigma^2}{2\tau^\alpha\Gamma(1-\alpha)}u^{1-\alpha}\hat{\hat{n}}(q,u), \tag{2.102}$$

and after carrying out the inverse Fourier-Laplace transform

$$\frac{\partial n(x,t)}{\partial t} = D\mathcal{L}^{-1}\left(u^{1-\alpha}\frac{\partial^2 \hat{n}(x,u)}{\partial x^2}\right) \tag{2.103}$$

where

$$D = \frac{\sigma^2}{2\tau^\alpha\Gamma(1-\alpha)}. \tag{2.104}$$

This can be simplified further by using a standard result in fractional calculus[33] [see Appendix, Eq. (2.167)],

$$u^{1-\alpha}\frac{\partial^2 \hat{n}(x,u)}{\partial x^2} = \mathcal{L}\left({}_0\mathcal{D}_t^{1-\alpha} n(x,t)\right) + \left({}_0\mathcal{D}_t^{-\alpha}\frac{\partial^2 n(x,t)}{\partial x^2}\bigg|_{t=0}\right). \tag{2.105}$$

In this equation ${}_0\mathcal{D}_t^{1-\alpha}$ denotes a Riemann–Liouville fractional derivative of order α and ${}_0\mathcal{D}_t^{-\alpha}$ denotes a fractional integral of order α. The fractional integral on the far right hand side of Eq. (2.105) can be shown to be zero under fairly general conditions[34] so that using Eq. (2.105) in Eq. (2.103) we obtain the celebrated fractional subdiffusion equation

$$\frac{\partial n(x,t)}{\partial t} = D\left({}_0\mathcal{D}_t^{1-\alpha}\frac{\partial^2 n(x,t)}{\partial x^2}\right). \tag{2.106}$$

This equation can be obtained phenomenologically by combining the continuity equation

$$\frac{\partial n}{\partial t} = -\frac{\partial q}{\partial x}$$

with an ad-hoc fractional Fick's law

$$q(x,t) = -D\left({}_0\mathcal{D}_t^{1-\alpha}\frac{\partial n(x,t)}{\partial x}\right).$$

The fractional integral in this expression provides a weighted average of the concentration gradient over the prior history.

The Green's solution for the subdiffusion equation can be written in closed form using Fox H functions[17] (see Table 2.2). The special case $\alpha = 1/2$ is more amenable to analysis since the solution in this case can be written in terms of Meijer G-functions

$$G(x,t) = \frac{1}{\sqrt{8\pi D t^{\frac{1}{2}}}} G^{3,0}_{0,3}\left[\frac{x^2}{16 D t^{\frac{1}{2}}}\,\middle|\, \begin{matrix} \text{---} \\ 0, \frac{1}{4}, \frac{1}{2} \end{matrix}\right] \tag{2.107}$$

that are included as special functions in packages such as Maple and Mathematica.

In general the Green's solution for linear fractional diffusion equations can be obtained using Fourier-Laplace transform methods. The first step is to carry out a Fourier transform in space and a Laplace transform in time using the known results for the Laplace transform of Riemann–Liouville fractional derivatives. The transformed solution is then obtained as the solution of an algebraic problem in Fourier-Laplace space. The next step is to carry out the inverse transforms. In fractional subdiffusion equations the inverse Fourier transform is straightforward. The inverse Laplace transform can be obtained by first expanding the Laplace transform as a series expansion in Fox H functions and then perform a term by term inverse Laplace transform. The advantage of using Fox H functions in this way is that derivatives of Fox H functions, and (inverse) Laplace transforms of Fox H functions can be evaluated using index shifting properties.[35] The review article by Metzler and Klafter[17] contains a useful summary of Fox H function properties including a computable alternating series for their evaluation.

We now consider the mean square displacement

$$\langle x^2(t)\rangle = \int_{-\infty}^{\infty} x^2 G(x,t)\, dx \tag{2.108}$$

which can be evaluated using the Fourier-Laplace representation

$$\langle x^2(t)\rangle = \mathcal{L}^{-1}\left(\lim_{q\to 0} -\frac{d^2}{dq^2}\,\hat{\hat{G}}(q,u)\right). \tag{2.109}$$

After rearranging Eq. (2.102) and using the result that $\hat{G}(q,0) = 1$ we have

$$\hat{\hat{G}}(q,u) = \frac{1}{u + q^2 D u^{1-\alpha}} \tag{2.110}$$

and then using Eq. (2.109)

$$\langle x^2(t)\rangle = \mathcal{L}^{-1}\left(2D_\alpha u^{-1-\alpha}\right) = \frac{2D}{\Gamma(1+\alpha)} t^\alpha. \tag{2.111}$$

2.2.3.4. *Subordinated diffusion*

The asymptotic subdiffusion that arises from a CTRW with power law waiting times can be considered as a subordinated Brownian motion stochastic process. If $B(t)$ denotes a Brownian motion stochastic process then a subordinated Brownian motion stochastic process $B(E(t))$ can be generated from a non-decreasing stochastic process $E(t)$ with values in $[0,\infty)$ which is independent of $B(t)$ and which starts at $E(0) = 0$. Meerschaert and Scheffler,[36] have shown that the limits of CTRWs with radially symmetric jumps and finite variance, as waiting times and jumps tend to zero, are processes of the form $B(E(t))$, where $E(t)$ is the generalized inverse of a strictly increasing Lévy process $S(t)$ on $[0,\infty)$.

In particular it is a straightforward exercise to show that if $n(x,t)$ is the Green's solution of the time fractional subdiffusion equation,

$$\frac{\partial n}{\partial t} = {}_0\mathcal{D}_t^{1-\alpha} \frac{\partial^2 n}{\partial x^2}$$

then

$$n(x,t) = \int_0^\infty n^\star(x,\tau)T(\tau,t)\,d\tau \tag{2.112}$$

where $n^\star(x,\tau)$ is the Green's solution of the standard diffusion equation

$$\frac{\partial n}{\partial \tau} = \frac{\partial^2 n}{\partial x^2}$$

and $T(\tau,t)$ is defined through the Laplace transform, with respect to t,[17]

$$\hat{T}(\tau,u) = u^{\alpha-1}e^{-\tau u^\alpha}. \tag{2.113}$$

The density $T(\tau,t)$ is related to the one-sided Lévy α-stable density $\ell_\alpha(z)$ through[37]

$$T(\tau,t) = \frac{t}{\alpha\tau^{\frac{\alpha+1}{\alpha}}}\ell_\alpha\left(\frac{t}{\tau^{\frac{1}{\alpha}}}\right). \tag{2.114}$$

Equation (2.112) defines a subordination process for $n(x,t)$ in terms of the operational time τ and the physical time t. The operational time is essentially the number of steps in the walk. In the standard random walk the number of steps is proportional to the physical time but in the CTRW with infinite mean waiting times the number of steps is a random variable. The solution of the time fractional diffusion equation at physical time t is a weighted average over the operational time of the solution of the standard diffusion equation.

2.2.3.5. *Lévy flights and fractional superdiffusion*

We now consider CTRWs with an exponential (Markovian) waiting time density but a Lévy step length density with power law asymptotics

$$\lambda(x) \sim \frac{A_\alpha}{\sigma_\alpha}|x|^{-1-\alpha}, \quad 1 < \alpha < 2. \tag{2.115}$$

The Lévy step length density enables walks on all spatial scales.

Our starting point is the Fourier-Laplace transformed master equation, Eq. (2.88), combined with the Laplace transform of the survival probability, Eq. (2.89), i.e.,

$$u\hat{\hat{n}}(q,u) = (1-\hat{\psi}(u))\hat{n}(q,0) + u\hat{\psi}(u)\,\hat{\lambda}(q)\hat{\hat{n}}(q,u). \tag{2.116}$$

The exponential waiting time density has a finite mean τ so that $\hat{\psi}(u) \sim 1-\tau u$ and then

$$u\hat{\hat{n}}(q,u) = \tau u\,\hat{n}(q,0) + \left(u-\tau u^2\right)\hat{\lambda}(q)\hat{\hat{n}}(q,u). \tag{2.117}$$

The Fourier transform of the Lévy step length density is given by[17]

$$\hat{\lambda}(q) = \exp(-\sigma^\alpha |q|^\alpha) \sim 1 - \sigma^\alpha |q|^\alpha. \tag{2.118}$$

The Lévy density can be expressed in terms of Fox H functions[38] through the inverse Fourier transform [see Appendix, Eq. (2.190)]. If we substitute Eq. (2.118) into Eq. (2.117) and retain leading order terms then we obtain

$$u\hat{\hat{n}}(q,u) - u\,\hat{n}(q,0) = -\frac{\sigma^\alpha |q|^\alpha}{\tau}\hat{\hat{n}}(q,u), \tag{2.119}$$

and then after inversion of the Laplace transform

$$\frac{\partial\,\hat{n}(q,t)}{\partial t} = -\frac{\sigma^\alpha |q|^\alpha}{\tau}\,\hat{n}(q,t). \tag{2.120}$$

It remains to invert the Fourier transform and this can be done using another standard result of fractional calculus

$$\mathcal{F}\left(\nabla^\alpha_{|x|} n(x,t)\right) = -|q|^\alpha\,\hat{n}(q,t) \tag{2.121}$$

where $\nabla^\mu_{|x|}$ is the Riesz fractional derivative [see Appendix, Eq. (2.173)] and $\mathcal{F}$ denotes the Fourier transform operator. The evolution equation for the probability density function is now given by

$$\frac{\partial n}{\partial t} = D\nabla^\alpha_{|x|} n \tag{2.122}$$

with the diffusion coefficient

$$D = \frac{\sigma^\alpha}{\tau}. \tag{2.123}$$

The solution, which can be expressed in terms of Fox H functions (see Table 2), has the asymptotic long-time behaviour[17]

$$n(x,t) \sim \frac{\sigma^\alpha t}{\tau |x|^{1+\alpha}}, \quad 1 < \alpha < 2. \tag{2.124}$$

The mean square displacement diverges in this model, i.e., $\langle x^2(t)\rangle \to \infty$. This is an unphysical result but it is partly ameliorated by a non-divergent pseudo mean square displacement,

$$\langle [x^2(t)] \rangle \sim t^{\frac{2}{\alpha}}, \tag{2.125}$$

which can be inferred from the finite fractional moment scaling[17]

$$\langle |x|^\delta \rangle \sim t^{\frac{\delta}{\alpha}}, \quad 0 < \delta < \alpha < 2. \tag{2.126}$$

The pseudo mean square displacement, Eq. (2.125) characterizes superdiffusion scaling for $1 < \alpha < 2$.

2.2.4. *Simulating random walks for fractional diffusion*

In this section we describe Monte Carlo methods to simulate random walk trajectories yielding probability density functions for standard diffusion, fractional Brownian motion, subdiffusion, and superdiffusion. Algebraic expressions for the probability density functions are summarized in Table 2.2.

Table 2.2. Probability density functions for standard and fractional diffusion equations.

Diffusion Process	Probability Density Function	
standard	$\frac{1}{\sqrt{4\pi Dt}}e^{-\frac{x^2}{4Dt}}$	Markovian Gaussian
fBm	$\frac{1}{\sqrt{4\pi Dt^\alpha}}e^{-\frac{x^2}{4Dt^\alpha}}$	Non-Markovian Gaussian
subdiffusion	$\frac{1}{\sqrt{4\pi Dt^\alpha}}H^{2,0}_{1,2}\left[\frac{x^2}{4Dt^\alpha}\,\middle\vert\,\begin{matrix}(1-\alpha/2,\alpha)\\(0,1)\,(1/2,1)\end{matrix}\right]$	Non-Markovian Non-Gaussian
superdiffusion	$\frac{1}{\alpha\lvert x\rvert}H^{1,1}_{2,2}\left[\frac{\lvert x\rvert}{(Dt)^{1/\alpha}}\,\middle\vert\,\begin{matrix}(1,1/\alpha)\,(1,1/2)\\(1,1)\,(1,1/2)\end{matrix}\right]$	Markovian Non-Gaussian

The general procedure for simulating a single trajectory is as follows:

(1) Set the starting position of the particle, x, and jump-time, t, to zero.
(2) Generate a random waiting-time, δt, and jump-length, δx, from appropriate waiting-time and jump-length densities, $\psi(t)$ and $\lambda(x)$ respectively.
(3) Update the position of the particle $x(t+\delta t) = x(t) + \delta x$.
(4) Update the jump-time $t = t + \delta t$ of the particle. For non-constant waiting-times (e.g. subdiffusion) both the position of the particle and its jump-time need to be stored.
(5) Repeat steps 1 to 4 until the new jump-time reaches or exceeds the required simulation run-time.

The probability density for finding the particle at a particular time and position can be constructed from an ensemble average over a large number of random walk simulations.

2.2.4.1. *Generation of waiting-times*

In the subdiffusive case we take the waiting-time density as the (shifted) Pareto law[39]

$$\psi(t) = \frac{\alpha/\tau}{(1+t/\tau)^{1+\alpha}}. \tag{2.127}$$

The parameters α and τ are the anomalous exponent and the characteristic time respectively. This probability density function has the asymptotic scaling

$$\psi(t) \sim \frac{\alpha}{\tau}\left(\frac{t}{\tau}\right)^{-1-\alpha} \tag{2.128}$$

for long times. A random waiting-time that satisfies the waiting-time density, Eq. (2.127) can be generated from a uniform distribution $\rho(r)\,dr = 1\,dr, \quad r \in [0,1]$ as follows:

$$\rho(r)\,dr = \rho(r(t))\frac{dr}{dt}\,dt = \psi(t)\,dt \tag{2.129}$$

but $\rho(r(t)) = 1$ so that

$$\frac{dr}{dt} = \psi(t). \tag{2.130}$$

The solution of Eq. (2.130), using Eq. (2.127), and the initial condition $r(0) = 0$ is given by

$$r(t) = 1 - \left(1 + \frac{t}{\tau}\right)^{-\alpha}. \tag{2.131}$$

We can now invert this equation to find the random waiting time $t = \delta t$ in terms of the random number r. This yields

$$\delta t = \tau\left((1-r)^{-\frac{1}{\alpha}} - 1\right) \tag{2.132}$$

where $r \in (0,1)$ is a uniform random number.

For the non-subdiffusive cases we take for simplicity a constant waiting time of $\delta t = \tau$ between jumps. The density for this case is simply

$$\psi(t) = \delta(t-\tau) \tag{2.133}$$

though the exponential density

$$\psi(t) = \frac{1}{\tau}e^{-\frac{t}{\tau}} \tag{2.134}$$

could also be used. In this latter case the generated random waiting-time is given by

$$\delta t = -\tau \ln(1 - r). \tag{2.135}$$

2.2.4.2. *Generation of jump-lengths*

In the case of superdiffusion we generate a jump-length from the Lévy α-stable probability density using the transformation method described in:[40,41]

$$\delta x = \sigma \left(\frac{-\ln u \cos \phi}{\cos\left((1-\alpha)\,\phi\right)} \right)^{1-\frac{1}{\alpha}} \frac{\sin(\alpha\phi)}{\cos\phi} \tag{2.136}$$

where $\phi = \pi(v - 1/2)$, σ is jump-length scale parameter, and u, $v \in (0,1)$ are two independent uniform random numbers.

For simplicity, the jumps in the non-superdiffusive cases are taken to the nearest-neighbour grid points only. For the standard diffusion and subdiffusive cases the particle, after waiting, has to jump either to the left or right a distance of Δx. The jump-length, for these cases, is generated from

$$\delta x = \begin{cases} \Delta x, & 0 \le r < \frac{1}{2} \\ -\Delta x, & \frac{1}{2} \le r < 1 \end{cases} \tag{2.137}$$

where $r \in (0,1)$ is uniform random number. The jump density in this case is

$$\lambda(x) = \frac{1}{2}\delta(x - \Delta x) + \frac{1}{2}\delta(x + \Delta x). \tag{2.138}$$

In the fractional Brownian case the particle may jump to the left or right or not jump at all. In this case Eq. (2.137) is modified to (where $0 < \alpha < 1$)

$$\delta x = \begin{cases} \Delta x, & 0 \le r < \alpha n^{\alpha-1} \\ -\Delta x, & \alpha n^{\alpha-1} \le r < 2\alpha n^{\alpha-1} \\ 0, & 2\alpha n^{\alpha-1} \le r < 1 \end{cases} \tag{2.139}$$

where n is the current step number for the time $t = n\tau$.

This random walk process leads to the probability density function for fractional Brownian motion at any given time t, but it is not an approximation of the whole trajectory of fractional Brownian motion. Methods for simulating trajectories of fractional Brownian motion have been described by Whitt.[42]

2.2.4.3. *Calculation of the mean-squared displacement*

To calculate the mean-squared displacement for the non-superdiffusive cases we simply evaluate the ensemble average of the particles position, $x(t)$, at each time-step $t_n = n\tau$. For simulations with a non-constant waiting time, this requires a bit of book-keeping as the particles do not necessarily jump at these times. However the position of the particle for a particular trajectory can be found from the stored jump-times noting the particles wait at their current location until the next jump-time. The mean-squared displacement is estimated using

$$\left\langle x^2(t_n)\right\rangle \simeq \frac{1}{M}\sum_{j}^{M}[x(t_n)]^2 \qquad (2.140)$$

where M is the number of trajectories averaged. This can be compared with the algebraic expressions for the mean square displacements for subdiffusion, Eq. (2.111), and standard diffusion ($\gamma = 1$) once the constant D is estimated. In the case of fractional Brownian motion, we can also compare with Eq. (2.111) but with denominator set to unity.

In the case of superdiffusion, where the mean-squared displacement diverges, we have computed the ensemble average

$$\left\langle |x|^\delta(t_n)\right\rangle \simeq \frac{1}{M}\sum_{j}^{M}[x(t_n)]^\delta \qquad 0<\delta<\alpha \qquad (2.141)$$

to compare with[17]

$$\left\langle |x|^\delta(t)\right\rangle = \frac{2}{\alpha}(Dt)^{\delta/\alpha}\frac{\Gamma(-\delta/\alpha)\Gamma(1+\delta)}{\Gamma(-\delta/2)\Gamma(1+\delta/2)}. \qquad (2.142)$$

2.2.4.4. *Probability density functions*

The waiting-times, δt, and step-lengths, δx, for simulating standard and fractional diffusion processes are listed in Table 2.3. The diffusion constants are also listed for the purposes of comparisons with the algebraic formulae in Table 2.2.

For the simulations presented here we take the relevant length scales to be $\Delta x = 1$ and $\sigma = 1$ and the waiting-time scales $\tau = 1$ for the non-subdiffusive simulations and $\tau = 0.1$ for the subdiffusive simulations. An

Table 2.3. Waiting times, step lengths and diffusion constants for simulating fractional diffusion random walks.

random walk	δt	δx	D
standard	τ	$\begin{array}{rll} \Delta x & \text{if} & 0 \le r < \frac{1}{2} \\ -\Delta x & \text{if} & \frac{1}{2} \le r < 1 \end{array}$	$\dfrac{\Delta x^2}{2\tau}$
fBm	τ	$\begin{array}{rll} \Delta x & \text{if} & 0 \le r < \alpha n^{\alpha-1} \\ -\Delta x & \text{if} & \alpha n^{\alpha-1} \le r < 2\alpha n^{\alpha-1} \\ 0 & \text{if} & 2\alpha n^{\alpha-1} \le r < 1 \end{array}$	$\dfrac{\Delta x^2}{\tau^\alpha}$
subdiffusion	$\tau\left((1-r)^{-\frac{1}{\alpha}} - 1\right)$	$\begin{array}{rll} \Delta x & \text{if} & 0 \le r < \frac{1}{2} \\ -\Delta x & \text{if} & \frac{1}{2} \le r < 1 \end{array}$	$\dfrac{\Delta x^2}{2\tau^\alpha \Gamma(1-\alpha)}$
superdiffusion	τ	$\Delta x \left(\dfrac{-\ln u \cos\phi}{\cos\left((1-\alpha)\phi\right)}\right)^{1-\frac{1}{\alpha}} \times \dfrac{\sin(\alpha\phi)}{\cos\phi}$	$\dfrac{\sigma^\alpha}{\tau}$

In the table, $r, u, v \in (0,1)$ are independent random numbers and $\phi = \pi(v - 1/2)$, $n = t/\tau$.

ensemble average of 100,000 trajectories were used to generate the simulation results for both the mean-squared displacement and probability densities shown in these notes. In the fractional Brownian motion and subdiffusion we took $\alpha = 1/2$. For the superdiffusive case we used $\alpha = 3/2$ and calculated the average Eq. (2.141) using $\delta = 3/4 = \alpha/2$. The results of the simulations are compared with algebraic results in Figs. 2.1–2.4. Note the log-log scales in the mean-squared displacement plots. The data values correspond to logarithms of the numbers shown on the axes. In each case the results of the simulations (open circles) agree with the theoretical results (solid lines).

2.2.5. *Fractional Fokker–Planck equations*

In the CTRWs described above we considered unbiased walks i.e., there was an equal probability to step left or right in a given step. It is possible to generalize the analysis to permit a bias, for example the step length density could be chosen to be a function of position to model the effects of CTRWs in a space varying force field. The biased CTRWs lead to fractional Fokker–Planck equations. In these notes we summarize key results that have been obtained and refer the reader to the original journal articles for details.

In the case of anomalous subdiffusion in an external force field $f(x,t)$

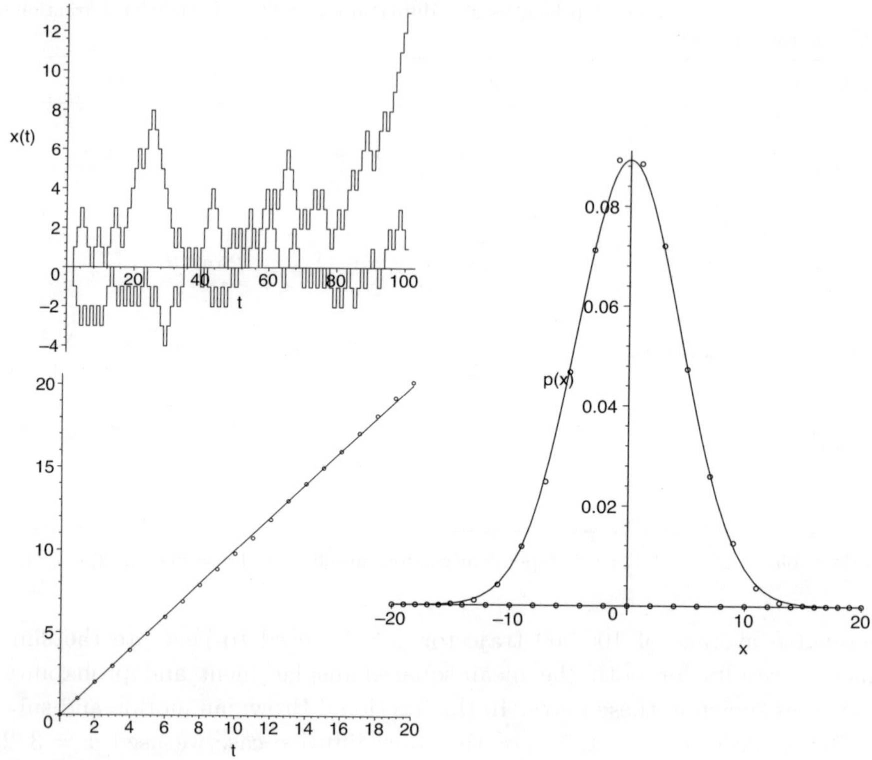

Fig. 2.1. Sample trajectories (top left), probability density function (right) and mean-squared displacement (lower left) for standard diffusion.

two fractional Fokker–Planck equations that have been considered are

$$\frac{\partial n(x,t)}{\partial t} = {}_0\mathcal{D}_t^{1-\alpha} D\nabla^2 n(x,t) - {}_0\mathcal{D}_t^{1-\alpha}\nabla\left(\frac{1}{\eta}f(x,t)n(x,t)\right) \qquad (2.143)$$

and

$$\frac{\partial n(x,t)}{\partial t} = {}_0\mathcal{D}_t^{1-\alpha} D\nabla^2 n(x,t) - \nabla\left(\frac{1}{\eta}f(x,t)\,{}_0\mathcal{D}_t^{1-\alpha}\nabla n(x,t)\right) \qquad (2.144)$$

where D is the diffusion coefficient for subdiffusion, Eq. (2.104), and the coefficents D and η are related through a generalized Einstein relation

$$D = \frac{k_B T}{m\eta}. \qquad (2.145)$$

Neither of the above equations have been derived from CTRWs in the general case $f = f(x,t)$. However in the case of subdiffusion in a time indepen-

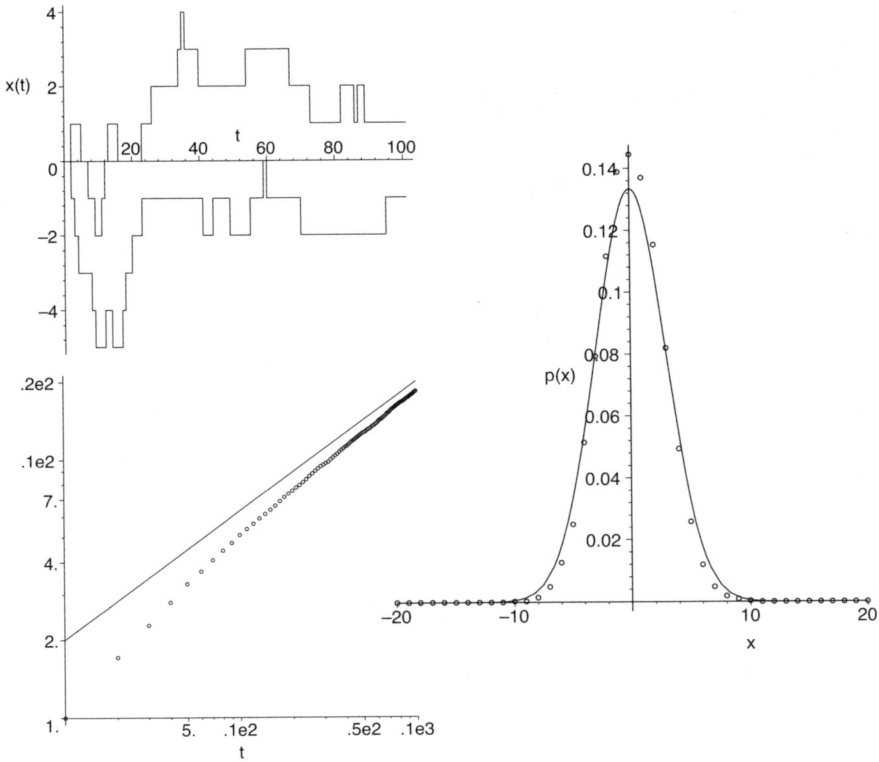

Fig. 2.2. Sample trajectories (top left), probability density function (right) and log-log mean-squared displacement (lower left) for fractional Brownian motion with $\alpha = 1/2$.

dent external force field $f = f(x)$ the fractional Fokker–Planck equation Eq. (2.143) has been derived from biased CTRWs[43] and in the case of subdiffusion in a space independent external force field $f = f(t)$ the second fractional Fokker–Planck equation Eq. (2.144) has been derived from a generalized master equation formulation of CTRWs.[44]

Given that both equations are equivalent in the case of time independent force fields this suggests that the second formulation might be preferred for generalizing to $f = f(x,t)$. Another argument in favour of this is that temporal variations in the external force field occur in physical time which is different to the operational time for subdiffusion whereas the first formulation produces a subordination over the same operational time scale. However if the force field is generated internally (e.g., by ionic concentration gradients or chemotaxis) then this subordination may be appropriate.

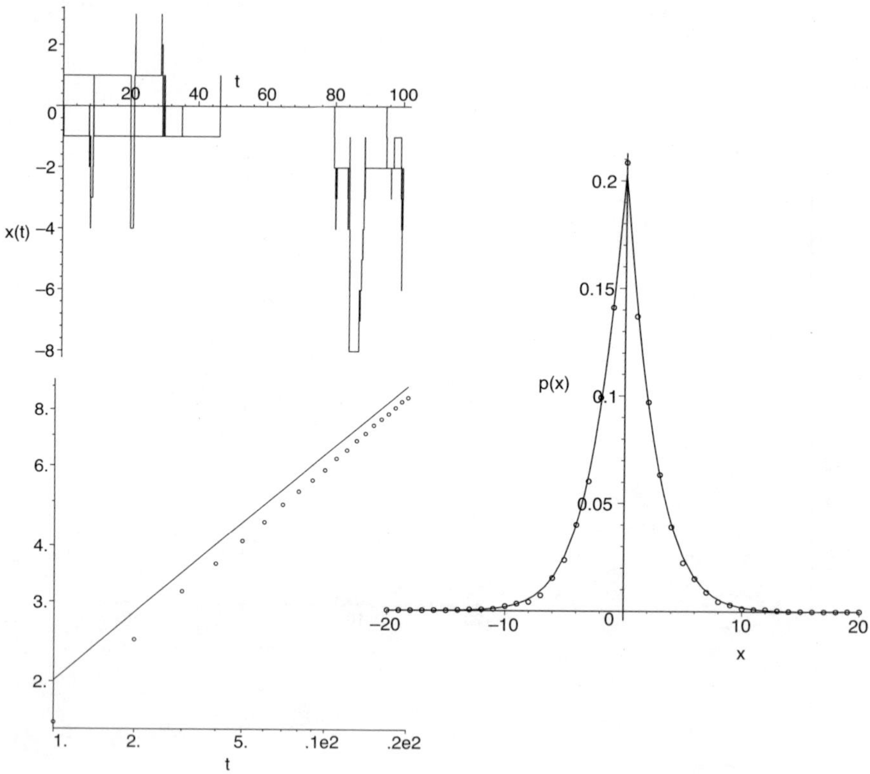

Fig. 2.3. Sample trajectories (top left), probability density function (right) and log-log mean-squared displacement (lower left) for fractional subdiffusion with $\alpha = 1/2$.

The derivation of a generalized fractional diffusion equation to describe fractional diffusion in an external (or internal) time and space varying force field is still an open problem.

2.2.6. *Fractional Reaction-Diffusion equations*

The CTRW formalism can also be extended to accommodate source or sink terms arising from reactions. These generalized CTRWs lead to fractional reaction-diffusion equations. Again, in these notes we simply summarize key results and refer the reader to the original journal articles for details.

In early CTRW formulations of fractional reaction-diffusion[45] a time fractional derivative was applied to the spatial diffusion term but not the reaction terms. However in other studies it was suggested that the time

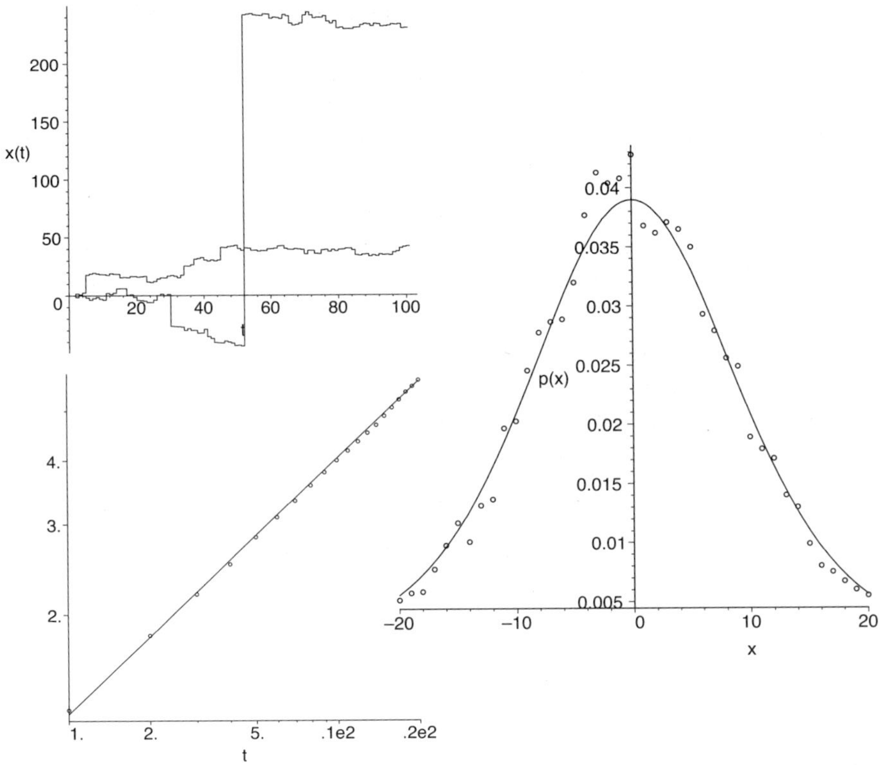

Fig. 2.4. Sample trajectories (top left), probability density function (right) and log-log pseudo mean-squared displacement with $\delta = 3/4$ (lower left) for fractional superdiffusion with $\alpha = 3/2$.

fractional derivative should operate equally on both terms.[39,46] This second formulation was motivated by considerations of subordination where reactions and diffusions are affected by the same operational time scales. More recently, at least in the case of linear reaction dynamics, it was shown[31,47] that neither approach properly describes subdiffusion with prescribed linear reaction kinetics. In the particular case where the reaction dynamics models exponential growth ($+k$) or decay ($-k$) during the CTRW waiting time intervals the CTRW master equation yields the balance equation[31]

$$n(x,t) = \Phi(t)e^{\pm kt}n(x,0) + \int_{-\infty}^{\infty}\int_0^t n(x',t')e^{\pm k(t-t')}\psi(t-t')\lambda(x-x')\,dt'\,dx' \tag{2.146}$$

and the governing fractional reaction diffusion equation is given by[31]

$$\frac{\partial n}{\partial t} = D\, e^{\pm kt}\, {}_0\mathcal{D}_t^{1-\alpha} \left(e^{\mp kt} \frac{\partial^2 n}{\partial x^2} \right) \pm kn. \qquad (2.147)$$

The above formalism has also been extended to multispecies subdiffusion with linear reaction kinetics.[48] Although some progress has been made in extending CTRWs to include nonlinear reaction kinetics[49] the derivation of general nonlinear fractional reaction diffusion equations is still an open problem. A possible generalization of the balance equation, Eq.(2.146), for nonlinear reactions is to replace the linear evolution operator $e^{\pm kt}$ in this equation with a nonlinear evolution operator.

2.2.7. *Fractional diffusion based models*

In addition to the fractional diffusion equations derived from CTRWs there are numerous other fractional diffusion equations that have been studied as models for physical, social or economic systems, with varying levels of justification. Examples include:
Space-time fractional Fokker–Planck equation[17]

$$\frac{\partial w}{\partial t} = \mathcal{D}_t^{1-\alpha} \left(\frac{\partial}{\partial x} \frac{V'(x)}{\eta} + K \nabla_{|x|^\mu} \right) w. \qquad (2.148)$$

Space-time fractional diffusion model for plasmas[50]

$$\mathcal{D}_t^{\beta} P = \chi \nabla_{|x|^\alpha} P. \qquad (2.149)$$

Fractional Black-Scholes model for option prices[51]

$$rV(x,t) = \frac{\partial V(x,t)}{\partial t} + \left(r + \sigma^\alpha \sec(\frac{\alpha\pi}{2}) \right) \frac{\partial V}{\partial x} - \sigma^\alpha \sec(\frac{\alpha\pi}{2}) \mathcal{D}_x^\alpha V. \qquad (2.150)$$

Fractional cable equation for nerve cells[52]

$$r_m c_m \frac{\partial V}{\partial t} = \frac{d r_m}{4 r_L(\gamma)} \mathcal{D}_t^{1-\gamma} \left(\frac{\partial^2 V}{\partial x^2} \right) - \mathcal{D}_t^{1-\kappa} (V - r_m i_e). \qquad (2.151)$$

2.2.8. *Power laws and fractional diffusion*

An average individual who seeks a friend twice his height would fail. On the other hand, one who has an average income will have no trouble in discovering a richer person with twice his income, and that richer person may, with a little diligence, locate a third party with twice his income, etc.
Elliot Montroll and Michael Shlesinger (1984)[30]

In the above sections we showed how CTRWs with asymptotic power law waiting time densities result in subdiffusion and CTRWs with asymptotic power law step length densities result in superdiffusion. We conclude these notes by commenting on possible origins of these power laws. As a first general remark, power laws $f(x)$ are scale invariant functions, i.e., $f(\lambda x) = \lambda^\alpha f(x)$ for some exponent α and all scale factors λ. Power law scaling is a characteristic feature of fractals, and power law distributions have been found to characterize numerous real world data sets[53] in which the complexity might be expected to extend over a large range of spatial or temporal scales.

A possible mechanism that has been suggested for power law waiting time densities in CTRWs[54] is that the random walker moves in an environment with an exponential distribution of trap binding energies

$$\rho(E) = \frac{1}{E_0} e^{-\frac{E}{E_0}} \tag{2.152}$$

with thermally activated trapping times

$$\tau = e^{\frac{E}{k_B T}}. \tag{2.153}$$

The waiting time density follows as

$$\begin{aligned}
\psi(\tau) d\tau &= \rho(E) \frac{dE}{d\tau}\, d\tau \\
&= \frac{1}{E_0} e^{-\frac{E}{E_0}} \left(\frac{k_B T}{\tau} \right) d\tau \\
&= \frac{1}{E_0} \tau^{-\frac{kT}{E_0}} \left(\frac{k_B T}{\tau} \right) d\tau \\
&= \left(\frac{k_B T}{E_0} \right) \tau^{-\frac{kT}{E_0} - 1}\, d\tau,
\end{aligned}$$

so that

$$\psi(t) = \alpha t^{-1-\alpha}. \tag{2.154}$$

Power law step length densities describe so called Lévy flights and they can be motivated by considering a generalized Central Limit Theorem.[55] In the standard Central Limit Theorem the normal distribution is the limiting stable law for the distribution of the normalized sum of random variables

$$\frac{X_1 + X_2 + \ldots X_N}{N^{\frac{1}{2}}}.$$

The proof of this is dependent on the X having a finite mean $\langle X \rangle$ and variance $\langle X^2 \rangle$. The probability density for the normalized sum of the random variables is the probability density for the position of the walker after N steps.

If the X do not have a finite variance then the sum of N steps also has infinite variance. It is then natural to seek a probability step length density with the same form as the probability density for the normalized sum of N steps in the limit of large N. This suggests a scale invariant function, and Paul Lévy was able to show that, if the variance is infinite, then the normalized sum

$$\frac{X_1 + X_2 + \dots X_N}{N^\alpha}$$

is governed by a symmetric stable law that does not decay exponentially as $|x| \to \infty$ but instead it has a power law tail $\sim C|x|^{-1-\alpha}$. The variance is infinite for $0 < \alpha < 2$. The mean is infinite $0 < \alpha < 1$ and this is unphysical so the range is restricted to $1 < \alpha < 2$. The solution of the space fractional diffusion equation, Eq. (2.122), is precisely the Lévy stable distribution, represented as a Fox H function in Table 2.2 [also see the Appendix, Eq. (2.190)].

2.3. Appendix: Introduction to fractional calculus

One can ask what would be a differential having as its exponent a fraction. Although this seems removed from Geometry ... it appears that one day these paradoxes will yield useful consequences.
Gottfried Leibniz (1695)

There are different possible ways to define fractional derivatives, all based on generalizing well known results in the ordinary calculus. Here we focus attention on the Riemann–Liouville definition although other definitions will be introduced through Fourier and Laplace transform involving fractional powers of the transform variables. Further details can be found in the excellent reference books by Oldham and Spanier (1974),[33] Miller and Ross (1993)[56] and Podlubny (1999).[57]

As a first introduction it is constructive to consider ordinary derivatives of power laws $f(x) = x^p$ then for integer $n > 0$

$$\begin{aligned}
\frac{d^n f}{dx^n} &= p(p-1)\dots(p-n-1)x^{p-n} \\
&= \frac{p!}{(p-n)!}x^{p-n} \\
&= \frac{\Gamma(p+1)x^{p-n}}{\Gamma(p-n+1)}
\end{aligned} \tag{2.155}$$

where we have used the definition of the Gamma function

$$\Gamma(\alpha+1) = \int_0^\infty e^{-t} t^\alpha \, dt \quad \forall \alpha \in \mathbb{R} \tag{2.156}$$

and the result that $n! = \Gamma(n+1), \quad n \in \mathbb{N}$. The result on the right hand side of Eq. (2.155) is well defined for $n \in \mathbb{R}^+$ and if n is non-integer this can be considered as a fractional derivative of a power law. An example is

$$\frac{d^{\frac{1}{2}}}{dx^{\frac{1}{2}}} x^p = \frac{\Gamma(p+1) x^{p-\frac{1}{2}}}{\Gamma(p+\frac{1}{2})}$$

A more general definition of a fractional derivative (that reproduces the above results for power laws) is the so called Riemann–Liouville fractional derivative which is in turn based on a Riemann–Liouville fractional integral.

2.3.1. *Riemann–Liouville fractional integral*

Consider the n fold integral ($n \in \mathbb{N}$)

$$\begin{aligned}\frac{d^{-n} f(x)}{dx^{-n}} &= \int_0^x \left(\int_0^{x_{n-1}} \cdots \left(\int_0^{x_2} \left(\int_0^{x_1} f(x_0) dx_0 \right) dx_1 \right) \cdots \right) dx_{n-1} \\ &= \frac{1}{\Gamma(n)} \int_0^x \frac{f(y)}{(x-y)^{-n+1}} dy \end{aligned} \tag{2.157}$$

where the compact expression on the right hand side is known as Cauchy's formula. This single integral is well defined for certain non-integer values of n which leads to the Riemann–Liouville definition of a fractional integral

$${}_0\mathcal{D}_x^{-q} = \frac{d^{-q} f(x)}{dx^{-q}} = \frac{1}{\Gamma(q)} \int_0^x \frac{f(y)}{(x-y)^{-q+1}} \, dy, \qquad q \in \mathbb{R}^+. \tag{2.158}$$

The integral is improper for $q < 1$ but converges for $0 < q < 1$. Note too that the integral diverges if $q \leq 0$ so the above formula will not work for a fractional derivative $\mathcal{D}_x^\alpha$ with $\alpha > 0$.

The formula for the fractional integral in Eq. (2.158) defines a weighted average of the function using a power law weighting function. A geometric interpretation of the fractional integral has recently been given by Podlubny.[58] Consider an auxiliary function

$$g(y) = \frac{1}{\Gamma(q+1)} \left(x^q - (x-y)^q \right) \tag{2.159}$$

and plot $g(y)$ versus y for $0 < y < x$. For each y along this curve construct a fence with a height $f(y)$. The standard integral $\int_0^x f(y) \, dy$ is the area of the projection of this fence onto the $(y, f(y))$ plane and the fractional integral ${}_0\mathcal{D}_x^{-q} f$ is the area of the projection of the fence onto the $(g(y), f(y))$ plane.

The Riemann–Liouville integral defined above is called a left-sided fractional integral. The right-sided Riemann–Liouville integral is defined as

$$ {}_x\mathcal{D}_a^{-q} = \frac{d^{-q} f(x)}{dx^{-q}} = \frac{1}{\Gamma(q)} \int_x^a \frac{f(y)}{(x-y)^{-q+1}}\, dy \qquad q \in \mathbb{R}^+. \tag{2.160} $$

Usually we will deal with a left-sided fractional integral and omit the zero subscript.

2.3.2. *Riemann–Liouville fractional derivative*

The Riemann–Liouville definition of a fractional derivative is the ordinary derivative of a fractional integral. Formally we define the Riemann–Liouville fractional derivative

$$ \mathcal{D}_x^q f(x) = \frac{d^q f(x)}{dx^q} = \frac{d^n}{dx^n}\left(\frac{d^{-(n-q)} f(x)}{dx^{-(n-q)}}\right) \qquad q \in \mathbb{R}^+, \qquad n = \lfloor q \rfloor + 1. \tag{2.161} $$

Examples

$$ \mathcal{D}_x^{\frac{1}{2}} x^p = \frac{d^{\frac{1}{2}} x^p}{dx^{\frac{1}{2}}} = \frac{d}{dx}\left(\frac{d^{-\frac{1}{2}} x^p}{dx^{-\frac{1}{2}}}\right) = \frac{d}{dx}\left(\frac{1}{\Gamma(\frac{1}{2})}\int_0^x \frac{y^p}{(x-y)^{\frac{1}{2}}} dy\right) = \frac{\Gamma(p+1) x^{p-\frac{1}{2}}}{\Gamma(p+\frac{1}{2})} $$

$$ \mathcal{D}_x^{\frac{1}{2}} e^x = \frac{x^{-\frac{1}{2}}}{\Gamma(\frac{1}{2})}\, {}_1F_1(1, \frac{1}{2}, x) $$

$$ \mathcal{D}_x^{\alpha}\,(\text{constant}) = \frac{x^{-\alpha}}{\Gamma(1-\alpha)}\,(\text{constant}) $$

$$ \mathcal{D}_x^{\alpha} x^p = \frac{\Gamma(p+1)}{\Gamma(p-\alpha+1)} x^{p-\alpha} $$

2.3.2.1. *Tautochrone problem*

One of the earliest applications of fractional calculus was in Abel's (1823) solution of the tautochrone problem (see for example Miller and Ross[56]); to find the shape of wire $x(y)$ such that the time of descent, τ, of a frictionless bead falling under gravity is a constant independent of the starting point. Conservation of energy for this problem yields

$$ \frac{1}{2} v^2 = g(h-y) \Rightarrow v = \sqrt{2g(h-y)}. $$

The velocity v can also be related to the arc length s and thus the shape of the wire by

$$v = \frac{ds}{dt} = \frac{\left(\sqrt{1 + \left(\frac{dx}{dy}\right)^2}\right) dy}{dt}.$$

After equating the two expressions for the velocity we now have

$$\sqrt{2g(h-y)} = \frac{\left(\sqrt{1 + \left(\frac{dx}{dy}\right)^2}\right) dy}{dt}.$$

This is separable and thus

$$\int_0^T \sqrt{2g}\, dt = \int_0^h \frac{\sqrt{1 + \left(\frac{dx}{dy}\right)^2}}{\sqrt{h-y}} dy = \int_0^h \frac{f(y)}{(h-y)^{\frac{1}{2}}} dy$$

where the shape of the wire $x(y)$ is governed by the differential equation

$$\frac{dx}{dy} = \sqrt{f^2(y) - 1}. \qquad (2.162)$$

The steps to find $f(y)$ now follow as

$$\begin{aligned} \sqrt{2g}T &= \int_0^h \frac{f(y)}{(h-y)^{\frac{1}{2}}} dy \\ &= \sqrt{\pi}\, \mathcal{D}_h^{-\frac{1}{2}} f(h) \\ \Rightarrow \mathcal{D}_h^{\frac{1}{2}} \sqrt{2g}T &= \sqrt{\pi}\, \mathcal{D}^{\frac{1}{2}}{}_h \mathcal{D}_h^{-\frac{1}{2}} f(h) = \sqrt{\pi} f(h) \\ \left(\frac{1}{\sqrt{\pi}\sqrt{h}}\right) \sqrt{2g}T &= \sqrt{\pi} f(h) \\ f(y) &= \frac{\sqrt{2g}}{\pi} \sqrt{\frac{T^2}{y}}. \end{aligned}$$

Of particular note in this application is that the fractional derivative w.r.t. h of the constant $\sqrt{2g}T$, yields a non-zero function of h.

The differential equation for the shape of the wire can now be written as

$$\frac{dx}{dy} = \sqrt{\frac{2gT^2}{\pi^2 y} - 1}, \qquad (2.163)$$

with parametric solution

$$x = a(\theta + \sin\theta)$$
$$y = a(1 - \cos(\theta))$$

which describes a cycloid. The cycloid is also the solution to the brachistochrone problem – the shape of the wire that results in the fastest path from the point of release.

2.3.3. *Basic properties of fractional calculus*

The Riemann–Liouville fractional derivative $D_x^q f(x)$ satisfies the following properties:

(i) $D_x^0 f(x) = f(x)$ identity property.
(ii) $D_x^q f(x) = f(x)$ is a standard derivative if $q \in \mathbb{N}$.
(iii) $D_x^q\left[af(x) + bg(x)\right] = a\, D_x^q f(x) + b\, D_x^q g(x)$ linearity property.
(iv) $D_x^\alpha[f(x)g(x)] = \sum_{m=0}^{\infty}\binom{\alpha}{m} D_x^m[f(x)]D_x^{\alpha-m}[g(x)]$ Leibniz product rule.

The Riemann–Liouville fractional integral $D_x^{-q} f(x)\quad q > 0$ satisfies the above properties together with

$D_x^{-q}\left(D_x^{-p} f(x)\right) = D_x^{-q-p} f(x)$ semi-group property.

2.3.4. *Fourier and Laplace transforms and fractional Calculus*

Here we use the notation; $\mathcal{L}$ to denote a Laplace transform with Laplace variable u; $\mathcal{F}$ to denote a Fourier transform with Fourier variable q; $\mathcal{D}_t^\alpha$ to denote a generic fractional derivative w.r.t. t of order α.

The Fourier transform pairs are

$$\hat{y}(q) = \int_{-\infty}^{+\infty} e^{iqx} y(x)\, dx, \qquad y(x) = \frac{1}{2\pi}\int_{-\infty}^{+\infty} e^{-iqx}\hat{y}(q)\, dx, \tag{2.164}$$

and the Laplace transform pairs are

$$\hat{y}(u) = \int_0^\infty e^{-ut} y(t)\, dt, \qquad y(t) = \int_{c-i\infty}^{c+i\infty} e^{ut}\hat{y}(u)\, du. \tag{2.165}$$

Some transform results for fractional derivatives are as follows:

(i) Riemann–Liouville

$$\mathcal{D}_t^{\alpha} y(t) = \frac{d}{dt}\left(\frac{1}{\Gamma(\alpha)}\int_0^t \frac{y(s)}{(t-s)^{1-\alpha}}\,ds\right) \quad 0<\alpha<1 \tag{2.166}$$

$$\mathcal{L}\left(\mathcal{D}_t^{\alpha} y(t)\right) = u^{\alpha}\hat{y}(u) - \left(\mathcal{D}_t^{\alpha-1} y(t)\right)\Big|_{t=0} \tag{2.167}$$

$$\mathcal{F}\left(\mathcal{D}_x^{\alpha} y(x)\right) = (iq)^{\alpha}\hat{y}(q) \tag{2.168}$$

(ii) Grunwald–Letnikov

$$\mathcal{D}_t^{\alpha} y(t) = \lim_{h\to 0}\frac{1}{h^{\alpha}}\sum_{j=0}^{\infty}(-1)^j \frac{\Gamma(\alpha+1)}{\Gamma(j+1)\Gamma(\alpha-j+1)} y(t-jh) \tag{2.169}$$

$$\mathcal{L}\left(\mathcal{D}_t^{\alpha} y(t)\right) = u^{\alpha}\hat{y}(u) \tag{2.170}$$

(iii) Caputo

$$\mathcal{D}_t^{\alpha} y(t) = \left(\frac{1}{\Gamma(1-\alpha)}\int_0^t \frac{\frac{d}{ds}y(s)}{(t-s)^{\alpha}}\,ds\right) \quad 0<\alpha<1 \tag{2.171}$$

$$\mathcal{L}\left(\mathcal{D}_t^{\alpha} y(t)\right) = u^{\alpha}\hat{y}(u) - \left(u^{\alpha-1}y(0)\right) \tag{2.172}$$

(iv) Riesz

$$\nabla_{|x|}^{\alpha} y(x) = -\frac{1}{2\cos(\frac{\pi\alpha}{2})}\left({}_{-\infty}\mathcal{D}_x^{\alpha} y(x) + {}_x\mathcal{D}_{\infty}^{\alpha} y(x)\right), \quad 1<\alpha<2 \tag{2.173}$$

where ${}_{-\infty}\mathcal{D}_x^{\alpha}$ and ${}_x\mathcal{D}_{\infty}^{\alpha}$ are left-sided and right-sided Riemann–Liouville fractional derivatives and

$$\mathcal{F}\left(\nabla_{|x|}^{\alpha} y(x)\right) = -|q|^{\alpha}\hat{y}(q). \tag{2.174}$$

2.3.5. *Special functions for fractional calculus*

Mittag–Leffler Function

$$E_{\alpha}(z) = \sum_{k=0}^{\infty}\frac{z^k}{\Gamma(\alpha k+1)} \quad \alpha>0 \tag{2.175}$$

$$\mathcal{L}\left(E_{\alpha}(-\left(\frac{t}{\tau}\right)^{\alpha})\right) = \frac{1}{u+\frac{u^{1-\alpha}}{\tau^{\alpha}}} \quad \alpha>0 \tag{2.176}$$

$$E_1(z) = \sum_{k=0}^{\infty} \frac{z^k}{\Gamma(k+1)} = \sum_{k=0}^{\infty} \frac{z^k}{k!} = e^z$$

Asymptotics

$$E_\alpha\left(-\left(\frac{t}{\tau}\right)^\alpha\right) \sim \exp\left(-\frac{1}{\Gamma(1+\alpha)}\left(\frac{t}{\tau}\right)^\alpha\right) \quad t \ll \tau, \quad 0 < \alpha < 1 \tag{2.177}$$

$$E_\alpha\left(-\left(\frac{t}{\tau}\right)^\alpha\right) \sim \frac{1}{\Gamma(1-\alpha)}\left(\frac{t}{\tau}\right)^{-\alpha} \quad t \gg \tau, \quad 0 < \alpha < 1 \tag{2.178}$$

Generalized Mittag–Leffler Function

$$E_{\alpha,\beta}(z) = \sum_{k=0}^{\infty} \frac{z^k}{\Gamma(\alpha k + \beta)} \qquad \alpha > 0, \quad \beta > 0 \tag{2.179}$$

$$E_{1,1}(z) = e^z \tag{2.180}$$

$$E_{1,2}(z) = \frac{e^z - 1}{z} \tag{2.181}$$

$$E_{2,2}(z) = \frac{\sinh(\sqrt{z})}{\sqrt{z}} \tag{2.182}$$

$$\mathcal{L}\left(t^{\alpha k+\beta-1} E^{(k)}_{\alpha,\beta}(\pm a t^\alpha)\right) = \frac{k! u^{\alpha-\beta}}{(u^\alpha \mp a)^{k+1}} \tag{2.183}$$

where $^{(k)}$ denotes the kth derivative with respect to z.

$$\mathcal{L}\left(t^{\frac{k-1}{2}} E^{(k)}_{\frac{1}{2},\frac{1}{2}}(\pm a\sqrt{t})\right) = \frac{k!}{(\sqrt{u} \mp a)^{k+1}} \tag{2.184}$$

Example: The fractional differential equation

$$\mathcal{D}_t^{\frac{1}{2}} y(t) = y(t)$$

with initial condition

$$\mathcal{D}_t^{-\frac{1}{2}} y(t)\Big|_{t=0} = C$$

has solution $t^{-\frac{1}{2}}E_{\frac{1}{2},\frac{1}{2}}(\sqrt{t})$. This can be shown as follows

$$\begin{aligned}
\mathcal{L}\left(\mathcal{D}_t^{\frac{1}{2}}y(t)\right) &= \mathcal{L}\left(y(t)\right) \\
u^{\frac{1}{2}}\hat{y}(u) - \mathcal{D}_t^{-\frac{1}{2}}y(t)\Big|_{t=0} &= \hat{y}(u) \\
\Rightarrow \hat{y}(u) &= \frac{C}{u^{\frac{1}{2}}-1} \\
\Rightarrow y(t) &= \mathcal{L}^{-1}\left(\frac{C}{u^{\frac{1}{2}}-1}\right) \\
&= t^{-\frac{1}{2}}E_{\frac{1}{2},\frac{1}{2}}(\sqrt{t})
\end{aligned}$$

Fox H Function.[17,35]

$$\begin{aligned}
H_{p,q}^{m,n}(z) &\equiv \frac{1}{2\pi i}\int_C \frac{\prod_{k=1}^{n}\Gamma(1-a_j+A_j\zeta)\prod_{j=1}^{m}\Gamma(b_j-B_j\zeta)}{\prod_{j=m+1}^{q}\Gamma(1-b_j+B_j\zeta)\prod_{j=n+1}^{p}\Gamma(a_j-A_j\zeta)} z^{\zeta}\, d\zeta \\
&= H_{p,q}^{m,n}\left[z\left|\begin{matrix}(a_1,A_1)\ \dots\ (a_p,A_p)\\(b_1,B_1)\ \dots\ (b_q,B_q)\end{matrix}\right.\right] \qquad (2.185)
\end{aligned}$$

Miscellaneous Results

(i)

$$H_{0,1}^{1,0}\left[z\left|\begin{matrix}-\\(b,1)\end{matrix}\right.\right] = z^b e^{-z} \qquad (2.186)$$

(ii)

$$H_{1,2}^{1,1}\left[z\left|\begin{matrix}(0,1)\\(0,1),(1-\beta,\alpha)\end{matrix}\right.\right] = E_{\alpha,\beta}(-z) \qquad (2.187)$$

(iii)

$$\begin{aligned}
&\mathcal{L}\left\{t^{\omega}H_{p,q}^{m,n}\left[zt^{-\sigma}\left|\begin{matrix}(a_p,A_p)\\(b_q,B_q)\end{matrix}\right.\right]\right\} \\
&= u^{-\omega-1}H_{p,q+1}^{m+1,n}\left[zu^{\sigma}\left|\begin{matrix}(a_p,A_p)\\(1+\omega,\sigma)\,(b_q,B_q)\end{matrix}\right.\right] \qquad (2.188)
\end{aligned}$$

(iv)

$$\begin{aligned}
&{}_0\mathcal{D}_z^{\nu}\left\{z^{\alpha}H_{p,q}^{m,n}\left[(az)^{\beta}\left|\begin{matrix}(a_p,A_p)\\(b_q,B_q)\end{matrix}\right.\right]\right\} \\
&= z^{\alpha-\nu}H_{p+1,q+1}^{m,n+1}\left[(az)^{\beta}\left|\begin{matrix}(-\alpha,\beta),\,(a_p,A_p)\\(b_q,B_q)\,(\nu-\alpha,\beta)\end{matrix}\right.\right] \qquad (2.189)
\end{aligned}$$

(v)

$$\mathcal{F}^{-1}\left(\exp(-D_\alpha t|q|^\alpha)\right) = \frac{1}{\alpha|x|} H_{2,2}^{1,1}\left[\frac{|x|}{(D_\alpha t)^{\frac{1}{\alpha}}} \middle| \begin{matrix} (1,\frac{1}{\alpha}),\,(1,\frac{1}{2}) \\ (1,1)\,(1,\frac{1}{2}) \end{matrix}\right] \tag{2.190}$$

The final result above defines the Lévy stable density in terms of Fox H functions.[38]

Acknowledgments

We would like to thank Claire Delides, a vacation scholar with us over the summer 2008, for a careful reading of these notes during preparation. We would also like to thank Professor Robert Dewar for bringing William Sutherland's work on Brownian motion to our attention.

References

1. R. Brown, Microscopical observations on the particles contained in the pollen of plants and on the general existence of active molecules in organic and inorganic bodies, *Edin. Phil. Journal*, July-September (1828) 358–371.
2. A. Einstein, On the movement of small particles suspended in a stationary liquid demanded by the molecular kinetic theory of heat, *Ann. d. Phys.* **17**, 549–560, (1905).
3. W. Sutherland, Dynamical theory of diffusion for non-electrolytes and the molecular mass of albumin, *Phil. Mag.* **9**, 781–785, (1905).
4. P. Langevin, Sur la thérie du mouvement brownien, *Comptes rendues Académie des Sciences (Paris)* **146**, 530–533, (1908). [English translation in *Am. J. Phys.*, **65**, 1079–1081, (1997).]
5. L. Boltzmann, Vorlesungen uber Gastheorie (Part I 1896, Part II 1898) [English translation by S.G. Brush, *Lectures on Gas Theory* Dover]
6. K. Pearson, The problem of the random walk, *Nature* **72**, 294, (1905).
7. Lord Rayleigh., The problem of the random walk, *Nature*, **72**, 318, (1905).
8. L. Bachelier, Théorie de la spéculation, *Annales Scientifiques de l'école Normale Supérieure* **3**(17), 21–86, (1900) . [English translation in M. Davis and A. Etheridge, *Louis Bachelier's Theory of Speculation: The Origins of Modern Finance*, (Princeton University Press, 2006).]
9. K. Pearson, Historical note on the origin of the normal curve of errors, *Biometrika*, **16**, 402–404, (1924) .
10. C.E. Weatherburn, *First Course in Mathematical Statistics*, (Cambridge University Press, 1949).
11. A. Fick, Über Diffusion, *Ann. Phys.* **94**, 59, (1855) ; *Poggendorffs Annalen* **94**, 59–86, (1855). [English translation: *Phil.Mag. S.4*, **10**, 30–39, (1855) .
12. L. Vlahos, H. Isliker, Y. Kominis, and K. Hizanidis, Normal and Anomalous Diffusion: A Tutorial, in T. Bountis (ed.) *Order and Chaos*, **Vol. 10**, (Patras University Press, 2008).
13. N. Wiener, Differential space, *Journal of Mathematics and Physics* **2**, 131–174, (1923). [Reprinted in *Selected Papers of Norbert Wiener* (MIT Press, 1964)]
14. G.E. Uhlenbeck, and L.S. Ornstein, "The Theory of Brownian Motion", *Phys. Rev.*, **36**, 823–841, (1930).
15. J. Perrin, *Annales de Chimie et de Physique* **8** (1909). [English translation: Brownian Movement and Molecular Reality, trans. F.Soddy (Taylor and Francis, London, 1910).]
16. J-P. Bouchard and A. Georges, Anomalous Diffusion in Disordered Media: Statistical Mechanics, Models and Applications, *Physics Reports*, **195**, Nos 4 & 5, 127–193, (1990) .
17. R. Metzler, and J. Klafter, The Random Walk's Guide to Anomalous Diffusion: A Fractional Dynamics Approach, *Phys. Reports* **339**, 1-77, (2000).
18. R. Metzler and J. Klafter, The Restaurant at the end of the Random Walk: Recent developments in the description of anomalous transport by fractional dynamics, *J. Phys. A: Math. Gen.*, **37**, R161–R208, (2004).

19. M.Z. Bazant (Ed.), *MIT Online Lecture Notes: Topics in Random Walks and Diffusion* `http://www-math.mit.edu/~bazant/teach/18.325/index.html`.
20. D. ben-Avraham and S. Havlin, *Diffusion and Reactions in Fractals and Disordered Systems*, (Cambridge University Press, 2000).
21. L.F. Richardson, Atmospheric diffusion shown on a distance-neighbour graph, *Proc. Roy. Soc. A*, **110**, 709–737, (1926) .
22. B. O'Shaughnessy and I. Procaccia, Diffusion on fractals, *Phys. Rev. A*, **32**, 3073–3083, (1985) .
23. S.A. Adelman, Fokker–Planck equations for simple non-Markovian systems, *J. Chem. Phys.* **64**, 124–130, (1976).
24. B.B. Mandelbrot and J.W. Van Ness, Fractional Brownian motions, fractional noises and applications, *SIAM Review* **10**, 422–437, (1968).
25. F. Mainardi and P. Pironi, The fractional Langevin equation: Brownian motion revisited, *Extracta Mathematicae* **11**, 140–154, (1996).
26. E. Lutz, Fractional Langevin equation, *Phys. Rev. E* **64**, 051106, (2001).
27. K.G. Wang, L.K. Dong, X.F. Wu, F.W. Zhu and T. Ko, Correlation effects, generalized Brownian motion and anomalous diffusion, *Physica A* **203**, 53–60, (1994).
28. E.W. Montroll and G.H. Weiss, Random walks on lattices. II *J. Math. Phys.* **6**, 167, (1965).
29. H. Scher and M. Lax, Stochastic Transport in a Disordered Solid. I. Theory, *Phys. Rev. B* **7**, 4491, (1973).
30. E.W. Montroll and M.F. Shlesinger, On the wonderful world of random walks, *Nonequilibrium Phenomena II: From Stochastics to Hydrodynamics*, Eds. J.L. Lebowitz and E.W. Montroll. (Elsevier Science Publishers, 1984).
31. B.I. Henry, T.A.M. Langlands and S.L. Wearne, Anomalous diffusion with linear reaction dynamics: From continuous time random walks to fractional reaction-diffusion equations, *Phys. Rev. E* **74**, 031116, (2006).
32. B. Berkowitz, A. Cortis, M. Dentz and H. Scher, Modelling non-Fickian transport in geological formations as a continuous time random walks, *Reviews of Geophysics* **44**, RG2003, (2006).
33. K.B. Oldham and J. Spanier, *The Fractional Calculus*, (Academic Press, 1974).
34. I. Podlubny, Physical interpretation of initial conditions for fractional differential equations with Riemann–Liouville fractional derivatives. *Rheol. Acta.* **45**, 765–771, (2006).
35. F. Mainardi, G. Pagnini and R.K. Saxena, Fox H functions in fractional diffusion, *J. Comput. Appl. Math.* **178**, 321–331, (2005).
36. M.M. Meerschaert and H.P. Scheffler, Triangular array limits for continuous time random walks, *Stochastic Processes and Their Applications* **118**, 1606–1633, (2008).
37. E. Barkai and R.J. Silbey, Fractional Kramers Equation, *Phys. Chem. B* **104**, 3866–3874, (2000).
38. W.R. Schneider, Stable distributions: Fox function representation and generalization, *Lecture Notes in Physics*, **262**, 497-511, (1986).
39. S.B. Yuste, L. Acedi and K.Lindenberg, Reaction front in an $A + B \to C$

reaction-subdiffusion process, *Phys. Rev. E* **69**, 036126, (2004).
40. D. Fulger, E. Scalas and G. Germano, Monte Carlo simulation of uncoupled continuous-time random walks yielding a stochastic solution of the space-time fractional diffusion equation, *Phys. Rev. E* **77**, 021122, (2008).
41. E.A. Abdel-Rehim and R. Gorenflo, Simulation of continuous time random walk of the space fractional diffusion equations, *J. Comput. Appl. Math.* **222**, 274–283, (2008).
42. W. Whitt, *Stochastic-process limits*, Springer Series in Operations Research (Springer-Verlag, 2002).
43. E. Barkai, R. Metzler and J. Klafter, From continuous time random walks to the fractional Fokker–Planck equation, *Phys. Rev. E* **61**, 132–138, (2000).
44. I.M. Sokolov and J. Klafter, Field-induced dispersion in subdiffusion, *Phys. Rev. Letts.* **97** (2006) 140602.
45. B.I. Henry and S.L. Wearne, Fractional reaction-diffusion, *Physica A* **276**, 448–455, (2000).
46. T.A.M. Langlands, B.I. Henry and S.L. Wearne, Turing pattern formation with fractional diffusion and fractional reactions, *J. Phys.: Condens. Matter* **19**, 065115, (2007).
47. I.M. Sokolov, M.G.M. Schmidt and F. Sagues, Reaction-subdiffusion equations, *Phys. Rev. E*, **73**, 031102, (2006).
48. T.A.M. Langlands, B.I. Henry and S.L. Wearne, Anomalous subdiffusion with multispecies linear reaction dynamics, *Phys. Rev. E* **77**, 021111, (2008).
49. A. Yadav and W. Horsthemke, Kinetic equations for reaction-subdiffusion systems: Derivation and stability analysis, *Phys. Rev. E* **74**, 066118, (2006).
50. D. del-Castillo-Negrete, B.A. Carreras, V.E. Lynch, Fractional diffusion in plasma turbulence, *Phys. of Plasmas* **11** (2004) 3854–3864.
51. A. Cartea, Fractional diffusion models of option prices with jumps, *University of London* Birkbeck Working Paper No. 0604 (2006)
52. B.I. Henry, T.A.M. Langlands and S.L. Wearne, Fractional cable models for spiny neuronal dendrites, *Phys. Rev. Letts.* **100**, 128103, (2008).
53. A. Clauset, C.R. Shalizi and M.E.J. Newman, Power law distributions in empirical data, *arXiv:0706.1062v1* 2007.
54. H. Scher and E.W. Montroll, Anomalous transit time dispersion in amorphous solids, *Phys. Rev. B.* **12**, 2455–2477, (1975).
55. M.F. Shlesinger, J. Klafter and G. Zumofen, Above, below and beyond Brownian motion, *Am. J. Phys.* **67**(12), 1253–1258, (1999).
56. K. Miller and B. Ross, *An Introduction to the Fractional Calculus and Fractional Differential Equations* (John Wiley & Sons, New York, 1993).
57. I. Podlubny, *Fractional Differential Equations* (Academic Press, New York, 1999).
58. I. Podlubny, Geometric and physical interpretations of fractional integration and differentiation, in A.Le Mehauté, J.A.T. Machado, J.C. Trigeassou and J. Sabatier (Eds) *Fractional Differentiation and its Applications*, (U Books, ISBN 3-86608-026-3, Germany, 2005).

Chapter 3

Space Plasmas and Fusion Plasmas as Complex Systems

R. O. Dendy[1,2]

[1]*Euratom/UKAEA Fusion Association, Culham Science Centre, Abingdon, Oxfordshire, OX14 3DB, U.K.*

[2]*Centre for Fusion, Space and Astrophysics, Department of Physics, Warwick University, Coventry, CV4 7AL, U.K.*

Contents

3.1. Introduction

Plasma physics[1] has proven exceptionally fruitful as a field of application for the concepts and techniques of complex systems science.[2–4] At first sight, this may appear paradoxical: the most accessible plasmas[5]—fusion experiments, Earth's magnetosphere, and the solar wind—possess high temperatures or low densities, implying that their constituent particles combine sparseness with random motion. This may appear unpromising as a basis for the deployment of complex systems science. Nevertheless plasmas in fact exhibit great diversity of collective nonlinear processes, operating over an exceptionally wide range of lengthscales and timescales, whose coupled interactions give rise to the observed phenomenology. The need to interpret, predict, and control this emergent phenomenology is central to the mission of plasma physicists, whether for fusion or in the geospace environment.

A handful of properties intrinsic to the plasma state give rise to its ability to self-organise on multiple lengthscales and timescales. First, being fully ionised, a plasma is permeated by electric and magnetic fields which can in principle act over distances of arbitrary length. The low inertia and consequent high mobility of the free electrons enables them to respond rapidly to any field, furthermore the instantaneous spatial distribution of the electrons and of any electric current arising from their motion act as sources of electric and magnetic field. In any volume of plasma , the distribution of charge, current, and electromagnetic field at any moment represents a self-consistent solution of the coupled nonlinear equations governing the dynamics of multiple charged particles together with Maxwell's equations. Given the enormous number of particles involved, which is of order Avogadro's number per gram of plasma, direct mathematical solution is impossible and physically motivated truncation—the construction of reduced equations—is called for.

This brings us to the second key feature of the plasma state. Plasmas typically coexist with magnetic fields, whose energy density is often comparable to the thermal or bulk flow kinetic energies of the plasma. Examples include Earth's magnetosphere, where large scale magnetic fields arise from the geodynamo or are convected from the solar corona by the solar wind, and fusion plasmas where, in the most successful tokamak[6] configuration, magnetic fields arise from internal plasma currents and from external current-carrying coils. It follows that magnetic fields affect the system's behaviour on all lengthscales and timescales from those of electron gyromotion to those characteristic of the entire system—these differ respectively by factors of 10^4 and 10^{10} in fusion experiments, for example. Phenomenology within specific intermediate ranges of lengthscales and timescales can be addressed mathematically using different levels of description obtained by truncating, or averaging over, the fullest kinetic description. For example, averaging over gyro-angles yields the gyrokinetic description, while averaging over lengthscales long compared to gyro-orbits yields the magnetohydrodynamic (MHD) fluid description. Importantly, however, phenomena that are primarily describable using a particular level of model are often strongly coupled to those operating at another level. For example, the stability of quasi-fluid MHD modes of oscillation can be critically affected by kinetic resonant wave-particle interactions.

In general, the overall phenomenology of a plasma system emerges from multiple couplings between processes on different lengthscales and timescales, creating innumerable feedback loops. For example, the overall energy confinement—the central emergent property for fusion applications—is governed in part by a self-consistent loop involving: temperature, density, and current profiles extending from the extremely hot core to the cold edge; collective instabilities which are locally triggered when profile gradients exceed certain limits, and which draw their free energy from the profiles; the saturated turbulence which is driven by these instabilities; and the energy transport resulting from this turbulence, which in turn reacts back on the temperature and current profiles. The lengthscales associated by the key coupled physical processes within the turbulence-and-transport loop extend from the thermal ion gyroradius to the system size. Fusion plasma confinement properties thus emerge from self organisation within a complex system. This fact was early recognised by one of the field's most distinguished founders,[7] but not always pursued energetically thereafter. For a recent review, see Ref. 8, which is drawn on for some of the subsequent material in this paper.

These considerations apply *a fortiori* to the next step fusion experiment, ITER,[9] where the plasma itself will assume control of key degrees of freedom hitherto governed externally. Specifically, plasma heating will be dominated by alpha-particles born at 3.5 MeV in fusion reactions between thermal (10–20 keV) deuterium and tritium nuclei. The local birth rate of fusion alpha-particles, and their associated collisional heating profile across the plasma, is determined by temperature and density profiles. These profiles will in turn respond to heating by alpha-particles, while also affecting the current profile through the temperature dependence of plasma conductivity, thereby linking to the transport-and turbulence loop outlined in the preceding paragraph. Significant—but not dominant—heating of the plasma by fusion alpha-particles has been demonstrated[10] in the JET experiment. Alpha-particles introduce additional key lengthscales and timescales into the system, due to their distinctive (because energy-dependent) orbital characteristics in the confining toroidal magnetic field. Resonance between characteristic frequencies of alpha-particle motion and of bulk MHD oscillations above 100 kHz can be an important channel for energy transfer within the plasma. From a complex systems perspective, this further illustrates the centrality of coupling between phenomena that, in isolation, arise at different levels of description.

It is clear from the foregoing outline that fusion plasmas, and their solar-terrestrial cousins, are complex systems in the strict sense. To develop this point, let us now focus on the generic properties of complex systems (represented in italic text hereafter), as for instance listed in the Preface of this book and in the first chapter, and examine their role in plasma physics.

Plasmas in fusion and space exhibit *emergence*, meaning that *some properties present at system level are not present at lower level*. This applies in various senses. First, we have already outlined how overall energy confinement is a property that emerges only at *system level* from the interplay of *coupled* physical processes operating across a *hierarchy* of lower levels, reaching down to single particle dynamics. Furthermore, as already described, each level of description (single fluid; two fluids—electrons and ions; kinetic ions and fluid electrons; gyrokinetic; and so on) within this hierarchy is determined by the characteristic lengthscale and timescale of whichever physical process dominates at that level. Plasmas are thus *multiscale*, and we have seen that the different levels of description and associated observed phenomenology *extend over several orders of magnitude, and distinct properties and functions are associated with different scales*. Returning to a second aspect of emergence, plasmas self organise persistent coherent macroscopic structures that only arise on lengthscales at, or just below, *system level*. Examples include magnetic islands[11] and zonal flows,[12] which *are not present at lower level*.

Plasmas are invariably *open*, in the sense that *energy and information are constantly being imported and exported across system boundaries*. The quest for fusion power from magnetically confined plasmas involves injecting energy at the 10 MW level into a gram of material occupying a volume of tens of cubic metres, for example using energetic neutral particle beams and radio frequency waves, with a view to overcoming natural loss from, for example, Bremsstrahlung and collisional and turbulent plasma energy transport processes. The role of information in plasmas, which is arguably underexplored and is highly topical, is briefly outlined in a separate paragraph below. The fact that such plasmas sustain, over seconds, the steepest steady-state temperature gradients known, while subject to energy fluxes of order 10 MWm^{-2}, is an instance of their ability to *adapt: in response to external or internal changes, the system can reorganise itself without breaking*.

Plasmas are *not completely predictable: unexpected behaviours can emerge—prediction becomes expectation*. For example, fusionplasma performance in future fusion experiments such as ITER is extrapolated using empirical dimensionless scaling laws in the absence of first principles predictions of global phenomenology. Key behaviours, such as enhanced confinement operating regimes[13] and edge localised modes (ELMs),[14] were not predicted. Transitions between confinement regimes typically have a *history: even a small change in circumstances can lead to large deviations in the future*, and reflect the existence of *multiple metastable states*. These transitions can occur spontaneously as plasma conditions evolve in time, or can be induced by careful sequencing of external drivers, notably auxiliary heating and fuelling. In both cases, history is crucial and there is an element of irreversibility.

Returning to the opening paragraph of this paper, we note that the typical very low density and high temperature of plasmas in magnetically confined fusion experiments, for example, implies a high degree of *disorder* at the lowest level of description, namely the self-consistent dynamics of charged particles and electromagnetic field. For this reason, *there is no compact and concise way to encode the whole information contained in the system.* On the one hand, particle-in-cell codes which implement this lowest-level description are best adapted to phenomena occurring on the fastest timescales and shortest lengthscales, for example in laser-plasma interactions where attosecond phenomenology is now being probed. On the other, as noted, higher level descriptions are reduced models; to construct these, information has deliberately been dropped.

Reference to the *information contained in the system* opens other doors. It appears that information theory may in future provide central unifying principles for complex systems science and plasma physics alike. Irrespective of their physical and mathematical embodiment, all complex systems have in common the creation, transmission, sharing and destruction of information. It is the ebb and flow, birth and death of information—a physical quantity—that underlies and enables the physical phenomenology. Quantifying the state and distribution of information within a complex system is thus crucial both to understanding its working, and to rigorously characterising its behaviour. We outline below some pioneering studies of mutual information[15–17] in the solar wind plasma upstream of Earth, using techniques tested[18] on standard complex systems models for the collective dynamics (i.e., flocking) of birds. While significant progress is being made, a fundamental question of great interest arises. Namely, the relation between: information evaluated at the bulk level of description—for example, mutual information used as a measure of nonlinear correlation between spatiotemporally separated but causally linked fluid flows; and information evaluated when finer structure—for example, in measurements of magnetic field fluctuations—is resolved in the same system.

In the remainder of this paper, we describe some results of applying a complex systems perspective to plasma physics. There are two main lines of attack, as follows.

First, complex systems science seeks to identify simple universal models[2,19] that capture the key physics of extended macroscopic systems, whose behaviour is governed by multiple nonlinear coupled processes that operate across a wide range of spatiotemporal scales. In such systems, energy release often occurs intermittently, in bursty events, and the phenomenology can exhibit scaling, that is, a significant degree of self-similarity. Within plasma physics, such systems include Earth's magnetosphere, the solar corona, and toroidal magnetic confinement experiments. Guided by broad understanding of the dominant plasma processes—for example, turbulent transport in tokamaks, or magnetic reconnection in some space and solar contexts—we construct minimalist complex systems models that yield relevant global behaviour. Examples outlined below include the sandpile approach to tokamaks[20–25] and to the magnetosphere,[26–33] and a reconnecting[34] multiple-loops model[35] for the solar coronal magnetic carpet.[36] Such models can address questions that are inaccessible to analytical treatment and are too demanding for computational resolution. These models are useful, and potentially valid, if they can replicate aspects of observed global phenomenology, or of event statistics, for which no explanation has been obtained from first principles including the underlying equations. For example, a simple sandpile model,[37] which implements critical-gradient-triggered avalanching transport associated with nearest-neighbour mode coupling, generates[23,25] some of the distinctive observed elements of tokamak confinement phenomenology such as ELMing and edge pedestals. The same sandpile model also generates distributions of energy-release events whose statistics resemble those observed in the auroral zone.[29] Similarly, a simple multiple-loops model,[35] which implements random photospheric footpoint motion combined with reconnection of intersecting oriented loops, generates global magnetic field structure resembling the solar coronal magnetic carpet, with power law distributions for energy-release events that are similar to those observed in the solar corona.[38–41]

Finding complex systems models whose behaviour resembles that of large-scale plasmas helps to identify the dominant physical processes that govern the observed phenomenology. This is otherwise difficult, given the many interacting plasma physics mechanisms, operating on diverse lengthscales and timescales and in nonlinear regimes, that combine together to produce the effects observed. Furthermore, such models may enable one to isolate a small set of control parameters - perhaps representing the combined effects of many experimental variables—whose values determine system behaviour. A simple model that captures the key physics can provide a means to both understand and control the overall system behaviour.

Second, complex systems science offers an array of statistical physics techniques that are specially adapted to capturing and quantifying the nonlinear features of macroscopic system behaviour. The review paper Ref. 42 describes how several techniques of nonlinear time series analysis were successfully applied to a variety of plasmas including astrophysical accretion discs,[43,44] the solar corona,[45] the solar wind and terrestrial ionosphere,[15,16] edge fluctuations in the MAST tokamak,[46] and ELM statistics in the JET tokamak.[47] In the present paper this aspect is brought up-to-date with further applications—some novel—to the MAST tokamak,[48] LHD stellarator,[49] and solar wind.[17,18,50]

3.2. Complex Systems Models for Global Tokamak Phenomenology

As noted by Kadomtsev[7] in 1992, for example, diffusive and Gaussian paradigms for the transport arising from turbulence in tokamak plasmas cannot account for all the confinement phenomenology observed. The first measurements of avalanching transport emerged later in the 1990s.[51,52] Analysis of edge plasma turbulence measurements[24,46,48,49,53–61] has consistently yielded non-Gaussian probability distribution functions (PDFs) that are long-tailed, pointing towards intermittency. Bursty transport is seen in a wide range of numerical simulations using different models;[62–69] for a recent review see Ref. 70. In parallel, there remains the question: why does the basic phenomenology of magnetically confined plasmas—enhanced confinement regimes, edge pedestals, ELMs, and so on—arise at all? And could its existence have been predicted by analogy with other physical systems?

These questions motivate the application to plasma phenomenology of the sandpile paradigm for rapid nonlocal nondiffusive transport events arising from critical gradient-triggered nearest-neighbour interactions. Consider, for example, the simple one-dimensional N-cell sandpile model of Ref. 23, further explored in Refs. 25 and 71, which incorporates other established models[19,72,73] as limiting cases. This is a centrally fuelled (at cell $n = 1$) model, whose distinctive feature is its rule for local redistribution of sand at a cell (say at $n = k$) when the critical gradient z_c is exceeded

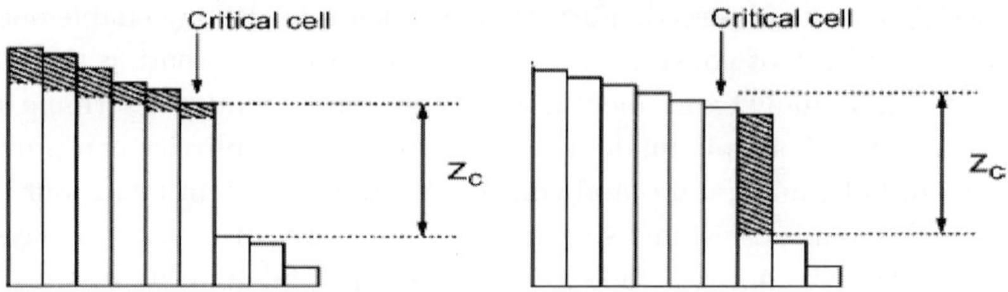

Fig. 3.1. Section of sandpile prior to flattening (Left), and after flattening (Right), for the case $L_f = 6$. The gradient at the critical cell initially exceeds z_c, so that sand from this cell and its five nearest leftward neighbours is transported downhill to its nearest rightward neighbouring cell. From Ref. 71, for which we see further discussion of boundary conditions.

there. The sandpile is flattened behind the unstable cell over a "fluidisation length" L_f, embracing the cells $n = k-(L_f-1), k-(L_f-2), ..., k$; and this sand is conservatively relocated to the cell at $n = k+1$; see Fig. 3.1. As in all sandpile models, the system is then iterated to stability, which generates an avalanche as redistribution is triggered sequentially across neighbouring cells, before it is fuelled again. The lengthscale L_f governs rapid redistribution, so that it is a proxy for turbulent vortex size, for example.

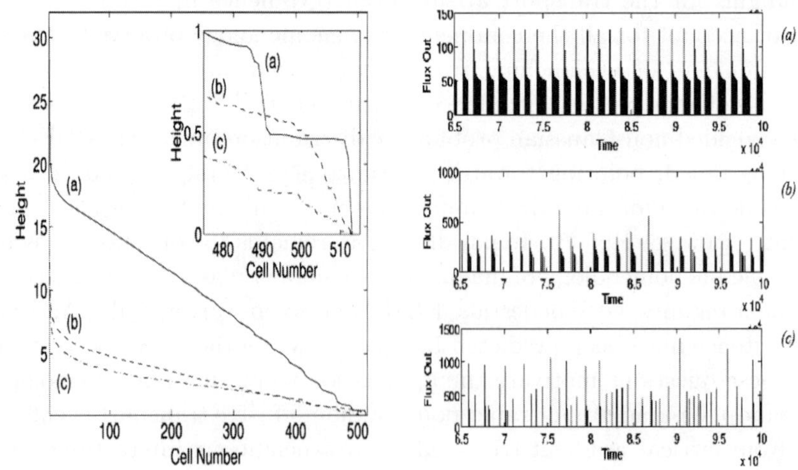

Fig. 3.2. Enhanced confinement, edge pedestals, and pulsed mass loss events in a sandpile, see Ref. 23. (Left): Time averaged height profiles of the 512 cell sandpile for $L = (a)50, (b)150, (c)250$. Inset: edge structure. (Right): Time series of external avalanches (MLEs) for the 512 cell sandpile for $L = (a)50, (b)150, (c)250$; these plots show magnitude of flux leaving the sandpile, versus time.

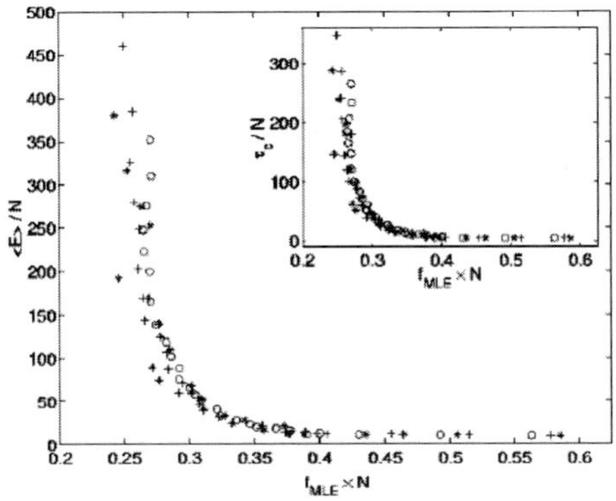

Fig. 3.3. Averaged stored energy versus frequency of mass loss events (MLEs) for the sandpile with number of cells N = 512, 4096, and 8192, see Ref. 23. Normalisation with respect to N shows robust scale invariance of this phenomenology. Inset: plot of confinement time versus MLE frequency shares features with measured correlation[74] of energy confinement with ELM frequency in some JET plasmas.

The normalised redistribution lengthscale L_f/N is the model's only control parameter, which governs different regimes of avalanche statistics and system dynamics. This sandpile model displays[23,25,71] phenomenology similar to several key features of tokamak confinement phenomenology: edge pedestals and enhanced confinement,[23] see Fig. 3.2 (Left); ELMing,[23] see Fig. 3.2 (Right); the dependence of ELM frequency on stored energy,[23] see Fig. 3.3; internal transport barriers;[25] and off-axis ECRH temperature profiles.[71] The existence of enhanced confinement and edge pedestals for this sandpile is complemented by the pulse-like time series for its external avalanches ("mass loss events", MLEs, see Fig. 3.2 (Right)), whose role mimics that of ELMs in tokamak plasmas. Not only does the character of the MLEs correlate with the confinement properties of the sandpile; there are also quantitative correlations. For example, Fig. 3.3 shows the scaling of the frequency of the MLEs with stored energy in the sandpile, which is similar in form to that obtained for the scaling of ELM frequency with stored energy in JET for certain plasmas, see Fig. 3.6 of Ref. 74. Figure 3.3 also confirms that the phenomenology is scale-invariant with respect to the number N of cells in the implementation of the sandpile model.

The emergence, from this paradigmatic complex system,[23] of counterparts to key aspects of tokamak confinement phenomenology appears significant. This sandpile algorithm provides a simple one-parameter model for studying generic nonlocal transport, conditioned by a critical gradient, in a macroscopic confinement system. Changing the value of the single control parameter L_f corresponds to altering the spatial range over which the transport process operates; different values of L_f would reflect different properties of the plasma turbulence underlying the transport. It follows that this small set of mathematical ingredients may be all that is required to generate the aspects of tokamak confinement phenomenology described above. In addition one may attach weight to the observations of avalanching transport in tokamaks and in largescale numerical simulations thereof, and therefore regard the avalanching transport that is built into sandpile algorithms as an additional point of contact with the physics of magnetically confined plasmas. It would then follow from the sandpile results that tokamak observations of avalanching transport are deeply linked to the existence of enhanced confinement and ELMs; and that ELMs may be a local manifestation of a global process. The existence of the single control parameter L_f, governing the confinement phenomenology and arising from the rapid transport, implies that it may be possible be combine the many experimental parameters into a very small number of underlying control parameters. We note also that large L_f, corresponding to low confinement [trace (c) of Fig. 3.2 (Left)], is known[23,73] to correspond to robust scale-free systems dynamics; whereas small L_f, corresponding to high confinement [trace (a) of Fig. 3.2 (Left)], does not give rise to self-similar dynamics. Pursuing this aspect of the analogy with the sandpile, the tokamak H-mode would correspond to low-dimensional behaviour, with the dynamics of the core only loosely coupled to quasi-oscillatory edge dynamics; while the tokamak L-mode would correspond to high-dimensional behaviour, with coupling extending across the system.

3.3. Complex Systems Modelling of Earth's Magnetosphere

At its most basic level, Earth's magnetosphere resembles a tokamak, in that it is a macroscopic magnetic confinement system for plasma set up by the interaction between a flow—the solar wind, instead of a toroidal current—and a magnetic field anchored in dense electrically conducting matter—the Earth's core, instead of copper or superconducting coils. Fusion and magnetospheric plasmas are driven-dissipative systems that combine energy injection, storage, and distinct release events; they exhibit structure and display phenomenology on a very broad range of lengthscales and timescales. Complex systems science should therefore provide insights into magnetospheric as well as fusion plasma physics.[33] Here we focus on applications of the sandpile models introduced above. Motivation for the sandpile approach to the magnetosphere arises both *a priori* and observationally, through analysis of the statistics of energy release events. For example, terrestrial magnetometer measurements of geomagnetic indices, which reflect energy deposition in the auroral zone arising from plasma transport of energy released in magnetic reconnection events in the distant magnetotail, can yield scale-free power law distributions[75,76] resembling those of avalanching processes in sandpiles, as was noted in Refs. 77 and 78 reviewed further in Refs. 79 and 80. These statistics appear to be intrinsic[81] to the internal plasma physics of the magnetosphere, rather than conditioned by the ultimate driver, the solar wind. The sandpile model of Ref. 78 yields internal

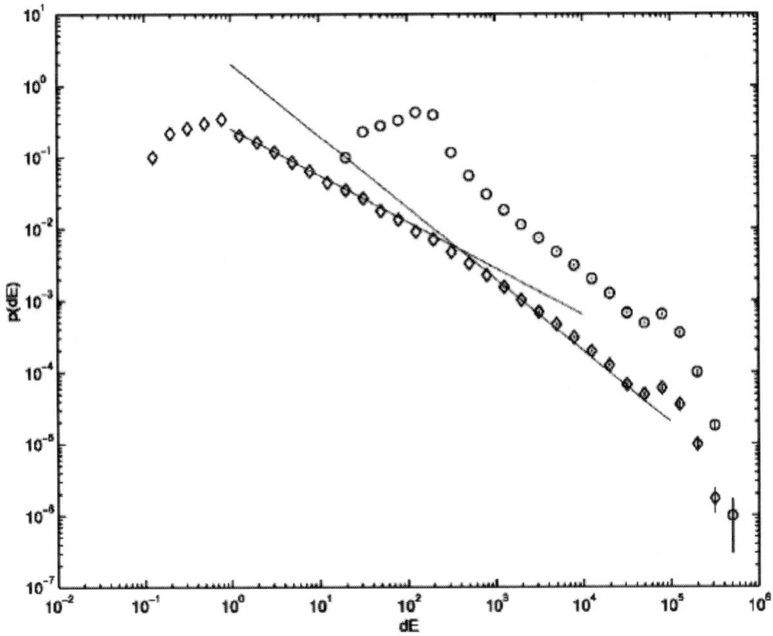

Fig. 3.4. Logarithmic plot of the measured distribution p(dE) of avalanche energy dE in a 5000 cell sandpile with $L_f = N$, for two central fuelling rates: slow (diamonds), and ten times faster (circles). Rollover at left reflects finite size of discrete fuelling events; power law slope is -1 above $dE = 10^3$; systemwide avalanches are visible at right, peaking near $dE = 10^5$. From Ref. 80.

avalanches (involving transport of sand within the system, but no mass loss) whose distribution is scale-free; and external avalanches (involving

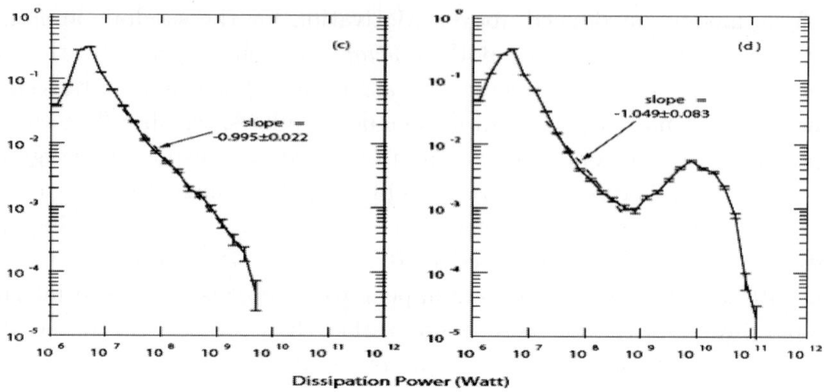

Fig. 3.5. Observed logarithmic frequency-magnitude plots of UV emission from Earth's auroral oval obtained by the POLAR UVI instrument. (Left) Quiet. (Right) Concurrent substorm activity. Comparison with Fig.6 suggests that the sandpile and magnetosphere share global phenomenology, with substorms driving systemwide avalanches. From Ref. 83.

transport of sand right across the system, resulting in some loss of sand) whose distribution has a well defined mean and hence an intrinsic scale. At the simplest level, this model could in principle encompass both intermittent local magnetospheric energy release events, arising for example from bursty bulk flows and pseudobreakups, as well as global energy release events, such as substorms. Furthermore the model is known to be robust[80,82] against substantial temporal fluctuations in the magnitude of the fuelling, and against different rates of fuelling. For example, Fig. 3.4 (from Ref. 80) plots the distribution of avalanche energy against the number of events for two types of central fuelling: slow (diamonds) and fast (circles). This is important for space and astrophysical applications where the driver may be highly variable, for example the solar wind, and differs from the classical self-organised criticality (SOC)[1] picture of slowly driven dynamics. The rollover at small energies in Fig. 3.4 reflects the minimum size of discrete energy packets with which the systems are fuelled; both distributions display a robust power law with slope -1 at higher energy; and the distinct distribution of systemwide avalanches with a well defined mean is visible at the highest energies.

The similarity between Fig. 3.4 and the observed distribution of certain magnetospheric energy release events, reproduced here as Fig. 3.5, was noted by Lui *et al.* in Ref. 83. Global snapshot images of ultraviolet emission across the entire auroral oval (see for example Fig. 3.3 of

Ref. 33) were obtained by the POLAR UVI instrument looking down on Earth. Analysis[83] of 9033 images taken during January 1997 gives rise to the magnitude-frequency plots of Fig. 3.5. The data are split into two categories: active (Fig. 3.5 (Right)), when there is concurrent substorm activity in the magnetotail, which gives rise to spatially extended UV emission spanning the auroral oval; and quiet (Fig. 3.5 (Left)), when substorm activity is absent, and auroral UV emission consists of more localised blobs. There is a straightforward mapping between the salient features of Fig. 3.5 (Right) and those of Fig. 3.4: the observed event distribution of UV emission from the auroral oval matches that from a sandpile that has blobby fuelling and displays both systemwide and internal avalanches. Identification of the systemwide avalanches with substorms, and of the internal avalanches with an ever-present process of auroral activity, is reinforced by Fig. 3.5 (Left). Taken at times when substorms are absent, Fig. 3.5(Left) has no population of large events with well-defined mean, unlike Fig. 3.5 (Right), but displays the same distribution of smaller scale events which are fitted by a power law. The fact that this slope is independent of the level of activity in the system suggests that it reflects continual underlying intermittent bursts associated with internal reconfiguration.

3.4. A Global Complex Systems Model for Aspects of Solar Coronal Plasma and Magnetic Fields

Magnetic loops are pushed out of the solar photosphere into the corona, giving rise to an inhomogeneous pattern of magnetic flux that is anchored to the conducting photosphere. Turbulent plasma flow on the photospheric surface drives the anchored flux loops into complex, stressed configurations.[84] When local magnetic field gradients become sufficiently steep, plasma instability allows the coronal magnetic field to change its topology via reconnection, suddenly releasing energy.[85] A cascade of reconnecting flux loops may be a mechanism underlying solar flares, with the larger flares originating in regions of strong coronal fields.

The range of lengthscales and timescales on which the underlying plasma processes operate, and on which the myriad loops arise and evolve, renders first principles modelling of the full system impossible. A complex systems model has therefore been proposed,[86] in which multiple directed loops evolve in space and time. It aims to capture the essentially topological nature of reconnection events in the solar coronal plasma, and handle a very large number of loops, while necessarily omitting the detailed physics of the reconnection process. In this multiple-loops dynamic model,[86] a pair of footpoints, having opposite polarity, anchors each directed loop to a two dimensional surface, representing the photosphere. These multiple magnetic loops undertake randomly driven motion at their footpoints, and can interact. Nearby footpoints of the same polarity aggregate; furthermore when loops intersect they can reconnect (see Fig. 3.6, Left) by exchanging footpoints, if this lowers the combined length of the pair. Depending on

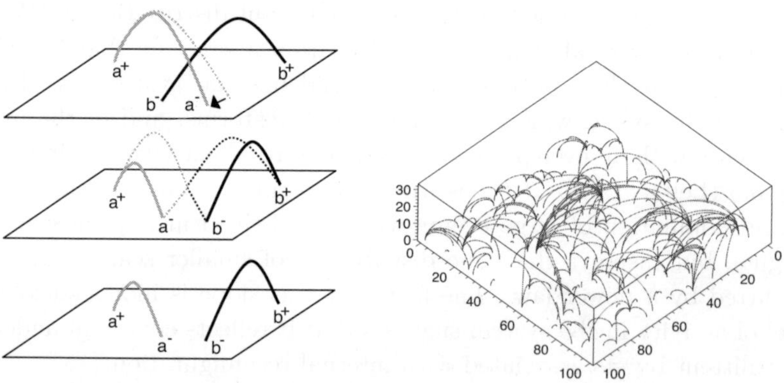

Fig. 3.6. (Left): Diagram showing the process of a reconnection event in the multiple-loops model of Ref. 86. In the top frame, loop *a* moves from its previous position (dashed line) and crosses loop *b*. Subsequently (middle frame) the loops exchange footpoints, and move to their final relaxed configuration (bottom frame). (Right): Snapshot of a configuration of multiple loops in steady state; see online Ref. 86 for a colour version.

the local density of neighbouring loops, and their configuration, a single reconnection can lead to further loop interactions, and thereby trigger a cascade of further reconnection. These cascades lower the overall length of loops in the system, and are identified with the energy release events occurring as solar flares. Loops are injected at small scales, so that, combined with random footpoint motion, an energetic and statistical steady state is achieved.

Numerical implementation[86] of this model gives rise to loop configurations (see Fig. 3.8, Right) that are qualitatively similar, in the steady state, to the magnetic carpet deduced from observations of the photospheric field.[87] The term magnetic carpet refers to the structure of field lines embedded in the coronal plasma, in regions that are not dominated by major structures such as flares, prominences, and active regions; it is characteristic of periods of low solar activity. The model yields a pattern of loops that dynamically forms a scale free network, where the number of loops emerging from each footpoint is distributed as a power law—a prediction which could be tested by solar magnetic field observations in future. A power law distribution of flare energies emerges from the model, and the power law index ≈ -3 obtained[86] is the same as that measured[44] for full-disc EUV/XUV solar irradiance during a low activity period in 1996; see Fig. 3.7, which compares Fig.4 of Ref. 86 with upper Fig. 2 of Ref. 44. The model outcomes resonate with the observed statistics of peak flare X-ray flux distributions,[88] of the energy released,[89] and of the quiescent time intervals between solar flares:[90] these are all characterised by power law distributions, with power law indices that are independent of the phase of the solar cycle.[91] The distribution of energy released is particularly striking, exhibiting scale-free behaviour over more than eight decades in energy. It is well known[92] that these statistics suggest that the solar corona may be in a state of SOC, sharing some common features with other intermittent scale free phenomena, including some in the geosciences and life sciences.[2,93] This complex sys-

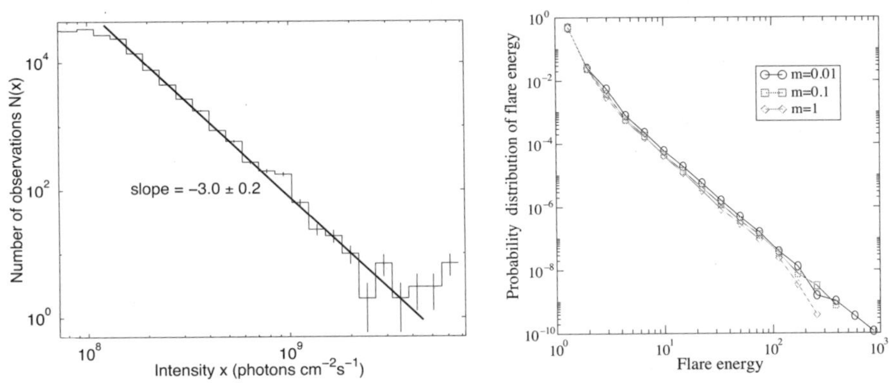

Fig. 3.7. Similarity between distributions of magnitudes of energy release events observed for the solar corona and obtained from the multiple loops model: a power law with index -3 in both cases. (Left): Logarithmic plot of number of observations N(x) of detrended full-disc solar flux intensity x, measured at 15s intervals by SOHO/SEM from January to June 1996, reproduced from Ref. 44. (Right): Logarithmic frequency-magnitude plot of energy release events occurring in the multiple loops model, reproduced from Ref. 86.

tems model[86] of the solar coronal magnetic field enables a global description that involves many more loops, and extends over greater lengthscales and timescales, than are accessible to analytical and computational techniques that start from the full equations. The striking and unpredictable outcome of the multiple loops reconnection model is the dynamical self-organisation of the magnetic field embedded in the coronal plasma, which gives rise to a power law distribution of solar flare energies, and which forms a scale-free network that qualitatively resembles the actual coronal magnetic field. This outcome is presumably independent of the specific details of the rules of the multiple loops model. From a complex systems perspective, the suggestion is that the solar coronal magnetic field behaves globally as it does because that is where it is led by the key physics elements captured in the multiple loops model.

3.5. Scaling Properties of Turbulent Fluctuations in Fusion Plasmas

The extent to which universal properties are displayed by edge plasma turbulence in toroidal magnetically confined plasmas is an important but unresolved question. Particularly interesting is the identification of generic features that may be shared by edge plasma turbulence in the three most promising confinement concepts: conventional tokamaks, spherical tokamaks, and stellarators. Generic features would arise in all three confinement systems, and would display universal characteristics; for example, their statistical properties, when rescaled with respect to the size of the device and other key bulk parameters, would be the same. Their identification thus requires quantitative comparison of the measured turbulence properties under different operating regimes for the different confinement systems, which in turn requires the application of modern techniques for the statistical analysis of nonlinear time series. As an example, we consider recent analyses[46,48,49] of probe measurements of the ion saturation current I_{sat} obtained from the plasma edge in the world-leading spherical tokamak MAST[46,48] and in the world's leading stellarator, the Large Helical Device.[49] In the latter case we address datasets previously described in Ref. 94.

We first examine scaling properties of the absolute moments $S_m = \langle|\delta x(t,\tau)|^m\rangle \propto \tau^{\zeta(m)}$ of fluctuations $\delta x(t,\tau) = \sum_{t'=t}^{t+\tau-1}$ with varying temporal scale τ, and obtain scaling exponents $\zeta(m)$. The results shown in Fig. 3.8, for example, demonstrate that the functional form of $\zeta(m)$, and the values of scaling exponents are robust quantitative discriminators of the plasma turbulence. In addition to a well defined discontinuity at 40μs, plots of $\log(S_m)$ versus $\log(\tau)$ demonstrate self-similar scaling $\zeta(m) = \alpha m$. The value of α is the same for all the MAST L-mode plasmas considered

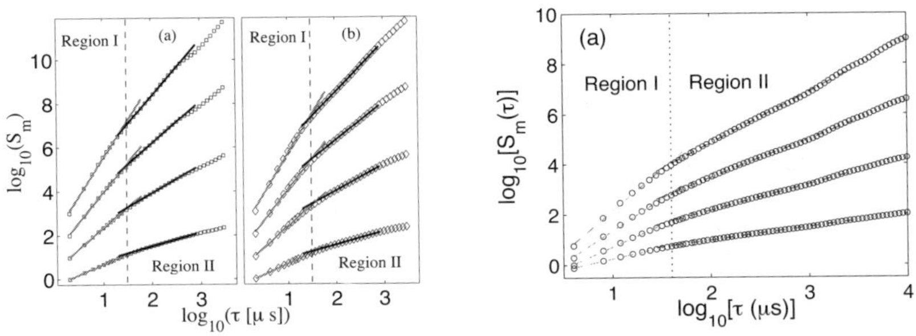

Fig. 3.8. Logarithmic plots of structure functions S_m of order $m = 1$ to 4 versus temporal scale τ for edge turbulence measurement in: (Left)MAST from Ref. 48; (Right) LHD from Ref. 49.

on timescales up to 40μs, suggesting universality in the character of these fluctuations. On longer timescales 40μs to 400μs, two distinct groups of scaling exponents are found, that exhibit weak dependence on magnetic field structure.[48]

Significant quantitative results[48,49] also arise from investigations of the probability density functions (PDFs) of the I_{sat} fluctuations. These can be strongly non-Gaussian, in particular they are sometimes long-tailed through having significantly more large events. The PDF of the fluctuations from the MAST plasmas considered, sampled on a timescale τ=2μs, is well fitted by an extremal[4] Fréchet distribution with index $a = 1.25$; see Fig. 3.9 (Left). For individual MAST plasmas, Fréchet distributions give the best fit for $\tau \leq 40\mu$s, and Gumbel for $\tau \geq 40\mu$s. This transition at 40μs, which may correspond to filamentary structures observed in optical imaging, is confirmed by the structure function scaling properties $S_m \sim \tau^{\zeta(m)}$ noted in Fig. 3.10 (Left). Figure 3.9 shows that the short timescale fluctuations in MAST, and those at one location in LHD, are both well fitted by extreme value distributions, whereas a simple Gaussian provides a good fit to the LHD distribution at a second nearby location. This sheds light on shared, and different, physical behaviour at two levels. There is a broad question: to what extent are the measured statistical properties similar? And there is a more technical question: given that the stellarator edge magnetic structure encompasses both regular and stochastic field line regions, do these affect local turbulence measurements and do they relate to spherical tokamak scenarios where the edge magnetic field is deliberately stochasticised? Experiments involving the latter process are currently under way in MAST.

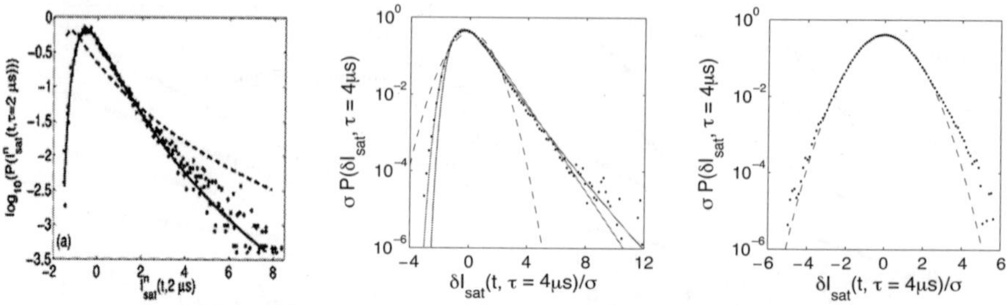

Fig. 3.9. Probability density functions for turbulent edge plasma fluctuations. (Left): Sampled on two-microsecond timescales in the MAST edge plasma. (Centre): Sampled at a particular edge location in LHD on four-microsecond timescales. (Right): Sampled in LHD simultaneously with the central panel, at a location only a few millimetres away. Fitted curves relate to extreme value distributions, see Refs. 48 and 49 for details.

3.6. Large Scale Astrophysical Objects as Complex Systems

In a striking demonstration of their wide applicability, extreme event distributions[4] have been shown[43,44] to fit the PDFs of X-ray signal intensities emitted from the astrophysical accretion disc plasmas in the microquasar GRS 1915+105 and the black hole X-ray binary Cygnus X-1; see Fig. 3.10. These accretion discs are formed by ambient plasma drawn towards, and orbiting around, stellar mass black holes, whose angular momentum prevents direct infall. Viscous outward transport of angular momentum within the disc enables matter at its inner edge to fall inwards, heating and radiating as it does so. Parenthetically, the question whether accretion discs may be in a state of self organised criticality is discussed in Ref. 95. As noted above,[4] extreme value distributions result from repeatedly selecting the maximum value from each of a large number of large samples—the brightest among many contemporaneous local flashes scattered across an extended object, for example. The fact that the tails of the PDFs in Fig. 3.10 can be fitted by extreme value distributions therefore suggests that the observed signals may be the brightest among multiple events occurring within each measurement window.

The observed signals are also found[43] to be self-similar. While self-similarity, like the non-Gaussian PDFs, is a strong indicator of highly-correlated processes such as turbulence, there are no *a priori* reasons to expect it in these contexts. Consider first the difference between the maximum and minimum values of a data time series $y(t)$ during an interval Δt; this defines its range $R(\Delta t)$ for that interval. Run a window of width Δt

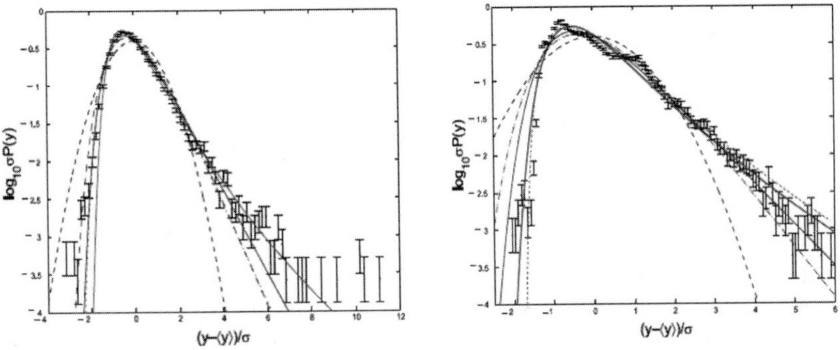

Fig. 3.10. Long-tailed probability density functions of the X-ray signal intensity[43] observed from (Left) Cygnus X-1 and (Right) GRS 1915+105 accretion disc plasmas, showing strong deviation from the Gaussian (dashed black lines). The coloured curves represent best fits of different classes of known extreme value distribution functions to these long-tailed datasets.

through the entire dataset, yielding a value of $R(\Delta t)$ at each step. From these, an ensemble-averaged value $< R(\Delta t) >$ is calculated, and this operation is then repeated for windows of different size Δt. If the time series $y(t)$ is self-similar, the ensemble-averaged value of the range scales with window size: $< R(\Delta t) >= c\Delta t^H$, where c and H are constants. This equation defines the Hurst exponent H which quantifies the self-similar growth of range. Figure 3.11 displays the results of this procedure for the Cygnus X-1 and GRS 1915+105 datasets whose PDFs are shown in Fig. 3.10, and for a set of edge turbulence measurements from the MAST tokamak. The well defined Hurst exponents demonstrate that the underlying nonlinear plasma physics gives rise to signals that are self-similar to a significant extent, implying turbulent processes with substantial correlation over long timescales—many days in the astrophysical cases. Furthermore, it is possible to quantify this self-similarity by means of the single model-independent quantity H. The value of H enables us to discriminate observationally between datasets, and it constrains models of these systems. For example, a numerical simulation of MAST edge turbulence under the conditions of plasma 6861 should generate a signal whose H-value matches that obtained in Fig. 3.11, and likewise for the astrophysical objects.

The differencing and rescaling technique[43] yields additional information on the timescales over which correlation persists in these instances of complex systems phenomenology within plasmas. The original data time series $y(t)$ is used to construct a differenced time series $Z(t, \tau) = y(t) - y(t - \tau)$.

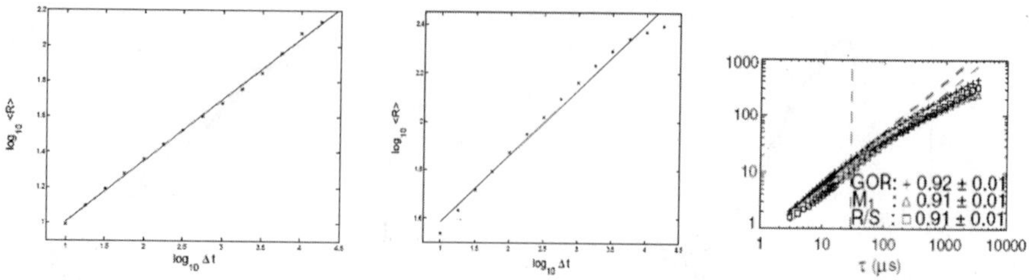

Fig. 3.11. Growth of range. Logarithmic plots of $< R(\Delta t) >$ versus Δt for: X-ray time series from Cygnus X-1(Left) and GRS 1915+105 (Centre),[43] where time steps are 90 minutes; and ion saturation current measurements in MAST L-mode plasma 6861 (Right),[46] where time units are microseconds, and several different algorithms for computing H are used. The value of the Hurst exponent is well defined in all cases: 0.35 ± 0.1, 0.27 ± 0.2, and 0.91 ± 0.01, respectively.

This time series gives the sequence of fluctuations in the data, over a timescale τ, at each time step. Many values of τ can be chosen, each of which generates a differenced time series $Z(t, \tau)$ describing fluctuations on the corresponding timescale. Thus a family of differenced time series $Z(t, \tau)$ is generated from a single data time series $y(t)$. Fluctuation amplitudes Z are typically small when τ is small, because the dataset does not have time to diverge, whereas larger-amplitude fluctuations Z occur for larger τ, during which interval the dataset has time to diverge substantially. In general, the distribution of fluctuations within the dataset on timescale τ is described by the probability density function $P(Z, \tau)$, known as the differenced distribution. For small τ, $P(Z, \tau)$ will typically be strongly peaked about $Z = 0$ and negligible for large Z, whereas $P(Z, \tau)$ will be significant at larger values of Z for larger τ.

If the fluctuations observed in the dataset are driven by a dominant underlying physical process which maintains correlation up to some timescale τ_c, and the fluctuations are self-similar, it follows that all members of the family of differenced distributions $P(Z, \tau : \tau < \tau_c)$ contain the same information. These distributions are, in essence, identical—they are simply stretched versions of each other. Mathematically, a simple rescaling operation applied to P and Z causes all the curves representing $P(Z, \tau)$ in this family to collapse onto a single curve, as follows. It will be found that the peak amplitudes $P(0, \tau)$ scale as $\tau^{-\alpha}$ for $\tau < \tau_c$, where the parameter α is inferred empirically. Rescaling $Z \to Z\tau^{-\alpha} = Z_s$ and $P \to P\tau^{\alpha} = P_s$ for each $P(Z, \tau)$ with $\tau < \tau_c$ creates rescaled PDFs which all collapse onto a single curve $P_s(Z_s)$ characteristic of the dominant physical process. To fix ideas, Fig. 3.12 shows differencing and rescaling in action for the simple example[8] of a Gaussian random walk $y(t)$.

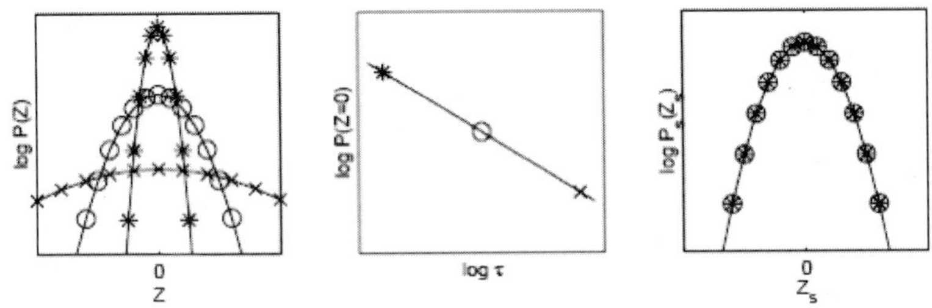

Fig. 3.12. Rescaling differenced distributions $P(Z)$. (Left) Unscaled PDFs of differenced series constructed from a Gaussian random walk,[43] showing curves for three different values of τ. (Centre) Power law scaling of $P(0, \tau)$ with τ, whose slope yields the scaling parameter α. (Right) Collapse of rescaled PDFs onto a single (Gaussian) curve for all τ.

If the procedure outlined above works for a given dataset, there emerge three new model-independent quantities characterising the dominant physical process: the correlation timescale τ_c up to which the differencing and rescaling operation succeeds, and beyond which it breaks down; the scaling parameter α; and the shape of the curve onto which the differenced PDFs collapse. As with the growth of range analysis above, there are no compelling reasons to expect the differencing and rescaling technique to operate successfully in astrophysical or fusion plasma objects. Nevertheless, on occasion it does so: see, for example, Figs. 3.13 and 3.14. Physically, Figure 3.13 establishes that the X-ray output of Cygnus X-1 varies in a way which is self-similar and correlated. It is controlled by a single non-Gaussian process, on timescales up to three years.[43] For the data from GRS 1915+105, differencing and rescaling also operates successfully,[43,44] but yields two well defined rescaling regimes with different values for the scaling parameter α, with correlations on timescales of days. This latter type of dual regime rescaling phenomenology is also found[46] for the MAST L-mode edge turbulence dataset analysed in the right-hand panel of Fig. 3.11, with correlations on timescales of tens of microseconds; see Fig. 3.14.

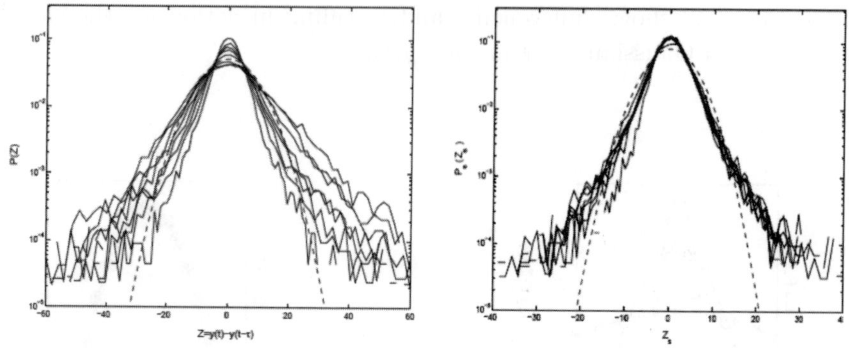

Fig. 3.13. (Left) Unscaled PDFs $P(Z, \tau)$ of differenced time series of X-ray signals from Cygnus X-1, plotted versus Z, with differencing parameter τ stepping up in half-integer powers of the 90 minutes timestep up to a maximum 10^4. Curves with lower $P(0)$ and broader tails correspond to higher values of τ. The scaling of $P(0)$ with τ yields the value of the parameter α used for rescaling. (Right) Rescaled PDFs $P_s(Z_s)$ plotted versus Z_s. The curves collapse onto a single characteristic distribution, which deviates substantially from a Gaussian (dashed line).[43]

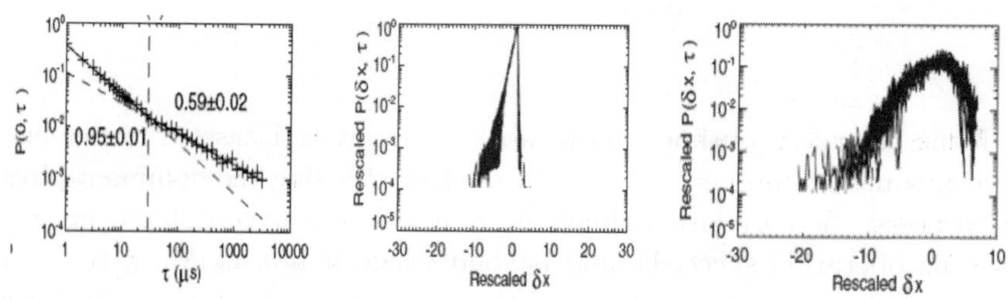

Fig. 3.14. (Left) The value of $P(0, \tau)$ plotted versus τ for the family of differenced distributions $P(\delta x, \tau)$ constructed from ion saturation current (x) measurements in MAST L-mode plasma 6861. The scaling parameter α is well defined within two sharply separated regions. (Centre) Collapse of rescaled PDFs for $\tau < 60$ μs onto a single curve, using slope 0.95 from left panel. (Right) Collapse of rescaled PDFs for $\tau > 60\mu$s onto a different single curve, using slope 0.59 from left panel.[46]

3.7. Solar Wind Plasma as a Complex System

The solar wind is a supersonic plasma flow which originates from the solar corona and propagates through interplanetary space, filling it until it reaches the local interstellar medium at the heliopause. It is evident from this description that the range of lengthscales and timescales on which solar wind physics unfolds is very extensive, furthermore solar wind fluctuations include both large scale episodic nonlinear perturbations and turbulence.

Once more, we are in complex systems territory. In the next section, we use the solar wind as a test bed for applying information theoretic techniques to a complex system. First, let us examine it from a more conventional complex systems perspective.

The solar wind provides unique opportunities for long duration *in situ* studies of magnetohydrodynamic turbulence in a plasma flowing supersonically with high magnetic Reynolds number $\sim 10^5$. Its spectral power density scales approximately as inverse frequency f^{-1} at lower frequencies (≤ 1 mHz); and as $f^{-5/3}$, reminiscent of Kolmogorov's inertial range, at higher frequencies (˜10-100 mHz). Both the $f^{-5/3}$ and f^{-1} fluctuations are often predominantly shear Alfvénic in character. The frequency at which the transition between power laws occurs (˜1-10 mHz) is observed to decline with increasing heliocentric distance in the plane of the ecliptic, and this extension of the $f^{-5/3}$ range at greater distances can be interpreted as evidence for an evolving turbulent cascade. The f^{-1} range is taken to reflect embedded solar coronal turbulence, convected with the solar wind, while the large-scale magnetic structure of the corona varies with the solar cycle and heliospheric latitude, creating variations in solar wind speed. Consider for example the work reported in Ref. 96. This concentrates on the quiet fast solar wind, where large transient events, such as those associated with coronal mass ejections, are absent. It exploits the unique out-of-ecliptic orbit of the *Ulysses* satellite, by focusing on its measurements of fluctuations in the three vector components of the magnetic field $\boldsymbol{B}$, taken above polar solar coronal holes at times of both minimum and maximum solar activity. By applying standard complex systems techniques—generalised structure function analysis, combined with extended self-similarity[97]—this work addresses key nonlinear plasma physics and MHD turbulence issues. These include: the extent to which the f^{-1} spectrum contains frozen information about coronal magnetic activity; the interplay between the coronal driver and the evolving inertial range turbulence; the degree to which the fluctuations exhibit self-similarity; and the dependence of the foregoing on heliocentric latitude and radial distance.

Extended self similarity (ESS) addresses[97] scaling in non-ideal situations where, for example, the effects of dissipation on small scales or of finite system size on large scales inhibit the development of self similar behaviour over a wide range—for example, the formation of a broad inertial range of turbulence. Recall the structure function definition $S_m(\tau) =< y(t+\tau) - y(t) >$ and that ideal scaling requires $S_m(\tau) \sim \tau^{\zeta(m)}$. ESS proceeds by replacing τ in the scaling expression by an initially unknown generalised timescale $g(\tau)$, such that $S_m(\tau) \sim [g(\tau)]^{\zeta(m)}$. It then follows that for structure functions of differing order p and q, $S_p(\tau) \sim [S_q(\tau)]^{\zeta(m)/\zeta(q)}$. Self similar behaviour is recaptured from the analysis of plots of $\log S_p(\tau)$ versus $\log S_q(\tau)$ for the datasets in question.

Figure 3.15, from Ref. 96, shows this technique in action for the solar wind. The straight line fits imply a global scaling $\zeta(2)/\zeta(3) \sim 0.75$ with

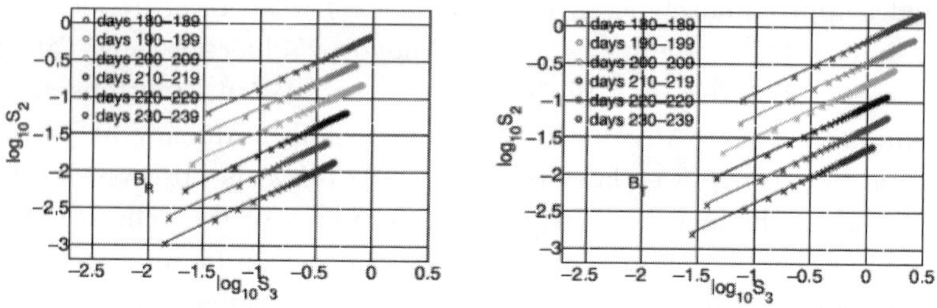

Fig. 3.15. Evidence from Ref. 96 for extended self similarity in magnetic field fluctuations in the solar wind, showing (Left) radial component and (Right) tangential component.The measurements are grouped into contiguous 10-day periods, the plots for each of which are offset in the y-direction for clarity. The straight lines show linear regression fits across the full range of sampling intervals, from 2 to 49 minutes.

one per cent accuracy. Comparison of scaling between the $f^{-5/3}$ and f^{-1} ranges, and its dependence on heliocentric latitude and distance, are among the more technical ways of exploiting such results described in Ref. 96. Here we simply note that a single robust quantitative measure of scaling has been extracted from the complex system presented by the solar wind plasma.

3.8. Information Theory and Plasma Physics

As we have described, the solar wind provides a natural laboratory for studying complex systems phenomenology in a magnetised plasma that shows, for example, a clear inertial range on timescales from minutes to hours, and whose magnetic Reynolds number is of order 10^5. Spacecraft located in the solar wind upstream of Earth's bow shock have the further advantage of sampling this high Mach number turbulent plasma flow far from any boundary layer. The signals obtained are often strongly nonlinear in character and are sparse, in the sense that observations are available at only a very small number of points in space, depending on the number of spacecraft that are simultaneously taking measurements of related plasma parameters in the upstream solar wind. In order to characterise the observed signals, and relate them to plasma models, it is therefore necessary to quantify spatial correlation within a strongly nonlinear system for which information is sparse. Studies of spatiotemporal correlation between coupled nonlinear signals in the solar wind and magnetosphere-ionosphere, reported in Refs. 15–17, have shown that the mutual Shannon information, and associated recurrence plot techniques, are helpful in this context. We refer to these papers, and to the review Ref. 8, for further information and for an account of how physically relevant timescales are extracted from these studies.

In outline, suppose we have two signals—that is, time series of data measurements—labelled A and B. These can be partitioned into an alphabet, meaning a discrete set which spans all the values which the signal can take. Then each measurement within A is a member of the set $\{a_1, ..a_i, ...a_n\}$, where a_1 and a_n are the minimum and maximum values that A is found to take. Within the discretised signal A, each value a_i is found to occur with a probability $P(a_i)$, and similarly for each element b_i in B, we find $P(b_i)$; for the joint probability of a_i in A and b_j in B we have $P(a_i, b_j)$. The rarer an element a_i, the more information its occurrence conveys, an observation that underlies Shannon's definition[98] $H(A) = -\sum P(a_i) \log_2 P(a_i)$ of the entropy of a signal. The mutual information $I(A, B)$ between two signals is then $H(A) + H(B) - H(A, B)$. Information is both a physical quantity and an intrinsically nonlinear measure of correlation. In these respects it differs fundamentally from the conventional linear cross-covariance, and thus offers a fresh perspective[15–17] on complex systems phenomenology in plasmas. Quantifying the mutual Shannon information shared between two causally linked but spatiotemporally separated plasma signals can identify key timescales, distinguish between plasma physics models for the propagation of perturbations, and measure the strength of the causal link.

As an example (for others, see Refs. 8,15,16), consider a recent information theoreticinformation theory!in plasma physics analysis[17] of data obtained during periods from 1998 onwards when the Wind, ACE and Cluster spacecraft were simultaneously in the upstream solar wind, and explored a range of spatial scales sufficient to determine correlation properties.[99] Nonlinear correlation is quantified by calculating the mutual information between measurements of magnetic field $\boldsymbol{B}$, flow velocity $\boldsymbol{v}$, and density ρ from spatially separated spacecraft. This enables us to compare the relative degree of correlation between different solar wind bulk parameters. The ordering of mutual information with respect to signal propagation relative to the background magnetic field direction is then related to current models and understanding of anisotropic solar wind plasma turbulence. These applications of mutual information have also required theoretical developments[18] that address both sparseness, and the use of timeseries of nonidentical quantities with different sampling rates.

Figure 3.16 shows plots[17] of normalised mutual information versus spatial separation between the ACE and WIND spacecraft, which took con-

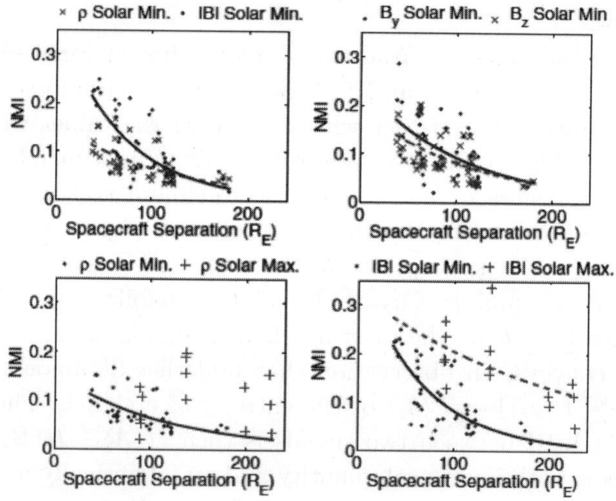

Fig. 3.16. Normalised mutual information between contemporaneous measurements of fluctuations in the solar wind by the ACE and WIND spacecraft, during periods of minimum and maximum solar activity. From Ref. 17.

temporaneous measurements in the upstream solar wind. These measurements were taken under a range of solar conditions, and it is clear from Fig. 3.16 that the spatial decline of shared information provides an effective measure of the extent of correlation of magnetic field strength, magnetic field components, and plasma density, and of the variation of this correlation with solar activity.

3.9. Conclusion

There are only two major conclusions from the foregoing. First, plasmas are typically complex systems in the strict sense. Second, the techniques of complex systems science yield fresh direct insight into the physics of plasmas. Since most of the visible matter in the universe is in the plasma state, these propositions imply a very broad scope for complex systems science.

Acknowledgments

It is a pleasure to acknowledge the contributions of many collaborators to the application of complex systems science to plasma physics, including Sandra Chapman, Joe Dewhurst, Ben Dudson, Jon Graves, John Greenhough, Per Helander, Bogdan Hnat, Keith Hopcraft, David Hughes, Ken McClements, Thomas March, James Merrifield, Ruth Nicol, Maya Paczuski, George Rowlands, Nick Watkins and Robert Wicks. This work was supported in part by Euratom (but the views and opinions expressed herein do not necessarily reflect those of the European Commission) and by the United Kingdom Engineering and Physical Sciences Research Council.

References

1. For a brief accessible introduction to the fundamentals of plasma physics, see for example Dendy R O 1990 *Plasma Dynamics* (Oxford: Oxford University Press)
2. Bak P 1996 *How Nature Works* (New York: Copernicus)
3. Badii R and Politi A 1999 *Complexity* (Cambridge: Cambridge University Press)
4. Sornette D 2000 *Critical Phenomena in Natural Sciences* (Heidelberg: Springer)
5. For chapter-length introductory descriptions of most manifestations of the plasma state, see Dendy R O (ed.) 1995 *Plasma Physics: An Introductory Course* (Cambridge: Cambridge University Press)
6. Wesson J 2004 *Tokamaks (3rd Edn)* (Oxford: Oxford University Press)
7. Kadomtsev B B 1992 *Plasma Phys. Control. Fusion* **34** 1931
8. Dendy R O, Chapman S C and Paczuski M 2007 *Plasma Phys. Control. Fusion* **49** A95
9. ITER Physics Basis Editors *et al.* 1999 *Nucl. Fusion* **39** 2137
10. Thomas P R *et al.* 1998 *Phys. Rev. Lett.***80** 5548
11. Shaing K C, Hegna C C, Callen J D and Houlberg W A 2003 *Nucl. Fusion* **43** 258
12. Diamond P H, Itoh S-I, Itoh K and Hahm T S 2005 *Plasma Phys. Control. Fusion* **47** R35
13. ITER Physics Expert Group on Confinement and Transport *et al.* 1999 *Nucl. Fusion* **39** 2175
14. Kamiya K *et al.* 2007 *Phys. Control. Fusion* **49** S43
15. March T K, Chapman S C and Dendy R O 2005 *Physica D* **200** 171

16. March T K, Chapman S C and Dendy R O 2005 *Geophys. Res. Lett.* **32** L04101
17. Wicks R T, Chapman S C and Dendy R O 2009 *Astrophys. J* **690** 734
18. Wicks R T, Chapman S C and Dendy R O 2007 *Phys. Rev. E* **75** 051125
19. Bak P, Tang C and K Wiesenfeld K 1987 *Phys. Rev. Lett.* **59**, 381
20. Newman D E, Carreras B A, Diamond P H and Hahm T S 1996 *Phys. Plasmas* **3**, 1858
21. Dendy R O and Helander P 1997 *Plasma Phys. Control. Fusion* **39**, 1947
22. Sanchez R, Newman D E and Carreras B A 2001 *Nucl. Fusion* **41** 247
23. Chapman S C, Dendy R O and Hnat B 2001 *Phys. Rev. Lett.* **86**, 2814
24. Graves J P, Dendy R O, Hopcraft K I and Jakeman E 2002 *Phys Plasmas* **9**, 1596
25. Chapman S C, Dendy R O and Hnat B 2003 *Plasma Phys. Control. Fusion* **45** 301
26. Consolini G 1997 *Cosmic Physics in the Year 2000* (Bologna, Italy: Società Italiana di Fisica) p 123
27. Chapman S C, Watkins N W, Dendy R O, Helander P and Rowlands G 1998 *Geophys. Res. Lett.* **25** 2397
28. Watkins N W, Chapman S C, Dendy R O, Helander P and Rowlands G 1999 *Geophys. Res. Lett.* **26** 2617
29. Lui A T Y, Chapman S C, Liou K, Newell P T, Meng C I, Brittnacher and Parks G K 2000 *Geophys. Res. Lett.* **27** 911
30. Watkins N W, Freeman M P, Chapman, S C and Dendy R O 2001 *J. Atmos. Solar-Terr. Phys.* **63** 1435
31. Chapman S C, Dendy R O and Hnat B 2001 *Phys. Plasmas* **8** 1969
32. Hnat B, Chapman S C, Rowlands G, Watkins N W and Freeman M P 2003 *Geophys. Res. Lett.* doi: 10.1029/2003GL018209
33. Chapman S C, Dendy R O and Watkins N W 2004 *Plasma Phys. Control. Fusion* **46** B157
34. Priest E R and Forbes T 2000 *Magnetic Reconnection* (Cambridge: Cambridge University Press)
35. Hughes D, Paczuski M, Dendy R O, Helander P and McClements K 2003 *Phys. Rev. Lett* **90** 131101
36. Schrijver C J *et al.* 1998 *Nature* **394** 152; see also `http://www.lmsal.com/carpet.htm`
37. Chapman S C 2000 *Phys. Rev. E,* **62** 1905
38. Dennis B R 1985 *Solar Phys.* **100** 465
39. Aschwanden M J *et al.* 2000 *Astrophys. J.* **535** 1047
40. Wheatland M J, Sturrock P A and McTiernan J M 1998 *Astrophys. J.* **509** 448
41. Charbonneau P, McIntosh S W, Liu H-L and Bogdan T J 2001 *Solar Phys.* **203** 321
42. Dendy R O and Chapman S C 2006 *Plasma Phys. Control. Fusion* **48** B313
43. Greenhough J, Chapman S C, Chaty S, Dendy R O and Rowlands G 2002 *Astron. Astrophys.* **385**, 693
44. Greenhough J, Chapman S C, Chaty S, Dendy R O and Rowlands G 2003

Mon. Not. R. Astr. Soc. **340**, 851
45. Greenhough J, Chapman S C, Dendy R O, Nakariakov V and Rowlands G, *Astron. Astrophys.* **409**, L17 (2003)
46. Dudson B D, Dendy R O, Kirk A, Meyer H and Counsell G C 2005 *Plasma Phys. Control. Fusion* **47** 885
47. Greenhough J, Chapman S C, Dendy R O and Ward D J 2003 *Plasma Phys. Control. Fusion* **45** 747
48. Hnat B, Dudson B H, Dendy RO, Counsell G C and Kirk A R 2008 *Nucl. Fusion* **48** 085009
49. Dewhurst J M, Hnat B, Ohno N, Dendy R O, Masuzaki S, Morisaki T and Komori A 2008 *Plasma Phys. Control. Fusion* **50** 095013
50. Nicol R M, Chapman S C and Dendy R O 2008 *Astrophys. J.* **679** 862
51. Rhodes T L, Moyer R A, Groebner R, Doyle E J, Peebles W A and Rettig C L 1999 *Phys. Lett. A* **253**, 181
52. Politzer P A 2000 *Phys. Rev. Lett.* **84**, 1192
53. Carreras B A, Hidalgo C, Sanchez E, Pedrosa M A, Balbin R *et al.* 1996 *Phys. Plasmas* **3** 2664
54. Callen J D and Kissick M W 1997 *Plasma Phys. Control. Fusion* **39** B173
55. Carreras B A, van Milligen B, Pedrosa M A, Balbin R, Hidalgo C *et al.* 1998 *Phys. Rev. Lett.* **80** 4438
56. Pedrosa M A, Hidalgo C, Carreras B A, Balbin R, Garcia-Cortes I *et al.* 1999 *Phys. Rev. Lett.* **82** 3621
57. Zaslavsky G M, Edelman M, Weitzner H, Carreras B, McKee G *et al.* 2000 *Phys. Plasmas* **7** 3691
58. Antar G Y, Counsell G, Yu Y, LaBombard B and Devynck P 2003 *Phys. Plasmas* **10** 419
59. van Milligen B P, Carreras B A and Sanchez R 2005 *Plasma Phys. Control. Fusion* **47** B743
60. Graves J P, Horacek J, Pitts R A and Hopcraft K I 2005 *Plasma Phys. Control. Fusion* **47** L1
61. Zweben S J, Boedo J A, Grulke O, Hidalgo C, LaBombard B, Maqueda R J, Scarin P and Terry J L 2007 *Plasma Phys. Control. Fusion* **49** S1
62. Garbet X and Waltz R 1998 *Phys. Plasmas* **5**, 2836
63. Sarazin Y and Ghendrih P 1998 *Phys. Plasmas* **5**, 4214
64. Beyer P, Sarazin Y, Garbet X, Ghendrih P and Benkadda S 1999 *Plasma Phys. Control. Fusion* **41**, A757
65. Li J and Kishimoto Y 2002 *Phys. Rev. Lett.* **89**, 115002
66. Tangri V, Das A, Kaw P and Singh R 2003 *Phys. Rev. Lett.* **91**, 025001
67. Villard L, Angelino P, Bottino A, Allfrey S J, Hatzky R *et al.* 2004 *Plasma Phys. Control. Fusion* **46**, B51
68. Garcia O E, Naulin V, Nielsen A H and Juul Rasmussen J 2004 *Phys. Rev. Lett.* **92** 165003
69. Russell D A, D'Ippolito D A, Myra J R, Nevins W M and Xu X Q 2004 *Phys. Rev. Lett.* **93** 265001
70. Scott B D 2007 *Plasma Phys. Control. Fusion* **49** S25
71. March T K, Chapman S C, Dendy R O and Merrifield J A 2004 *Phys. Plasmas*

11, 659
72. Dendy R O and Helander P 1998 *Phys. Rev. E* **57**, 3641
73. Chapman S C 2000 *Phys. Rev. E,* **62**, 1905
74. Fishpool G M 1998 *Nucl. Fusion* **38**, 1373
75. Tsurutani B *et al* 1990 *Geophys. Res. Lett.* **17**, 279
76. Takalo J and Timonen J 1998 *Geophys. Res. Lett.* **25** 2101
77. Consolini G 1997 *Cosmic Physics in the Year 2000* (Bologna, Italy: Società Italiana di Fisica) p 123
78. Chapman S C, Watkins N W, Dendy R O, Helander P and Rowlands G 1998 *Geophys. Res. Lett.* **25** 2397
79. Watkins N W, Freeman M P, Chapman, S C and Dendy R O 2001 *J. Atmos. Solar-Terr. Phys.* **63** 1435
80. Chapman S C, Dendy R O and Hnat B 2001 *Phys. Plasmas* **8** 1969
81. Hnat B, Chapman S C, Rowlands G, Watkins N W and Freeman M P 2003 *Geophys. Res. Lett.* doi 10.1029/2003GL018209
82. Watkins N W, Chapman S C, Dendy R O, Helander P and Rowlands G 1999 *Geophys. Res. Lett.* **26** 2617
83. Lui A T Y, Chapman S C, Liou K, Newell P T, Meng C I, Brittnacher and Parks G K 2000 *Geophys. Res. Lett.* **27** 911
84. Parker E N 1983 *Astrophys. J.* **264** 642; 1988 *Astrophys. J.* **330** 474; 1994 *Spontaneous Current Sheets in Magnetic Fields* (New York: Oxford University Press)
85. Priest E R and Forbes T 2000 *Magnetic Reconnection* (Cambridge: Cambridge University Press)
86. Hughes D, Paczuski M, Dendy R O, Helander P and McClements K 2003 *Phys. Rev. Lett* **90** 131101
87. Schrijver C J *et al.* 1998 *Nature* **394** 152; see also http://www.lmsal.com/carpet.htm
88. Dennis B R 1985 *Solar Phys.* **100** 465
89. Aschwanden M J *et al.* 2000 *Astrophys. J.* **535** 1047
90. Wheatland M J, Sturrock P A and McTiernan J M 1998 *Astrophys. J.* **509** 448
91. Charbonneau P, McIntosh S W, Liu H-L and Bogdan T J 2001 *Solar Phys.* **203** 321
92. Lu E T and Hamilton R J 1991 *Astrophys. J.* **380** L89
93. Turcotte D L 1999 *Rep. Prog. Phys.* **62**, 1377
94. Ohno N, Masuzaki S, Miyoshi H, Takamura S, Budaev V P, Morisaki T, Ohyabu N and Komori A 2006 *Contrib. Plasma Phys.* **46** 692
95. Dendy R O, Helander P and Tagger M 1998 *Astron. Astrophys.* **337** 962
96. Nicol R M, Chapman S C and Dendy R O 2008 *Astrophys. J.* **679** 862
97. Benzi R, Ciliberto S, Trippicione R, Baudet C, Massaioli F and Succi S 1993 *Phys. Rev. E* **48** 29
98. Shannon C E 1948 *Bell Syst. Tech. J.* **27** 379
99. Matthaeus W H, Dasso S, Weygand J M, Milano L J, Smith C W and Kivelson M G 2005 *Phys. Rev. Lett.* **95** 231101

Chapter 4

Bayesian Data Analysis

Michael S. Wheatland
School of Physics,
University of Sydney,
NSW 2006
m.wheatland@physics.usyd.edu.au

Bayesian methods provide a systematic approach to inference and data analysis in science. This chapter presents a tutorial on Bayesian analysis, with emphasis on the relationship to conventional methods. An application to solar flare prediction is then described.

Contents

4.1. Scientific Inference

Inference is the process of going from observed effects to underlying causes, and is the inverse process to deduction. Whereas deduction is exact, inference is imprecise, and necessarily probabilistic. Inference is the basis of science: we are always faced with observations we would like to explain in terms of underlying physical causes.

Bayesian inference is an approach to the problem based on an identity in conditional probability (Bayes's theorem). Notable Bayesians have included Pierre-Simon Laplace (who inferred the mass of Saturn from contemporary observations using Bayesian methods, and obtained a value consistent with modern estimates), the economist John Maynard Keynes, and the applied mathematician and geophysicist Harold Jeffreys. Bayesian inference has at times been controversial, because of its incorporation of subjective prior information into the process of inference. Historically the Bayesian approach was referred to as "subjective probability." In recent decades there has been wider acceptance and application of Bayesian methods in a range of disciplines, driven by a recognition of the utility and power of the methods. Increases in computational speed and the use of Markov chain Monte Carlo methods have also played a part in this adoption.

This chapter presents an overview of Bayesian methods in Section 4.2, and then an example of their application to the problem of solar flare prediction in Section 4.3. Section 4.2.1 presents Bayes's theorem and explains its use for inference. Sections 4.2.2 and 4.2.3 describe basic approaches to parameter estimation and hypothesis testing in the Bayesian method, and Section 4.2.4 illustrates these approaches in application to a simple example: coin tossing. Section 4.2.5 gives a brief account of a relatively recent development, Markov chain Monte Carlo (MCMC) methods. Sections 4.2.6 and 4.2.7 discuss the relationship between Bayesian and classical methods of parameter estimation and hypothesis testing. Section 4.3.1 provides background on the problem of solar flare prediction, and Section 4.3.2 describes properties of flare statistics. A Bayesian approach to prediction exploiting these statistics is then presented in Section 4.3.3, and is illustrated in application to whole-Sun prediction of soft X-ray flares in Section 4.3.4.

4.2. A Tutorial on Bayesian Methods

4.2.1. *Bayes's theorem*

Consider two propositions, X and Y (these may be thought of as statements that are either true or false). The probability that both are true may be written

$$\begin{aligned} P(X, Y) &= P(X|Y) \times P(Y) \\ &= P(Y|X) \times P(X), \end{aligned} \tag{4.1}$$

where $P(X|Y)$ is the probability X is true, given that Y is true (a conditional probability).

The Reverend Thomas Bayes[1] applied Eq. (4.1) to inference by identifying one of the propositions with a hypothesis or model (labelled H), and the other with available data (labelled D), and writing the equation in the form

$$P(H|D) = \frac{P(D|H) \times P(H)}{P(D)}. \tag{4.2}$$

In many cases it is sufficient to omit the evidence term and use the statement of proportionality rather than equality:

$$P(H|D) \propto P(D|H) \times P(H), \tag{4.3}$$

and then the requirement that the probabilities sum to unity over all possible hypotheses:

$$\sum_i P(H_i|D) = 1 \tag{4.4}$$

is used to determine the missing factor.

The terms in Eq. (4.2) are given names: $P(H|D)$ is called the "posterior" probability, $P(D|H)$ is the "likelihood," $P(H)$ is the "prior" probability, and $P(D)$ is sometimes called the "evidence." Eqs. (4.2) or (4.3) may be interpreted as statements of how an initial estimate of the probability of a hypothesis (the prior) is modified by new information (the likelihood), to give an updated estimate of the probability of a hypothesis (the posterior). Eq. (4.1) is a fact about conditional probability. However, in the application to inference there is some ambiguity because of the subjectivity inherent in the choice of the prior. The probability that one person assigns to a hypothesis being true, a priori, may not match that of another person.

4.2.2. *Bayesian parameter estimation*

In inference there are two basic problems: parameter estimation, i.e. deciding the best values for the parameters of a given model, and hypothesis testing or model selection, i.e. deciding between competing models. First we consider parameter estimation.

The basic approach is to express the model H in terms of model parameters, labelled $\boldsymbol{\theta} = [\theta_1, \theta_2, ..., \theta_N]$. The functional form of the likelihood $P(D|\boldsymbol{\theta})$ must be identified in terms of these parameters, based on the model, and possibly details of how the data were obtained (the likelihood may incorporate observational uncertainties). A prior $P(\boldsymbol{\theta})$ also needs to be chosen, based on existing knowledge. Bayes's theorem is then applied in the form

$$P(\boldsymbol{\theta}|D) \propto P(D|\boldsymbol{\theta})P(\boldsymbol{\theta}) \tag{4.5}$$

to give the posterior as a function of the model parameters.

In many cases the posterior will have a single maximum, as a function of the model parameters, and the parameter values corresponding to the maximum provide "best estimates" for the parameters. The width of the posterior in the vicinity of the maximum (how localized the maximum is) provides an estimate of the uncertainties in the best estimates.

If the interest is with only one parameter, say θ_1, then it is possible to integrate over the other parameters, to produce a univariate posterior:

$$P(\theta_1|D) = \int P(\boldsymbol{\theta}|D)d\theta_2 d\theta_3 ... d\theta_N. \tag{4.6}$$

This process of integrating over unwanted or "nuisance" parameters is called "marginalization."

In the Bayesian method, the posterior is taken to provide complete information about parameters, and methods of obtaining best estimates of parameters from the posterior are of secondary importance. Correspondingly, there are many ways to obtain best estimates. Expected values are often used. The expected value of a function $f(\theta_1)$ is

$$E\left[f(\theta_1)\right] = \int f(\theta_1)P(\theta_1|D)d\theta_1, \tag{4.7}$$

or

$$E\left[f(\theta_1)\right] = \int f(\theta_1)P(\boldsymbol{\theta}|D)d\boldsymbol{\theta} \tag{4.8}$$

in the multi-dimensional case. Expected values of powers of θ_1 provide means and standard deviations which may be used as best estimates and uncertainties:

$$\begin{aligned} \theta_{1,\mathrm{est}} &= E\left[\theta_1\right], \\ \sigma^2_{1,\mathrm{est}} &= E\left[\theta_1^2\right] - \left(E\left[\theta_1\right]\right)^2. \end{aligned} \tag{4.9}$$

Alternatively, the location of the maximum of the posterior (the mode) is often used as a best estimate:

$$\left.\frac{d}{d\theta_1}P(\theta_1|D)\right|_{\theta_{1,\mathrm{est}}} = 0. \tag{4.10}$$

If the posterior function is approximately Gaussian, then the following formula may be used to estimate the associated uncertainty:

$$\sigma_{1,\text{est}}^{-2} = - \left. \frac{d^2}{d\theta_1^2} \ln P(\theta_1|D) \right|_{\theta_{1,\text{est}}} . \tag{4.11}$$

The right hand side of Eq. (4.11) is the coefficient of $\frac{1}{2}(\theta - \theta_{1,\text{est}})^2$ in the Taylor expansion of $-\ln P(\theta_1|D)$ around $\theta_{1,\text{est}}$. If the posterior is Gaussian this is equal to σ^{-2}, where σ is the usual width (standard deviation). The formula uses only the behaviour of the posterior function at the peak, so it is important to check that the global behaviour is approximately Gaussian, to ensure the estimate is meaningful.

Fig. 4.1 illustrates these two approaches to obtaining best estimates and uncertainties.

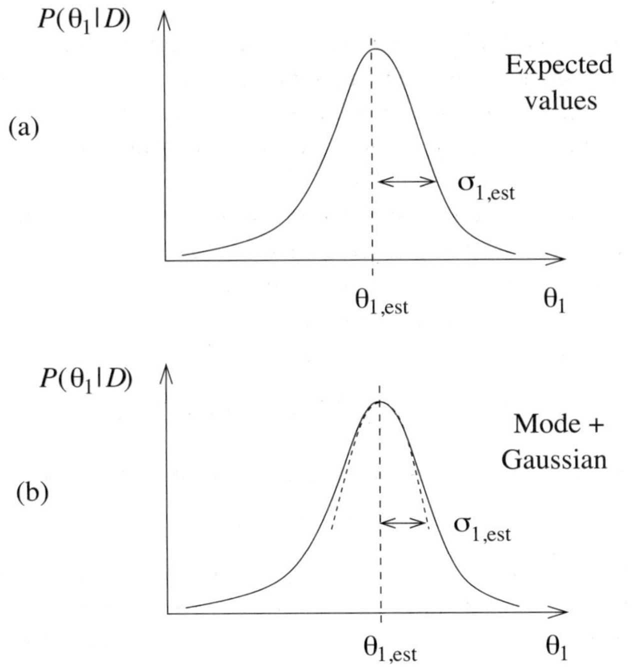

Fig. 4.1. Best estimates based on expected values [panel (a)], and on the mode and local Gaussian behaviour [panel (b)].

4.2.3. *Bayesian hypothesis testing*

Bayesian hypothesis testing involves taking ratios of Bayes's theorem. For two competing hypotheses, H_1 and H_2, we have

$$\frac{P(H_1|D)}{P(H_2|D)} = \frac{P(D|H_1)}{P(D|H_2)}\frac{P(H_1)}{P(H_2)}. \tag{4.12}$$

It should be noted that the common evidence term $P(D)$ in the two statements of Bayes's theorem has cancelled, and plays no further role. The ratio of posteriors $O_{12} = P(H_1|D)/P(H_2|D)$ is called the "odds ratio", and is equal to the ratio of the likelihoods, modulated by the ratio of the priors. For exclusive hypotheses $[P(H_1|D) + P(H_2|D) = 1]$ it is possible to assign an absolute probability for a model, e.g. $P(H_1|D) = O_{12}/(1 + O_{12})$. More generally there may be many competing hypotheses, and it is necessary to order the relative probabilities.

4.2.4. *Is this coin fair?*

To illustrate the methods of Bayesian parameter estimation and hypothesis testing, we consider a simple example often used in text books:[2] coin tossing. Suppose that you have a coin, and that you would like to determine, on the basis of tossing the coin, whether it is fair. For example, in 10 tosses of the coin you observe two heads. Is the coin fair?

As a problem in Bayesian parameter estimation, we can consider trying to infer the "bias" θ of the coin, which we define as the probability of obtaining a head in a single toss. For a fair coin, $\theta = \frac{1}{2}$.

If r heads are observed in n tosses of the coin, the likelihood of this data D is given by the binomial distribution

$$P(D|\theta) = \frac{n!}{r!(n-r)!}\theta^r(1-\theta)^{n-r}, \tag{4.13}$$

although all we really need is the statement of proportionality:

$$P(D|\theta) \propto \theta^r(1-\theta)^{n-r}. \tag{4.14}$$

In Bayesian inference it is necessary to choose a prior, describing the state of knowledge or ignorance about parameters in the absence of data. If you were suspicious about the coin before you started tossing it, you might consider a uniform prior, assigning equal probability to all possible values of θ:

$$P(\theta) = \begin{cases} 1 \text{ if } 0 \le \theta \le 1, \\ 0 \text{ otherwise.} \end{cases} \tag{4.15}$$

Alternatively, if you were somewhat confident of fairness, but still wanted to admit other possibilities, you might consider a Gaussian prior, peaked about one half:

$$P(\theta) \propto \begin{cases} \exp\left[-\frac{1}{2}(\theta - \frac{1}{2})^2/\sigma^2\right] & \text{if } 0 \leq \theta \leq 1, \\ 0 & \text{otherwise,} \end{cases} \tag{4.16}$$

with a width σ chosen to reflect your suspicion about the coin.

The posterior for the problem is then given by the product of the likelihood and the chosen prior:

$$P(\theta|D) \propto \begin{cases} \theta^r(1-\theta)^{n-r}P(\theta) & \text{if } 0 \leq \theta \leq 1, \\ 0 & \text{otherwise,} \end{cases} \tag{4.17}$$

and the normalization condition $\int_0^1 P(H|D)dH = 1$ is used to determine the constant of proportionality.

Fig. 4.2 illustrates the evaluation of this posterior, for both choices of the prior. Each panel shows a probability distribution function (PDF) for θ. Panel (a) illustrates the two priors, with the uniform prior shown by the solid curve and the Gaussian prior (with $\theta = 0.2$) shown by the dashed curve. Panel (b) shows the corresponding posterior distributions for the observation of two heads in ten tosses of the coin. Panel (c) shows the corresponding posterior distributions for the observation of 28 heads in 100 tosses of the coin.

After 10 tosses of the coin [panel (b) in Fig. 4.2], the posterior distributions are quite broad. The value $\theta = 0$ is ruled out, because heads have been observed. Values of θ above about 0.7 are unlikely, but only $\theta = 1$ is strictly impossible (since tails have been observed). The two choices of prior lead to somewhat different posterior distributions. On the basis of these results, it is hard to make a very definitive estimate of θ, and the prior plays an important role.

After 100 tosses of the coin [panel (c) in Fig. 4.2], the posterior distributions are much narrower. Values of θ less than about 0.1 and larger than about 0.5 are very unlikely, based on the data. The two choices of prior lead to quite similar posterior distributions. On the basis of these results, more definitive estimates of θ may be made, and the role of the prior is less important.

To illustrate more quantitative parameter estimation, we can consider the case of the uniform prior, which is straightforward to evaluate analytically. The normalisation of the posterior is achieved using the Eulerian

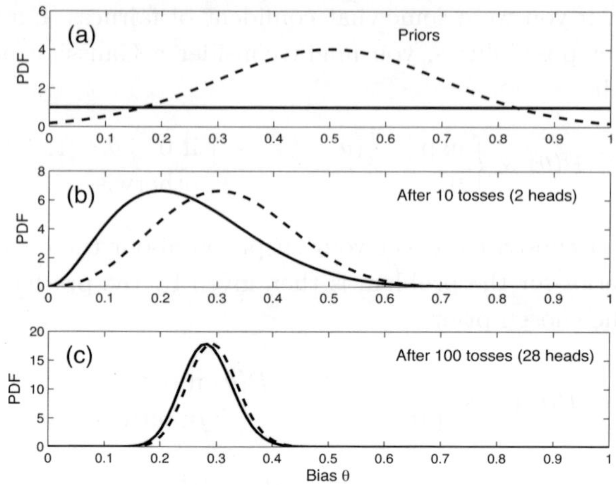

Fig. 4.2. Bayesian inference applied to coin tossing. Panel (a) shows two possible choices for the prior distribution of the probability θ of a coin landing heads: a uniform prior (solid), and a Gaussian prior (dashed). Panel (b) shows the corresponding posterior distributions based on the observation of two heads in ten tosses. Panel (c) shows the corresponding posterior distributions based on 28 heads in 100 tosses.

integral

$$\int_0^1 \theta^r (1-\theta)^{n-r} d\theta = \frac{r!(n-r)!}{(n+1)!}, \tag{4.18}$$

so that

$$P(\theta|D) = \frac{(n+1)!}{r!(n-r)!}\theta^r (1-\theta)^{n-r}. \tag{4.19}$$

Evaluating Eq. (4.9) for this distribution gives

$$\theta_{\text{est}} = \frac{r+1}{n+2}, \qquad \sigma_{\text{est}}^2 = \frac{\theta_{\text{est}}(1-\theta_{\text{est}})}{n+3}. \tag{4.20}$$

(The result for θ_{est} is known as Laplace's rule of succession, and was famously used by Laplace to estimate the probability that the Sun will rise tomorrow.[3]) For the case of two heads in 10 tosses, we have $\theta_{\text{est}} \approx 0.27$ and $\sigma_{\text{est}} \approx 0.12$, which is suggestive of a departure from fairness but not definitive. For 28 heads in 100 tosses we have $\theta_{\text{est}} \approx 0.28$ and $\sigma_{\text{est}} \approx 0.045$, which is becoming quite definitive. Alternatively, we can use Eqs. (4.10) and (4.11), leading to

$$\theta_{\text{est}} = \frac{r}{n}, \qquad \sigma^2_{\text{est}} = \frac{\theta_{\text{est}}(1-\theta_{\text{est}})}{n}. \tag{4.21}$$

For $r = 2$ and $n = 10$ we have $\theta_{\text{est}} = 0.2$ and $\sigma_{\text{est}} \approx 0.13$, and for $r = 28$ and $n = 100$ we have $\theta_{\text{est}} = 0.28$ and $\sigma_{\text{est}} \approx 0.045$.

As an example of a hypothesis test, we consider the question of whether, on the basis of the data, the coin is more likely to be heads biased ($\theta > 1/2$), or tails-biased ($\theta < 1/2$). For a uniform prior, the odds ratio of the two models may be evaluated analytically:

$$\begin{aligned} O_{ht}(r,n) &= \frac{\int_{\frac{1}{2}}^{1} \theta^r (1-\theta)^{n-r} d\theta}{\int_0^{\frac{1}{2}} \theta^r (1-\theta)^{n-r} d\theta} \\ &= \frac{I_{\frac{1}{2}}(n-r+1, r+1)}{1 - I_{\frac{1}{2}}(n-r+1, r+1)} \end{aligned} \tag{4.22}$$

where $I_x(a,b)$ is the incomplete Beta function. [This example does not correspond exactly to Eq. (4.12) because here we have integrated each posterior over the relevant values of the model parameter θ.] Evaluating this expression for the examples of interest gives $O_{ht}(2,10) = 67/1981 \approx 3.4 \times 10^{-2}$, and $O_{ht}(28,100) \approx 4.3 \times 10^{-6}$. For the case of 10 tosses, the coin is more likely to be tails-biased, although the result is not definitive, but for 100 tosses the tails bias is very strongly favoured.

4.2.5. *Markov chain Monte Carlo (MCMC)*

Normalisation and calculation of expected values involves evaluating integrals, for example of the form of Eq. (4.7), which may be multi-dimensional. Until recently, this presented a practical problem for Bayesian inference. However, "Markov chain Monte Carlo" (MCMC) methods now provide a powerful, general solution to the problem. Here we present only the basic idea, which is particularly simple. (For details, see e.g. Ref.[4])

If a sample $\{\theta_{1i}, i = 1, 2, ..., n\}$ of random variables from a probability distribution $P(\theta_1|D)$ is available, then an estimate of an expected value may be constructed via a sum:

$$\begin{aligned} E\left[f(\theta_1)\right] &= \int f(\theta_1) P(\theta_1|D) d\theta_1, \\ &\approx \frac{1}{n}\sum_i f(\theta_{1i}). \end{aligned} \tag{4.23}$$

Markov chain Monte Carlo methods provide ways to generate appropriate sets $\{\theta_{11}, \theta_{12}, ...\}$, using only uniformly-distributed random variables, which are simple to generate (approximately) on a computer, and evaluations of the function $P(\theta_1|D)$. The methods produce Markov chains (sequences of random numbers, such that each number depends only on the previous number) with the property that, after an initial "burn-in" period of non-stationarity, the Markov Chain becomes stationary, and then approximates a sequence of samples from $P(\theta_1|D)$. A number of different MCMC algorithms are commonly used, including the Metropolis, Metropolis-Hastings, and Gibbs sampler methods.

4.2.6. *Relationship to maximum likelihood and least squares*

"Maximum likelihood" and "least squares" are methods commonly used for parameter estimation, which are closely related to the Bayesian approach. We briefly discuss the relationship, and the approximations and assumptions being made these methods.

Consider a model involving parameters $\boldsymbol{\theta} = [\theta_1, \theta_2, ..., \theta_N]$, and a set of data $\mathbf{D} = [D_1, D_2, ..., D_M]$. Bayes's theorem may be stated

$$P(\boldsymbol{\theta}|\mathbf{D}) \propto P(\mathbf{D}|\boldsymbol{\theta})P(\boldsymbol{\theta}). \tag{4.24}$$

Assuming a uniform prior gives

$$P(\boldsymbol{\theta}|\mathbf{D}) \propto P(\mathbf{D}|\boldsymbol{\theta}), \tag{4.25}$$

i.e. the posterior is proportional to the likelihood. The "maximum likelihood estimate" $\boldsymbol{\theta}_{\mathrm{ML}} = [\theta_{\mathrm{ML}1}, \theta_{\mathrm{ML}2}, ..., \theta_{\mathrm{ML}N}]$ is the set of model parameters which maximizes the likelihood, i.e. satisfies

$$\left.\frac{\partial}{\partial \theta_i} P(\mathbf{D}|\boldsymbol{\theta})\right|_{\theta_{\mathrm{ML}i}} = 0, \tag{4.26}$$

for $i = 1, 2, ..., N$. In conventional statistical inference, the only justification for this estimate is that it makes the observed data most probable, or most likely.[5] However, given Eq. (4.25), this estimate also maximizes the posterior, i.e. makes the model most probable. Hence we see that the maximum likelihood estimate is the Bayesian modal estimate, assuming a uniform prior.

Assuming that the data points are independent, we have

$$P(\mathbf{D}|\boldsymbol{\theta}) = P(D_1|\boldsymbol{\theta})P(D_2|\boldsymbol{\theta}) \ldots P(D_M|\boldsymbol{\theta}). \tag{4.27}$$

If the model gives data values $\mathbf{F}(\boldsymbol{\theta}) = [F_1(\boldsymbol{\theta}), F_2(\boldsymbol{\theta}), ..., F_M(\boldsymbol{\theta})]$ in the absence of observational errors (uncertainties), and the errors are assumed to be Gaussian distributed, then the likelihood of each datum is

$$P(D_i|\boldsymbol{\theta}) = \frac{1}{\sqrt{2\pi}\sigma_i} \exp\left\{ -\frac{[F_i(\boldsymbol{\theta}) - D_i]^2}{2\sigma_i^2} \right\} \tag{4.28}$$

for $i = 1, 2, ..., M$, where the $\sigma_i = \sigma_i[F_i(\boldsymbol{\theta})]$ are uncertainties which depend on the model data values in a specified way.[a] Combining Eqs. (4.27) and (4.28), the overall likelihood is

$$P(\mathbf{D}|\boldsymbol{\theta}) \propto \exp\left[-\frac{1}{2}\chi^2(\boldsymbol{\theta}) \right], \tag{4.29}$$

where

$$\chi^2(\boldsymbol{\theta}) = \sum_{i=1}^{M} \frac{[F_i(\boldsymbol{\theta}) - D_i]^2}{\sigma_i^2}. \tag{4.30}$$

The quantity $\chi^2 = \chi^2(\boldsymbol{\theta})$ is usually called "chi-square," or the "chi-square statistic." The log-likelihood is

$$\ln P(\mathbf{D}|\boldsymbol{\theta}) = \text{const} - \frac{1}{2}\chi^2(\boldsymbol{\theta}), \tag{4.31}$$

and clearly the likelihood/log-likelihood is a maximum when chi-square is a minimum. The estimate for the model parameters obtained by minimizing chi-square, which we label $\boldsymbol{\theta}_{\text{LS}}$, is usually called the "least squares" estimate.

From this derivation, we see that the least squares estimate is also a Bayesian estimate, subject to additional of assumptions and approximations. In principle these assumptions may be relaxed in the Bayesian approach. For example, it is possible to incorporate errors other than Gaussian errors, which is sometimes appropriate. Also, a non-uniform prior may be introduced to reflect prior knowledge about the model parameters. For example, if one parameter represents energy, than the prior may be used to enforce the requirement of non-negativity of this parameter. More generally, the prior may be used to "bias" certain areas of the parameter space, if it is known a priori that certain values are more likely to be correct. This process has no counterpart in classical methods.

[a]The Gaussian or "normal" distribution is often appropriate to describe observational uncertainties. Some insight into the almost ubiquitous success of the Gaussian to describe errors is provided by the central limit theorem,[5] which states (roughly) that the sum of a large number of independent random variables from a variety of distributions is normally distributed. For a more detailed explanation, see Ref.[6]

Bayesian methods are also distinct from classical methods in other specific ways. The Bayesian approach provides a posterior distribution, rather than the limited information afforded by best estimates and uncertainties. This may be of particular use if the posterior distribution has an unusual shape (for example is multi-modal, or otherwise departs significantly from a Gaussian). The posterior distribution contains the totality of information available from inference, and this may be scrutinized in different ways. Bayesian methods also place a fundamentally different emphasis on the roles of data and of models. Classical methods work with the likelihood, which presupposes a model, and assesses the probability of the data given the model. The model is essentially treated as being perfect, and the data imperfect. From the Bayesian perspective, the roles are reversed. The posterior assesses the probability of the model given the data, so the data is presupposed, or perfect, and the model imperfect. For many scientists this may appear to be a more natural perspective: science involves the construction and refinement of models based on available observations.

4.2.7. *Classical hypothesis testing*

Classical hypothesis testing is quite different to the Bayesian approach presented in Section 4.2.3. The classical method involves the choice of a "statistic," and here we consider the use of chi-square.

A large value of $\chi^2_{\rm LS} = \chi^2(\boldsymbol{\theta}_{\rm LS})$ (where $\boldsymbol{\theta}_{\rm LS}$ is the least squares estimate, obtained as explained in Section 4.2.6), may be an indication that something is wrong. One possibility is that the model is incorrect. The "chi-square test" involves calculating $P_d(\chi^2 > \chi^2_{\rm LS})$, the probability of obtaining a larger value of χ^2 than $\chi^2_{\rm LS}$, for the given data, assuming that the model is correct. The quantity $P_d(\chi^2 > \chi^2_{\rm LS})$ is called the significance, and depends on $d = M - N$, the "number of degrees of freedom." It is straightforward to calculate this quantity based on the likelihood defined by Eqs. (4.27)-(4.30).[7] The calculation evaluates the probability of getting data that departs further from the (fixed) model than the data that was observed. The classical approach to hypothesis testing then involves "rejecting" the model if the significance is too small, say less than 1%.

A variety of criticisms of this procedure have been raised. First, it is not possible to accept a model, only to reject it. Given a suitably aberrant set of data, any model will be rejected. As Harold Jeffreys stated, "There has not been a single date in the history of the law of gravitation when a modern significance test would not have rejected all laws and left us with no law."[8] A related criticism is that the method does not consider alternative hypotheses. Finally, there is a degree of arbitrariness in the choice of the statistic, and also in the choice of a significance level for rejection. The Bayesian method explicitly deals with these problems. If a hypothesis is generally accepted, then the prior should reflect this, and a test based on a single set of aberrant data will not lead to the model being rejected. The Bayesian method forces consideration of competing hypotheses, and does not involve the arbitrary choice of a statistic, or of a significance level.

4.3. An Application to Solar Flare Prediction

4.3.1. *Background*

Solar flares are magnetic explosions in the ionised outer atmosphere of the Sun, the solar corona. Flares occur in and around sunspots, where intense magnetic fields penetrate the visible surface of the Sun, and thread the overlying coronal plasma. During a flare some of the energy stored in the magnetic field is released and appears in accelerated particles, radiation, heating, and bulk motion. The flare mechanism is accepted to be magnetic reconnection, a process involving a change in connectivity of magnetic field lines, but many aspects of the process remain poorly understood. Flares occur suddenly, and there are no (known) infallible indicators that a flare is about to occur. Hence flare prediction is probabilistic.

Large flares strongly influence our local "space weather." They can lead, for example, to enhanced populations of energetic particles in the Earth's magnetosphere (the region magnetically connected to the Earth), and these particles can damage satellite electronics, and pose radiation risks to astronauts and to passengers on polar aircraft flights. The space weather effects of large flares motivate a need for accurate solar flare prediction.

A variety of properties of active regions are correlated with flare occurrence. For example, certain sunspot classifications,[9] qualitative measures of magnetic complexity,[10] and moments of quantitative photospheric magnetic field maps[11,12] provide flare predictors, of varying reliability. Operational flare forecasters refer to the tendency of a region which has produced large flares in the past to produce large flares in the future as "persistence," and this provides one of the most reliable predictors for large flare occurrence in 24-hour forecasts.[13] The US National Oceanic and Atmospheric Administration (NOAA) uses an "expert system" based on sunspot classification and other properties of active regions to assign probabilities for the occurrence of large flares.[9]

Flares are commonly classified by their peak flux at X-ray wavelengths, in the 1-8 Å band measured by the Geostationary Observational Environmental (GOES) satellites. Moderate size flares correspond to "M-class" events, with a peak flux in the range $\geq 10^{-5}\,\mathrm{W\,m^{-2}}$ to $\geq 10^{-4}\,\mathrm{W\,m^{-2}}$. Large flares correspond to "X-class" events, with peak flux above $\geq 10^{-4}\,\mathrm{W\,m^{-2}}$. The NOAA predictions assign corresponding probabilities ϵ_{M} and ϵ_{X} for the occurrence of at least one event with peak flux above these levels within 24 hours.

Existing methods of flare prediction are not very accurate. One measure of success of probabilistic event forecasts is provided by the "skill score," defined as

$$\mathrm{SS}(f,x) = 1 - \frac{\mathrm{MSE}(f,x)}{\mathrm{MSE}(\langle x\rangle,x)}, \tag{4.32}$$

where f denotes the forecast value, x denotes the observation (a one or a zero, according to whether an event did or did not occur, respectively), $\langle ...\rangle$ denotes an average over the forecasts, and

$$\mathrm{MSE}(f,x) = \langle (f-x)^2\rangle \tag{4.33}$$

denotes the mean-square error. The skill score quantifies the improvement of the forecasts over a prediction of the average in every case. The maximum of the skill score is one, representing perfect prediction, and negative values of the skill score indicate predictions worse than forecasting the average. The NOAA published statistics describing the success of its forecasts for 1986–2006[b] The skill score for one-day forecasting of X-class flares is positive for only 7 of the 21 years.

4.3.2. *Flare statistics*

Flare occurrence follows a power-law frequency-size distribution, where "size" denotes some measure of the flare magnitude, for example the peak flux in X-rays measured by GOES. In other words, the number of events per unit time and per unit size S, denoted $N(S)$, obeys

$$N(S) = \lambda_1(\gamma-1)S_1^{\gamma-1}S^{-\gamma}, \tag{4.34}$$

where λ_1 is the total rate of events above size S_1 oberved, and $\gamma \approx 1.5-2$ is a constant, which depends on the specific choice of S. Although the distribution is typically constructed based on all flaring active regions present

[b]See http://www.swpc.noaa.gov/forecast_verification/.

on the Sun over some period of time, it also appears to hold in individual active regions.[14] The appearance of this power law in flare occurrence motivated the avalanche model for flares, in which the energy release mechanism consists of a sequence of elementary events which trigger one another, and in which the system is in a self-organised critical state.[15,16]

Flare occurrence in time may be modelled as a Poisson process.[17,18] For intervals in which the mean rate of flaring λ does not vary greatly, the distribution of waiting times τ is then

$$P(\tau) = \lambda \exp(-\lambda\tau). \tag{4.35}$$

Over longer time scales, the rate will vary with time, and the distribution is more complex.

Fig. 4.3 illustrates these properties of flare statistics. Panel (a) shows a schematic of a sequence of events, as size versus time, and also illustrates a waiting time τ. Panel (b) shows the power-law frequency-size distribution, and panel (c) shows the Poisson waiting-time distribution.

4.3.3. *Event statistics method of prediction*

Given the relative success of persistence as a flare predictor, and the simple statistical rules describing flare occurrence, it is worthwhile to consider methods of prediction relying on flare statistics alone. Refs.[19] and[20] develop such an approach, using the Bayesian method.

The basic idea is as follows. If S_1 is the size of a "small" event (chosen such that small events are well observed), and S_2 is the size of a "big" event (which you would like to predict), then the power-law frequency-size distribution Eq. (4.34) implies that the rates λ_1 and λ_2 of events above the two sizes are related according to

$$\lambda_2 = \lambda_1 \left(S_1/S_2\right)^{\gamma-1}. \tag{4.36}$$

Eq. (4.36) allows estimation of the rate of big events even if none have been observed. Given this estimate, the probability of at least one big event in a time T_P is

$$\begin{aligned} \epsilon &= 1 - \exp(-\lambda_2 T_P) \\ &= 1 - \exp\left[-\lambda_1 \left(S_1/S_2\right)^{\gamma-1} T_P\right], \end{aligned} \tag{4.37}$$

using Eq. (4.35). If M events are involved in the estimation of the rate λ_1, then it follows that $\sigma_\epsilon/\epsilon \approx M^{-1/2}$.[19] Hence the prediction becomes accurate if many small events are observed.

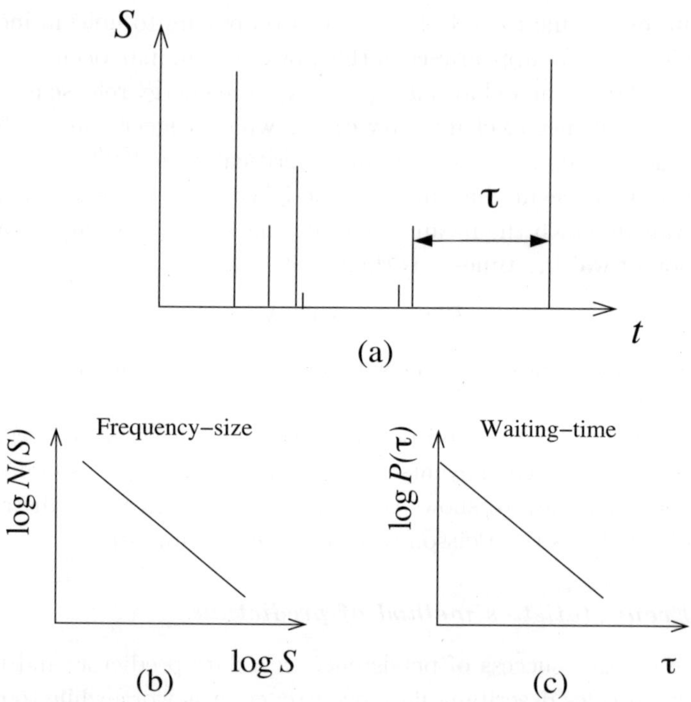

Fig. 4.3. Schematic illustration of flare statistics. Panel (a): flare events, showing size S versus time, and indicating a waiting time τ. Panel (b): frequency-size distribution. Panel (c): waiting-time distribution.

The Bayesian aspect of the method concerns the estimation of γ and λ_1 from the observed data. Specifically, if the data D consists of events $s_1, s_2, ..., s_M$ at times $t_1 < t_2 < ... < t_M$, then the problem is to calculate posterior distributions $P_\gamma(\gamma|D)$ and $P_1(\lambda_1|D)$. Given these, the posterior distribution for λ_2 is

$$P_2(\lambda_2|D) = \int_1^\infty d\gamma \int_0^\infty d\lambda_1 P_1(\lambda_1|D) P_\gamma(\gamma|D) \times \delta\left[\lambda_2 - \lambda_1 (S_1/S_2)^{\gamma-1}\right], \tag{4.38}$$

using Eq. (4.36). Finally, the posterior distribution for ϵ is obtained using

$$P_\epsilon(\epsilon|D) = P_2\left[\lambda_2(\epsilon)|D\right] \left|\frac{d\lambda_2}{d\epsilon}\right|, \tag{4.39}$$

where $\lambda_2(\epsilon) = -\ln(1-\epsilon)/T_P$, from Eq. (4.37).

The inference of the power-law index γ follows from Eq. (4.34), which implies the likelihood:[21]

$$P(D|\gamma) \propto \prod_{i=1}^{M} (\gamma - 1)(s_i/S_1)^{-\gamma}. \tag{4.40}$$

A uniform prior is used. If $M \gg 1$, the posterior/likelihood is sharply peaked in the vicinity of the modal/maximum likelihood estimate

$$\gamma_{\mathrm{ML}} = \frac{M}{\ln \pi} + 1, \quad \text{where} \quad \pi = \prod_{i=1}^{M} \frac{s_i}{S_1}. \tag{4.41}$$

In this case a suitable approximation to the posterior in Eq. (4.38) is provided by $P_\gamma(\gamma|D) = \delta(\gamma - \gamma_{\mathrm{ML}})$.

The inference of the rate λ_1 of small events is complicated by the time variation of the rate. The procedure used is to estimate the rate at the time the prediction is made using the "Bayesian blocks" procedure from Ref.[22] This procedure is a Bayesian change-point algorithm for decomposing a point process into a piecewise-constant Poisson process by iterative comparison of one- versus two-rate Poisson models.

Fig. 4.4 illustrates the procedure. Panel (a) shows a sequence of data, consisting of point events on a time line, during an observation interval T. The prediction interval T_P is also shown. The Bayesian blocks procedure compares the relative probability of one- and two-rate models for the observation interval T, for all choices of change point corresponding to an event time. If a two-rate model is more probable, then the data in each of the two chosen intervals is used for comparison of one- and two-rate models, and these intervals may be further sub-divided. An interval for which the one-rate model is more probable is a Bayesian block. The procedure continues iteratively in this way, until a sequence of Bayesian blocks is decided on, as shown in panel (b). The data D' in the last block, consisting of M' events in time T', then supplies a likelihood for the current data given the rate λ_1:

$$P_1(D'|\lambda_1) \propto \lambda_1^{M'} \mathrm{e}^{-\lambda_1 T'}, \tag{4.42}$$

based on the assumption of Poisson occurrence. The prior may be taken to be uniform,[19] or a prior may be constructed based on the rates in the other blocks.[20]

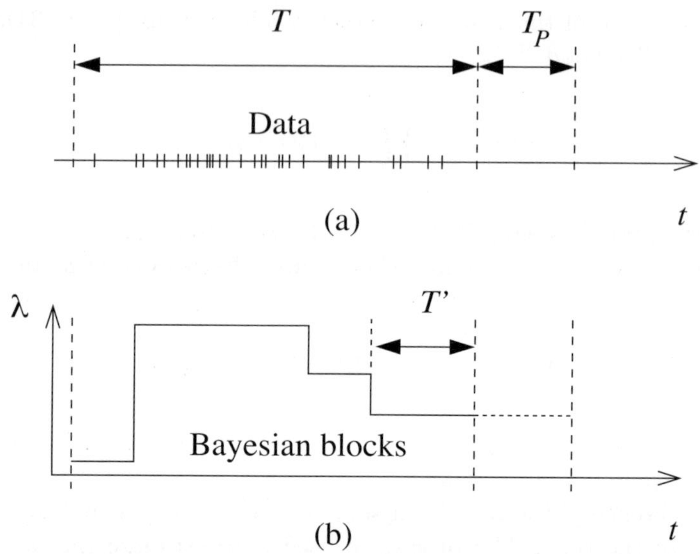

Fig. 4.4. Schematic illustration of Bayesian blocks determination of current rate. Panel (a): data, consisting of point events in time line during an observation interval T. The prediction interval T_P is also shown. Panel (b): Bayesian blocks decomposition of the rate λ, and identification of the most recent interval T' when the rate is approximately constant.

4.3.4. *Whole-Sun prediction of GOES flares*

To illustrate the method, we consider whole-Sun prediction of GOES soft X-ray flares, as described in detail in Ref.[20]

The largest soft X-ray flare of the modern era occurred on 4 November 2003, and saturated the GOES detectors at X28 (a peak flux in the 1-8 Å GOES band of $2.8 \times 10^{-3}\,\mathrm{W\,m^{-2}}$), although it was later estimated to be as large as X45.[23] It is interesting to consider applying the method for that day, following.[20]

The data D consists of one year of events prior to the day from the whole Sun, above peak flux $S_1 = 4 \times 10^{-6}\,\mathrm{W\,m^{-2}}$ (corresponding to a GOES C4 event). This gives 480 events. Probabilities ϵ_{MX} and ϵ_{X}, for the occurrence of at least one event in the range M to X, and at least one X-class event, respectively, were inferred for the 24 hours of 4 November 2003.

Fig. 4.5 illustrates the Bayesian blocks procedure in application to the data. Panel (a) shows the 480 events plotted as peak flux versus time. Panel (b) shows the Bayesian blocks decomposition: there are 13 blocks, and the last block has a duration of $T' = 15$ days and contains $M' = 104$

events.

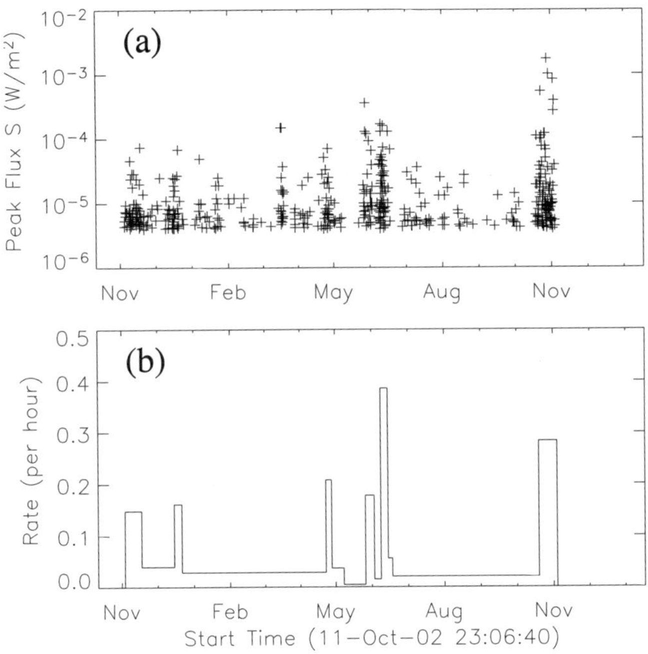

Fig. 4.5. Bayesian blocks applied to one year of GOES events prior to 4 November 2003.

Fig. 4.6 shows the posteriors for the predictions. The solid curve corresponds to $\epsilon_{\rm MX}$, and the dashed curve corresponds to $\epsilon_{\rm X}$. The best estimates (using expected values) are shown by short vertical lines at the bottom, and are $\epsilon_{\rm MX} \approx 0.73 \pm 0.03$, and $\epsilon_{\rm X} \approx 0.19 \pm 0.02$. These values are quite high, reflecting the recent high rate of flaring on the Sun. However, the estimates also highlight the limitations of probabilistic forecasting: the prediction for X-class events is only 20%, yet the largest flare of the last three decades is about to occur. (Incidentally, three M-class events were also recorded on 4 November 2003.)

The whole-Sun implementation of the method was tested on the GOES record for 1976-2003.[20] For each day a prediction was made based on one year of data prior to the day, following the procedure outlined for 4 November 2003. Comparison was then made with whether or not events occurred

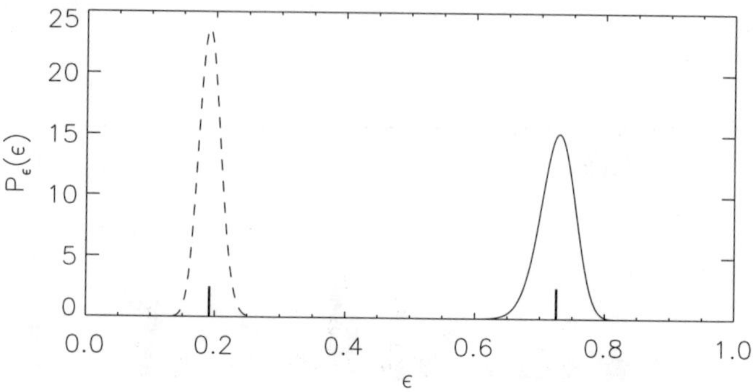

Fig. 4.6. Posterior distributions for predictions for 4 November 2003. The solid curve is the posterior for $\epsilon_{\rm MX}$, the probability of getting at least one flare in the range M to X, and the dashed curve is the posterior for $\epsilon_{\rm X}$, the probability of getting at least one X flare. The short vertical lines at the bottom indicate best estimates, using expected values.

on each day, and the success of the method was evaluated statistically. Table 4.1 provides statistics for the predictions for 1987-2003, for which years NOAA predictions are also available. The mean-square errors [see Eq. (4.33)], and the skill scores [see Eq. (4.32)] are listed. The event statistics method achieves very comparable results to the NOAA method, and even performs somewhat better, in terms of the skill score, for prediction of X-class flares.

Table 4.1. Comparison with NOAA predictions, for 1987-2003.

	Event statistics		NOAA	
	M-X	X	M-X	X
MSE(f,x)	0.143	0.031	0.139	0.032
SS(f,x)	0.258	0.078	0.262	-0.006

4.4. Summary

This chapter presents a tutorial on Bayesian methods, and an example of application to solar flare prediction. The emphasis has been on the basic principles, and on the relationship to conventional methods. For more details on Bayesian approaches, I recommend Refs. 2,4,6, and 24.

References

1. T. Bayes, An essay towards solving a problem in the doctrine of chances, *Phil. Trans. Roy. Soc.* **53**, 370–418, (1763).
2. D. Sivia, *Data Analysis, A Bayesian Tutorial.* (Oxford University Press, Oxford, 1996).
3. P.-S. Laplace, *A Philosophical Essay on Probabilities (1819; Translated from the 6th French edition by Frederick Wilson Truscott and Frederick Lincoln Emory).* (Dover Publications, New York, 1951).
4. W. Gilks, S. Richardson, and D. Spiegelhalter, *Markov Chain Monte Carlo in Practice.* (Chapman and Hall/CRC, Boca Raton, 1996).
5. J. Rice, *Mathematical Statistics and Data Analysis.* (Wadsworth and Brooks, Pacific Grove, California, 1988).
6. E. Jaynes, *Probability Theory, The Logic of Science.* (Cambridge University Press, Cambridge, 2003).
7. J. Mathews and R. Walker, *Mathematical Methods of Physics.* (Addison-Wesley, Redwood City, California, 1970), 2nd edition.
8. H. Jeffreys, *Theory of Probability.* (Clarendon Press, Oxford, 1961), 3rd edition.
9. P. McIntosh, The classification of sunspot groups, *Solar Phys.* **125**, 251–267, (1990).
10. I. Sammis, F. Tang, and H. Zirin, The dependence of large flare occurrence on the magnetic structure of sunspots, *Astrophys. J.* **540**, 583–587, (2000).
11. K. Leka and G. Barnes, Photospheric magnetic field properties of flaring versus flare-quiet active regions. IV. A statistically significant sample, *Astrophys. J.* **656**, 1173–1186, (2007).
12. G. Barnes, K. Leka, E. Schumer, and D. Della-Rose, Probabilistic forecasting of solar flares from vector magnetogram data, *Space Weather.* **5**, S09002, (2007).
13. D. Neidig, P. Wiborg, and P. Seagraves. The role of persistence in the 24-hour flare forecast. In eds. R. Thompson, D. Cole, P. Wilkinson, M. Shea, D. Smart, and G. Heckman, *Solar-Terrestrial Predictions: Workshop Proceedings, Leura, Australia, 16-20 October, 1989*, pp. 541–545, Boulder, Colorado, (1990).
14. M. Wheatland, Flare frequency-size distributions for individual active regions, *Astrophys. J.* **532**, 1209–1214, (2000).
15. E. Lu and R. Hamilton, Avalanches and the distribution of solar flares, *Astrophys. J.* **380**, L89–L92, (1991).
16. P. P. Charbonneau, S. McIntosh, H.-L. Liu, and T. Bogdan, Avalanche models for solar flares (Invited review), *Solar Phys.* **203**, 321–353, (2001).
17. M. Wheatland, Rates of flaring in individual active regions, *Solar Phys.* **203**, 87–106, (2001).
18. M. Wheatland and Y. Litvinenko, Understanding solar flare waiting-time distributions, *Solar Phys.* **211**, 255–274, (2002).
19. M. Wheatland, A Bayesian approach to solar flare prediction, *Astrophys. J.* **609**, 1134–1139, (2004).

20. M. Wheatland, A statistical solar flare forecast method, *Space Weather.* **3**, S07003, (2005).
21. T. Bai, Variability of the occurrence frequency of solar flares as a function of peak hard x-ray rate, *Astrophys. J.* **404**, 805–809, (1993).
22. J. Scargle, Studies in astronomical time series analysis. V. Bayesian blocks, a new method to analyze structure in photon counting data, *Astrophys. J.* **504**, 405–418, (1998).
23. N. Thomson, C. Rodger, and R. Dowden, Ionosphere gives size of greatest solar flare, *Geophys. Res. Lett.* **31**, L06803, (2004).
24. P. Gregory, *Bayesian Logical Data Analysis for the Physical Sciences, A Comparative Approach with Mathematica Support.* (Cambridge University Press, Cambridge, 2005).

Chapter 5

Inverse Problems and Complexity in Earth System Science

I. G. Enting
MASCOS
139 Barry St.
The University of Melbourne
Vic 3010, Australia

Developing a science of the combined atmosphere-hydrosphere-biosphere-lithosphere system is essential as humanity progresses through the so-called Anthropocene — the period with a discernable human imprint on the physical earth system. Developing the science through a classical 'reductionist' approach, often based on controlled experiment, is not possible if the system involved is the whole planet. This chapter explores how analysis of earth system can be developed through a range of inversion calculations, particularly what is known as data assimilation. Earth system models are calibrated and used for interpretation of processes by having the model track the observed changes from natural biogeophysical variability and humanity's current 'uncontrolled experiment' involving massive releases of greenhouse gases. The carbon cycle is considered in detail as an illustration of the principles underlying a more holistic analysis. Some strategies are described for calibration and data assimilation in terrestrial carbon models.

Contents

5.1. Complex Systems

5.1.1. *Prologue*

Increasingly, there has been an appreciation that the science of global change must confront the issue of complexity in the earth system. At the most obvious level is the turbulent nature of the geophysical fluid dynamics of the atmosphere and oceans.[1] Potential instability in the chemical composition of the atmosphere noted in theoretical cases for methane[2] and most dramatically in the ozone hole.[3] The complexity of ecosystems has been variously noted.[4,5] Falkowski et al.[6] have noted aspects of the carbon cycle that reflect complex system behaviour such as the CO_2-climate link over glacial-interglacial cycles. The book by Bunde et al.[7] places the behaviour of the earth system in a complex systems context, comparing it to complex behaviour in financial and physiological systems.

This chapter considers aspects of analysing the earth systems in complex systems terms, emphasising the carbon cycle, reflecting both my own expertise and the essential role of carbon dioxide as a driver of climate change.

5.1.2. *Reinventing reductionism*

Complex systems science is frequently promoted as an alternative (or extension) to a reductionist approach.[8] The latter is based on the view that an understanding of the behaviour of a whole system can be obtained from an understanding of the behaviour and interactions of its parts. Conceptually one understands ecosystems in terms of organisms understood in terms of cells understood in terms of biochemistry understood in terms of chemistry understood in terms of atoms understood in terms of sub-atomic 'particles' (and/or fields). However, as emphasised by Cohen and Stewart,[9] to the extent that the reductionist paradigm is ever implemented in practice, what is done is to explain each level in terms of particular, carefully selected, 'features' of the behaviour of the components. Further, as discussed by Ellis,[10] such 'explanations' still fail to capture important aspects of causality.

One or both of two limitations can prevent this reductionist vision being achieved:

i the computational task of analysing the behaviour is impractical;
ii it is not possible to perform the controlled experiments that are needed in order to determine the behaviour of the component parts.

Frequently, aspects (i) and (ii) appear as complementary difficulties whose relative importance changes depending on the space and/or time scales on which the component parts are considered. If the components are considered on small scales (e.g. laboratory scales) then controlled experiments are possible, but assembling such descriptions into a global-scale representation is computationally impossible. Conversely, if components are considered on scales little smaller than the Earth as a whole, then it is not possible to perform controlled experiments to determine the behaviour of the components.

A further difficulty with reductionist approaches is that:

iii a reductionist approach can direct thinking into following the causal channels associated with reductionist analysis with the risk of ignoring influences from the broader context.

Point (iii) is largely a matter of how reductionist analyses are used. In many real-world problems, it will be necessary to synthesise the information obtained by reductionist analyses into a 'systems view' of the problem. As an example from an application area beyond the scope of the present chapter, McMichael[11] shows the need for multiple perspectives in analysing the occurrence of disease.

The textbook description of reductionism implies the ability, in principle, of explaining systems in terms of their underlying component systems, all the way down to the 'theory of everything' envisaged as the ultimate goal of theoretical physics. E.O. Wilson, who describes reductionism as 'the search strategy employed to find points of entry into otherwise impenetrably complex systems,' goes on to characterise reductionism as aiming at 'consilience' — total consistency between different levels of description.[12]

Wilson also emphasised that 'reduction' is also to be regarded as a precursor of 'synthesis' that re-assembles the results from the reductionist decomposition.

As noted above, the classic description of reductionism is a myth since in practice, reductionism involves explaining systems in terms of carefully selected **features** of the behaviour of the components.[9] This, it must be emphasised, is **a good thing**. Much of science would be impractical if one had to re-evaluate ecology in terms of a chain of re-evaluated physiology from chemistry from each new variation in theoretical physics. As long as the key **features** remain, such re-evaluation is not needed.

In these terms we can recognise that much the most successful applications of traditional reductionist science have been in terms of features that are linearly additive. In particular, mass, energy, chemical species etc, are subject to additive conservation laws. As noted by John Finnigan (personal communication), much of the relevant mathematics is the mathematics of Courant and Hilbert.[13]

In contrast, complex systems are those where the features are non-linear and algorithmic. The relevant mathematical tools are such things as network analysis, agent-based modelling, cellular automata[14] etc.

5.1.3. *The modelling spectrum*

The 'synthesis' envisaged by Wilson[12] as following on from the 'reduction' is essentially the process of modelling. The various aspects of modelling can be usefully compared in terms of the modelling spectrum introduced by Karplus.[15] The spectrum was described as running from black-box models (characterised as being highly empirical, and only representing relations between inputs and outputs) through to white box models (characterised as having relations between inputs and outputs defined through processes involving internal states of the system expressed in mechanistic terms). Black box models are generally statistical and white-box models (glass-box models would have been a better term) are generally deterministic. In between lies modelling represented by various shades of grey. Although Karplus envisaged different parts of the spectrum as being occupied by models from different fields of study, I argued that various forms of carbon cycle modelling spanned much of the spectrum.[16]

Examples of the steps in the spectrum of carbon cycle modelling, running from white box to black box are:

curve-fitting: Purely statistical fits include regression-type analyses of

CO_2 trends and cycles, correlation studies relating CO_2 and ENSO and empirical fits of transfer relations connecting concentrations to emissions. These apply at the globally aggregated level. Statistical approaches can also apply in spatially disaggregated forms as in the use of empirical relations between vegetation and climate.[17]

constant airborne fraction: This is a specific functional relation between anthropogenic emissions and concentrations of CO_2: the airborne fraction is the ratio of growth-rate to emissions. The assumption of a constant airborne fraction is exact for the case of a linear system with exponentially-growing emissions. In some cases the term 'airborne fraction' is used to denote the ratio of CO_2 growth rate to fossil emissions. However, as suggested by Oeschger et al.,[18] it is better to refer to this as the 'apparent airborne fraction.'

response functions: These are a characterisation of 'transfer relations' between forcing and response as in equation (5.1) below. They can be obtained as purely statistical fits, but frequently they are obtained from empirical fits to the behaviour of mechanistic models. Spatially-disaggregated response function representations are possible,[19] but seem to have seldom been used.

lumped models: Lumped models use coarse aggregations of processes with the aim of capturing important features while neglecting minor details. In carbon cycle models, and biogeochemical models in general, this is often a representation in terms of aggregated 'pools' of constituent, the important feature being conservation of mass of the various elements and their isotopes.

lumped geographically explicit: Lumped models can be used in a spatially-explicit mode. Indeed components of earth system models often take this form.

earth system: Earth system models, built around General Circulation Models (GCMs) for atmosphere and Ocean represent the white-box end of the modelling spectrum for the earth system.

A significant problem is that as one moves away from purely statistical approaches and towards the white-box end of the spectrum, 'standard' statistical techniques for determining uncertainties become less directly applicable and statistical concepts reflecting system complexity may be required. Since the carbon cycle (and more generally the earth system) involves many interacting components, some modelling may involve combining models from different parts of the spectrum.

Characteristics	Carbon Cycle	Climate System
Black box		
Empirical	Fitting CO_2 trends	Fitting temperature trends
Stochastic	Airborne fraction	
Grey box		
	Response function	Response function
	Box model	Energy balance model
White box		
Deterministic		Atmos/ocean GCM
Reductionist	Spatially resolved	
Mechanistic		Earth system model

In contrast to Rutherford's "All science is either physics or stamp collecting," the concept of a spectrum of modelling argues for a continuum of approaches across science. In introducing the concept of the modelling spectrum Karplus[15] considers that different parts of the spectrum would come from different disciplines. Enting[16] noted that different aspects of carbon modelling could cover much of the spectrum. This can be extended to using models from different parts of the spectrum to model the same aspect of the carbon cycle (or other system). One of the requirements of the modelling spectrum is that of statistical consistency between the different levels of description. The progression along the spectrum from black box towards white box models corresponds to increasing the amount reductionism in the modelling.

In discussing environmental modelling, Young et al.[20] emphasise the positive aspects of stochastic modelling. Their discussion is complicated by the fact that they fail to acknowledge that 'stochastic modelling' can mean two different things: analysing models formulated in terms of stochastic processes and/or using stochastic techniques to analyse models. Most commonly, stochastic analysis techniques are used to analyse stochastic models and deterministic techniques are used to analyse deterministic models. However this correspondence is by no means universal. Stochastic models can be analysed by deterministic techniques (especially when closed form expressions are available for the probability distributions involved) and stochastic techniques (e.g. Monte Carlo integration) can be applied to purely deterministic systems. The present discussion is concerned with the stochastic vs. deterministic aspects of model formulation and not with the relative merits of computational techniques.

5.1.4. *A dichotomy*

To clarify some of the issues associated with considering modelling as a spectrum, we reduce the spectrum to a dichotomy that we term 'curve fitting' and 'process modelling.' The basis of these approaches are:

Curve Fitting —

- Build model on the basis of observed correlations between putative 'inputs' and 'outputs.'

Process modelling —

- Build model on the basis of 'mechanistic' relations between system components.

The main arguments, for and against these approaches, can be summarised in terms of trade-offs associated with moving in opposite directions along the modelling spectrum:

	Curve fitting	Process model
For	Formalism for analysing uncertainties	Wider range of validity
Against	No reason to assume validity beyond domain used for calibration	Vulnerable to neglect of processes — 'Kelvin error'

The term 'Kelvin error' refers to the risk of missing a process from the modelling, taking its name from Lord Kelvin's massive underestimates of the ages of the earth and sun due to neglect of nuclear processes.

The dichotomy has been discussed by Young and Garnier[21] in connection with their analysis of CO_2 stabilisation calculations of the type performed for the IPCC Radiative Forcing Report[22] and discussed in section 5.3.3 below. However, this discussion, promoting the benefits of their 'data-based' approach, has two related flaws. Firstly, in spite of their emphasis on data-based approaches, they ignore one of the most important data sets, the distributions of natural and anthropogenic ^{14}C, from which much of the understanding of the carbon cycle had been achieved prior to the availability of ice-core data. Secondly, when they note that mechanistic modelling tends to reflect modellers' understanding, Young and Garnier fail to note that such understanding is usually based on incorporating a much wider body of observational data. It is this aspect, of being broadly data-based, and including using such observationally-based principles as conservation of mass, that gives mechanistic models the possibility of applicability beyond the types of data for which they were calibrated.

Figure 5.1, from my recent book,[23] gives a schematic of how the inclusion of additional process in climate models over successive decades has increased the domain of validity. My book noted that the modelling dichotomy is playing a large role in public discussion of human-induced climate change. A large consensus view, drawing heavily on process-based climate models is challenged in public debate by claims based on extrapolation of various trends in climate data and putative proxies, with the claim that indirect curve-fitting approaches are more reliable than process models as predictors of future climate change. These 'debates' echo the characterisation by Christie[24] of earlier 'debates' over ozone depletion that although "*scientific debate is now closed with a clear consensus behind the orthodox views of the ozone hole ... a public and political debate continues in some quarters based largely on the same flawed and outdated arguments.*"

The long-term success of process-based models of climate change is emphasised by the 1977 prediction *about 1°C by 2000 AD (25% increase in carbon dioxide) and about 3°C by 2050 AD (doubling of atmospheric carbon dioxide) with an uncertainty factor of roughly two* — in a report[25] endorsed by WMO Panel of Experts on Climate Change as "*a very useful statement of knowledge of anthropogenic influences on climate.*" In contrast to this prediction (which has been borne out so far) is the claim that "*Scientists in the 1970's predicted an ice age*" based mainly on an article in *Newsweek*[26] that didn't actually name any scientists and was based on extrapolating trends (mainly from the northern hemisphere). In promoting this piece of pseudo-history, the self-styled 'greenhouse sceptics' ignore the fact that the 'prediction' was based on the type of 'curve-fitting' that many of them now favour.

There is, of course, no justification for abandoning 'curve fitting' approaches, simply on the grounds of current politicised mis-use or a failed prediction of doubtful provenance. What is needed is a consistent reconciliation of both approaches. The ideal is to capture the best aspects of both approaches, using statistical techniques to identify information content of deterministic models.[27]

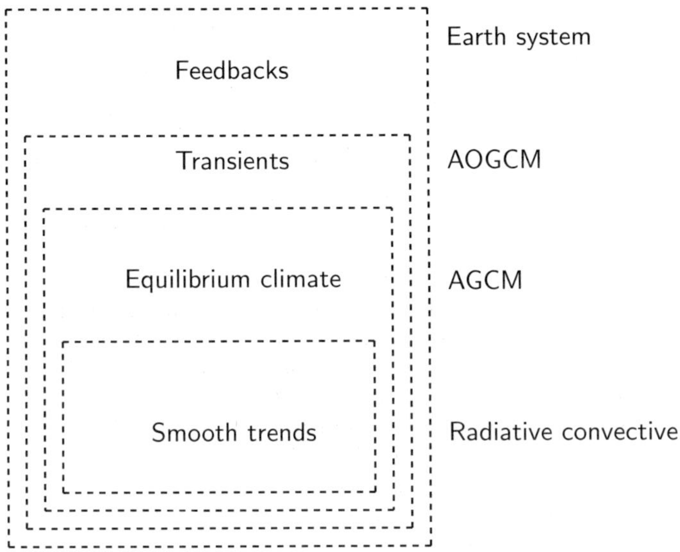

Fig. 5.1. Schematic of increasing sophistication in climate models (model types on right) showing increases in the domain of validity. (from[23]).

5.1.5. *A big picture?*

Regardless of the semantic question of whether chaotic behaviour is a *sufficient* condition to indicate complexity, many undoubtedly complex systems exhibit chaotic behaviour (or chaotic states). While the state of the atmosphere is a classic example (leading to the 'butterfly effect' metaphor[28]), sensitive dependence on initial conditions is found even in systems that appear much simpler — e.g. the three-body problem in celestial mechanics.[29]

As often been noted, many of the concepts that are being used to characterise complex systems have their origins in statistical physics (see section 5.2.3). The concept of a thermodynamic 'phase change' is seen as an example of a more general 'catastrophe.'

Particularly influential has been the sandpile model.[30,31] Notionally this is a model of avalanches on a pile of sand that gains grains from a steady source and loses grains as avalanches cross an outer boundary. Compared to models of thermodynamic states, this model has the surprising characteristic that it achieves a critical state (characterised by power-law behaviour of avalanche size) without the adjustment of a control parameter. The phenomenon is termed self-organised criticality. Sizes of particular avalanches have a sensitive dependence on the initial state.

Sensitive dependence on initial conditions is essentially the norm in the 'historical' sciences, almost by definition. The sandpile model has been used by various authors[32,33] as a paradigm for contingent phenomena from earthquakes to stock market crashes to world wars.

An 'admission of defeat' in the face of contingency is in the final words from Gould[34]

> *"... — why do humans exist? A major part of the answer, touching those aspects that science can treat at all, must be because* Pikaia *survived the Burgess decimation. This response does not cite a single law of nature; it embodies no statement about predictable evolutionary pathways, no calculation of probabilities based on general rules of anatomy or ecology. The survival of* Pikaia *was a contingency of 'just history.' I do not think that any 'higher' answer can be given and I cannot imagine that any resolution could be more fascinating."*

In this context, *Pikaia* refers to a minor species in the fauna of the Burgess shale from the early Cambrian. *Pikaia* is the only chordate (the phylum that includes vertebrates) present. Earlier, Gould is quite explicit that this sort of reference to the survival of *Pikaia* is a shorthand meaning the survival of one particular species (the one that was our ancestor) of those (apparently few) chordates existing at that time. The survival of an early chordate is of course just one many critical events leading to human history, including several mass extinctions and possibly reduction of human ancestors to a few thousand individuals as a consequence of the Toba eruption.

In contrast to the theme of contingency presented by Gould,[34] Conway Morris[35] suggests that there are constraints (from biology, chemistry and physics) that restrict evolution to a limited range of possibilities. He cites the case of South America where, during the period in which it was separate from other land masses, evolution proceeded in comparable directions to the rest of the world.

Clearly, two important questions for systems dominated by chaos (or other forms of contingency) are:

- How far ahead can we make specific predictions? (e.g. How many days ahead can we predict the weather?)
- How well can we determine the range of the space over which chaotic variability will range? (How well can we predict how the climate will respond to changed boundary conditions?).

These questions recur in inversions of atmospheric CO_2, as exemplified by the different approaches proposed by Prinn[36] and Enting,[37] the former noting the chaotic nature of atmospheric transport and seeking to use this as part of the data fitting, the latter looking for the long-term averages that do not depend on tracking a specific state. Further work remains to be done on establishing the domains of validity (and optimality) of the two approaches.

The issues in inversion mirror those associated with prediction, defining the question:

- how does the resolution of useful information from the past degrade as one looks further back?
- how much past data can be usefully assimilated?
- what are the appropriate 'spin-up' times for data assimilation in various types of system?

5.2. The Earth System

5.2.1. *Earth*

The earth system is seen as comprising the atmosphere, oceans, lithosphere, cryosphere and the biota. This evolves through the dynamics of interactions within and between the components as well is responding to external perturbations from the Sun, the rest of the solar system and the galaxy beyond.

The evolution of the atmosphere can be regarded as having passed through a sequence rather different atmospheres. Following the original accretion of matter, with the original gases largely lost, volcanic outgassing released a range of cases including H_2O, CO_2, SO_2, CH_4, H_2S and CO. With the advent of photosynthesis came a drawdown of CO_2, and slow build-up of oxygen (in part from H_2O, with H_2 lost to space). After several billion years there was sufficient oxygen from photosynthesis to oxidise surface minerals and oceanic iron. Subsequently oxygen accumulated in atmosphere.

The lithosphere acts as a control, with long-term influence on the atmosphere through processes such as silicate weathering, as new mountains are raised. This leads to sequestration of carbon through deposition of carbonate sediments. The positions of the continents shape the global climate through constraints on energy transfer, particularly through the oceans. The existence of large amounts of liquid water is the defining feature of the planet, to the extent that many have commented that 'Earth' is a misnomer. Like the atmosphere, the oceans provide a medium for the transport of chemical species and energy. The latent heat associated with the change of state of water is a major influence on the local to global energy balance. The oceans, or possibly their margins, provided the setting for the origin of life.

The fundamental distinction among living organisms is whether or not the cells have nuclei. The more primitive *prokaryota* (lacking nuclei) comprise *bacteria* which have range of chemical lifestyles, including photosynthesis (e.g. cyanobacteria) and *archea* which are genetically distinct and have even more diverse chemical lifestyles, many in extreme environments. The *eukaryota* (those with nucleated cells) require free oxygen at about 0.5% or more and evolved at least 2 billion years ago from symbiosis of prokaryotic organisms. The eukaryota comprise: protists (amoeba etc.), fungi, plants and animals. The expansion of life from the oceans to land represented a major shift in the environment of the earth.

Williams[38] sees the evolution of life as showing a directional trend, reflecting the increasingly oxidising atmosphere. As well as the minimum noted above for nucleated cells, other processes, essential for higher organisms, were identified as requiring even higher oxygen concentrations.

The cryosphere has played a varying role in the earth system. Over the last million years, a sequence of glacial/interglacial alternations has seen the high latitude ice-sheets expand and contract. Similar periods have occurred in the past, possibly including a 'snowball earth' (or at least 'slushball earth') period when extensive ice coverage maintaining a high reflectivity perpetuated cool conditions — possibly until build up of volcanic CO_2 lead to a greenhouse warming.

5.2.2. *Questions*

An important stimulus for the development of an integrative earth science was the unexpected discovery of the ozone hole.[3] Crutzen and Stoermer[39] have suggested that the present time could be characterised as being in a new geological epoch – the Anthropocence – a period with a discernable human imprint on the physical earth system. Apart from the direct concerns about warming and rising sea-levels, increasing CO_2 is giving a direct global-scale chemical imprint through lowering the pH of the oceans — a drop of 0.1 so far (i.e. a 26% increase in hydrogen ion concentration), with serious concerns about the impact on marine ecosystems.[40,41]

Some of the global-scale concerns have been: anthropogenic climate change; ozone depletion and the ozone hole; instabilities and tipping points, such as the type of abrupt natural change that occurred in the Younger Dryas; and nuclear winter.

This diversity of concerns suggests the questions:

- Is there any organising principle for a unified science of the earth system?
- Can studies of complexity contribute to understanding the earth system?

5.2.3. *Implications of complexity for earth system science*

The review of complexity and climate by Rind[1] noted that the climate system exhibits many of the classic types of complex behaviour but posed the question: "*Of what advantage is it to know this?*" A partial answer may be that studies of complex systems may at least teach us what sort of questions can be sensibly asked regarding global change. In other words studies of global change need to become part of the general question of: *How do we analyse complex systems?* However, there is a converse to Rind's question: *How can studies of Earth System Science contribute to the more general understanding of complex systems?* One such case has been the ideas of chaos arising from the work of Lorenz[29,42] as one of a number of independent discoveries of this phenomenon.[29] The premise of this chapter (and my earlier report[43]) is that the practice of data assimilation and similar inverse problems is one in which the experience in the Earth Sciences has much to offer the more general study of complex systems. In the background is the key question: Can one legitimately claim the existence of a field of *Complex System Science* — of which *Complex Earth System Science* might be a part? Sets of review articles, suggesting that there is a new emerging science, have been published by *Science* and *Nature* with overviews by Gallagher and Appenzeller[8] and Ziemelis[44] respectively.

Bak[32] takes the view that there can be a theory of complexity. He identifies some of its characteristics as:

- at most it can (and should) explain why there is variability — it will be unable to explain particular variations;
- it must be abstract;
- it must be statistical.

Bak argues that much of physical theory (i.e. statistical mechanics, quantum mechanics and chaotic dynamics) already has such characteristics. Bak proposes the sandpile model,[30,31] which exhibits self-organised criticality, as a paradigm for the behaviour of a multitude of complex systems. In a non-technical review, Buchanan[33] goes further and states that "*the ubiquity of the critical state may well be considered the first really solid discovery of complexity theory.*"

Expressing a somewhat different view, Kadanoff[45] notes that:

> "*At one time, many people believed that the study of complexity could give rise to a new science. In this science, as in others, there would be general laws with specific situations being understandable as the inevitable working out of these laws of nature. Up to now we have not found any such laws. Instead, studies of specific complex situations, for example the Rayleigh-Bénard cell, have taught us lessons — homilies — about the behaviour of systems with many independently varying degrees of freedom. These general ideas have broad applicability, but their use requires care and judgement. Our experience with complex systems encourages us to expect richly structured behaviour, with abrupt changes in space and time and some scaling properties. We have found quite a bit of self-organization and have learned to watch out for surprises and big events. So even though there is apparently no science of complexity, there is much science to be learned by studying complex systems.*"

The perception of complexity can be highly context-dependent. For example:

- Atmospheric dynamics tend to regard chaos as an inherent characteristic of the atmosphere, and so no longer see 'chaos' as a new paradigm. Rind[1] (citing Lorenz in an unjustified identification of chaos and complexity) suggests that *"the very concept of complexity originally arose in concert with atmospheric processes."*
- In contrast, a number of authors regard fully-developed chaos as too simple to count as complexity and identify complexity as a special state 'on the edge of chaos.'
- Many physicists would see low-order catastrophes as: (a) a long-known concept (e.g. the van der Waals equation from 1873[46]); and (b) inadequate as descriptions of reality, giving a poor description of actual liquid-gas critical points.
- Fractal behaviour (power-law statistics) has sometimes been taken as a characteristic of complex systems. The concept of fractals (self-similar distributions) was defined by Mandelbrot[47] who identified such be-

haviour in many human and natural phenomena. Power-law behaviour arises in cooperative phase transitions and which can be analysed using the techniques of statistical mechanics. Fractal behaviour has also been found in other aspects of condensed-matter physics.[48] As noted above, Buchanan[33] attached great significance to the wide-spread occurrence of power-law behaviour. However, given the diversity of critical behaviour, power-law behaviour can have only a limited role as a diagnostic.

- A number of workers have proposed algorithmic or computational definitions of complexity (e.g. by emphasising the importance of universal computation[49]). Such a definition would virtually rule out any general laws of complexity of the type discussed by Kadanoff:[45] if a system is described by generic laws then, under algorithmic definitions, it should not be regarded as complex.

Some discussions of complexity note that the behaviour of the complex system is influenced much more by the nature of the connections between components than by the detailed features of the components themselves. Such considerations are influenced by statistical physics where the phenomenon is known as universality and can be 'explained' by the renormalisation group.[50] For more general complex systems, however, there are not established criteria for the validity of universality or of the applicability of the renormalisation group. Establishing such criteria would be an important part of any general theory of complexity by allowing simplifications in the complex calculations. In the absence of such developments, this chapter proposes that inversion techniques may provide a viable alternative to the controlled experiments.

In the first chapter of this volume, Tomaso Aste and Tiziana Di Matteo note the characteristics of complex systems (or complex states) as being:

- emergent;
- open;
- adaptive;
- historically contingent;
- not fully predictable;
- multi-scaled;
- disordered;
- having multiple states.

An important concept for complex systems is that of the 'emergent possible' as described by Kauffman.[51] The idea is that as a system has complex components emerge, it becomes possible to have other emergent properties build on the initial components. Kauffman sees that idea as having wide validity in areas ranging from chemical evolution of the earth and its biota, through to innovation in cultural and economic systems. Newly emergent properties open the way for additional levels of causality as discussed by Ellis.[10]

With this characterisation, we can confront the question: What type of complex system is the earth system? A comprehensive survey is beyond the scope of this chapter, but the following sections capture some specific examples. Questions that need to be considered in complex systems terms include both big-picture questions such as: "is the interaction of the biosphere and the physical earth system, regarded as a complex adaptive system, such that the adaptation can be appropriately described as the 'geophysiology' of the Gaia hypothesis?" (see section 5.5) and more specific questions such as "what defines the limits (in temperature and greenhouse gas concentrations) of the last millions years of glacial-interglacial oscillations?"

5.2.4. *Inverse problems*

My own definition of an inverse problem[52] is one in which the chain of mathematical calculation is the opposite to that of real-world causality. Such calculations can be difficult because many real-world processes act to dissipate small-scale features in which case the inverse problem is trying to reconstruct features from information degraded by noise. Mathematicians tend to take such difficulties, technically called ill-conditioning, as the defining characteristic of inverse problems. Although many of the key ideas of inverse problems were first developed in the solid earth sciences[53,54] there is as extensive literature on applications in a range of fields of science such as astronomy,[55] oceanography[56] and remote sensing.[57]

While the study of the earth as a complex system is placing new emphasis on inverse problems, the principles are not dramatically new. In a very real sense, inversion is what science does, as in the celebrated comment by Charles Darwin that "*no observation is of any use unless it is for or against some theory*." What is new in the complex systems environment is greater mathematical and computational sophistication. The qualitative change is the need for more careful statistical analysis to separate multiple influences on observed behaviour. I have previously suggested[43] that data assimilation where analysis and synthesis proceed in alternation (see section 5.2.5) should be regarded as a paradigm for looking at complex systems when naive reductionism becomes difficult or impossible. Data assimilation has potential far beyond being an operational tool for numerical weather prediction.

Inverse problems include calibration and deconvolution as particular cases. A simple illustration is given by the relation between CO_2 concentration, $C(t)$, given anthropogenic sources, $S(t)$, and a linearised response, $R(t)$, that gives the amount of a CO_2 perturbation that remains in the atmosphere after time t.

$$C(t) - C(t = 0) = \int_0^t R(t - t')\, S(t')\, dt' = \int_0^t S(t - t'')\, R(t'')\, dt'' \qquad (5.1)$$

The change of variables shows that there is a formal mathematical equivalence between the problems of 'calibration' (deduce $R(t)$ given $C(t)$ and $S(t)$) and 'deconvolution' (deduce $S(t)$ given $C(t)$ and $R(t)$). However the nature of the inversion is very different because of the different statistical characteristics of the two problems. In the calibration problem, one is usually seeking a smooth function, often by the process of estimating model parameters. In contrast, in the deconvolution problem, one is usually seeking all the variability that can be validly deduced from the observational data.

The extreme sensitivity to model and data errors makes a careful statistical analysis particularly important in inversion calculations. Tarantola's books[58,59] present generic inversion techniques in a fully statistical manner, specifically using Bayesian statistics. Similarly, in meteorological data assimilation, Kalnay[60] emphasises the statistical basis of assimilations, as does my book on trace gas inversions.[52] An analysis of the relation between inverse problems and statistics is given in more formal terms by Evans and Stark.[61]

Tarantola[58] asserts that the Bayes posterior distribution, $\Pr_{\mathrm{posterior}}(\mathbf{x}|\mathbf{z}) \propto \Pr_{\mathrm{prior}}(\mathbf{x}) \Pr(\mathbf{z}|\mathbf{x})$ is **the** answer to the generic inverse problem of inferring parameters, $\mathbf{x}$ from data $\mathbf{z}$. His later book,[59] emphasises Monte Carlo techniques as providing sets of realisations from this distribution. More commonly, one wants a specific parameter estimate with uncertainties, derived from some principle such as maximum of the likelihood $L(\mathbf{x}|\mathbf{z}) = \Pr(\mathbf{z}|\mathbf{x})$ (or maximum of posterior distribution in the Bayesian case). For multivariate normal distributions with linear relations $\mathbf{G}\,\mathbf{x} = \mathbf{z} + \mathit{noise}$ between observations $\mathbf{z}$, and parameters $\mathbf{x}$, these are particularly simple with the respective negatives of the logarithms being given, to within an additive constant, by quadratic forms:

$$\Theta = -\ln[L(\mathbf{x}|\mathbf{z})] = [\mathbf{G}\,\mathbf{x} - \mathbf{z}]^{\mathsf{T}}\mathbf{X}[\mathbf{G}\,\mathbf{x} - \mathbf{z}] \tag{5.2}$$

where $\mathbf{X}$ is the inverse covariance matrix of the observations $\mathbf{z}$, or in the Bayesian case

$$\Theta = -\ln[\Pr(\mathbf{x}|\mathbf{z})] = [\mathbf{G}\,\mathbf{x} - \mathbf{z}]^{\mathsf{T}}\mathbf{X}[\mathbf{G}\,\mathbf{x} - \mathbf{z}] + [\mathbf{x} - \mathbf{x}_0]^{\mathsf{T}}\mathbf{W}[\mathbf{x} - \mathbf{x}_0] \tag{5.3}$$

where $\mathbf{W}$ is the inverse covariance matrix prior distribution of parameters with mean $\mathbf{x}_0$.

This separation of data and priors is a specific case of fitting multiple (independent) data sets with

$$\Theta = -\ln[\Pr(\mathbf{x}|\mathbf{z})] = \sum_i [\mathbf{G}_i\,\mathbf{x} - \mathbf{z}_i]^{\mathsf{T}}\mathbf{X}_i[\mathbf{G}_i\,\mathbf{x} - \mathbf{z}_i] + [\mathbf{x} - \mathbf{x}_0]^{\mathsf{T}}\mathbf{W}[\mathbf{x} - \mathbf{x}_0] \tag{5.4}$$

The maximum likelihood (or maximum of posterior distribution) estimates are obtained from linear relations

$$\hat{\mathbf{x}} = \mathbf{V}^{-1}[\mathbf{W}\mathbf{x}_0 + \sum_i (\mathbf{G}_i)^{\mathsf{T}}\mathbf{X}_i\mathbf{z}_i] \tag{5.5}$$

where $\mathbf{V}$ is the inverse covariance of these estimates and is given by

$$\mathbf{V} = \mathbf{W} + \sum_i (\mathbf{G}_i)^{\mathsf{T}}\mathbf{X}_i\mathbf{G}_i \tag{5.6}$$

5.2.5. *Data assimilation in meteorology*

In addressing the complexity of the earth system, my technical report[43] proposed that the experience in analysis of the Earth System and its components, in particular data assimilation for weather forecasting, can provide useful examples for the analysis of more general complex systems. In order to demonstrate this assertion, a number of components of the Earth System are considered. In each case, complex behaviour was identified and a number of associated inverse problems are reviewed. In systems characterised by high levels of contingency, inversion techniques, using observations as boundary conditions can provide an alternative to reductionist approaches based on controlled experiments.

Weather forecasting is a task undertaken daily (and often several times a day) by weather services around the world. The primary tool is an atmospheric general circulation model whose time-evolution provides the forecast. The chaotic nature of atmospheric models means that the model

results progressively diverge from the behaviour of the real atmosphere. The consequences of this are:
(a) forecasts are limited to times of order 1 week;
(b) the model must be continually matched to observations at the start of each forecast calculation.

This matching process is termed *data assimilation.* A classic description is given in the book by Daley[62] and the book by Kalnay[60] describes more recent practice.

Enting[52] compared meteorological data assimilation to other inversion problems, especially the trace gas inversion described in Section 5.3.2. In particular, he cited Lorenc[63] as identifying the special features of data assimilation in meteorology as being:

- a large pre-defined parameter basis, i.e. the variables of the forecast model;
- a set of observations that were far too few to determine these basis components;
- a large amount of prior information comes from ongoing operation of the forecasting system;

and also noted the additional issue that the observed quantities are generally not the dynamical variables used in the models.

Increasingly, assimilation schemes for weather forecasting also need to incorporate information about the land surface, including vegetation, soil-moisture and snow-cover. This is needed so that the forecast model can predict the evolution of surface properties (and their consequent influence on the atmosphere). This imposes an extra degree of difficulty because the time-scales involved in surface processes are longer than those of the atmosphere — a difficulty known as the 'spin-up problem'.[62]

5.3. Carbon

5.3.1. *The carbon cycle*

The volume *The Global Carbon Cycle*, edited by Field and Raupach[64] gives a recent overview of the carbon cycle. The cycle links the atmospheric, oceanic, geological and biospheric components of the earth system through a diverse range of carbon exchange processes acting on a wide range of spatial and temporal scales. In some cases, smaller-scales can be aggregated into larger scales. Sundquist[65] describes the carbon cycle in terms of such a hierarchy of scales and processes. Such an aggregation is possible in linear (or near-linear) systems. For the case of the carbon cycle, conservation of carbon mass provides a linear constraint. However non-linearities in the carbon cycle, and in particular the inter-dependence of carbon and climate, serve to couple different space and time scales, leading to the possibility of emergent complex behaviour, most notably abrupt climate change.

Some of the levels described by Sundquist are:

- atmosphere/biosphere/ocean-surface, for modelling times scales up to a few years;
- add a deep ocean component to atmosphere/biosphere/ocean-surface, for modelling decades to millennia;
- including ocean carbonate sediments for multi-millennia time scales;
- additional geological processes for longer time-scales.

Immediate human concerns cover time-scales of decades to centuries and so analyses are mainly concerned with the first two levels. In this case, in equation (5.1), $R(\infty) \neq 0$, i.e. for any perturbation, some CO_2, (of order 10 to 15%) remains in the atmosphere for thousands of years until excess carbon is removed from the atmosphere-ocean-biosphere system by geological processes. When considering the longer time-scales, the atmosphere/biosphere/ocean-surface system can be considered as in equilibrium and represented as a single reservoir.

The multiplicity of processes involved in the carbon cycle means that a wide range of observations will be needed in order to improve our understanding. Canadell et al.[66] have given an overview of of the various types of data available. The interpretation of such multiple data streams has sometimes been termed 'multiple constraints' analysis.[67] While no new principles seem to be required for such analyses (with the combination of disparate data being described by (5.4) or more generally by (5.11)), careful statistical analysis is important in order to ensure an appropriate relative weighing of the data streams. A qualitative description of important statistical issues of carbon cycle data is given by Raupach et al.[68]

5.3.2. *Estimating carbon fluxes*

The 'classic' CO_2 inversion problem, described in my book,[52] is the estimation of surface fluxes for surface concentration data. This was put on a quantitative basis, including uncertainty analysis, with the introduction of Bayesian synthesis inversion. This estimated fluxes $\mathbf{x}$ from concentrations $\mathbf{z}$ using (5.5, 5.6) where the $\mathbf{G}$ was implemented by calculating a set of responses to specific basis fluxes, giving a coarse discretisation of $\mathbf{G}$ as an explicit matrix. The 'multiple constraints' were introduced via the 'priors.'

The disadvantage of these techniques is the need to perform multiple runs in order to generate an explicit matrix representation of $\mathbf{G}$. In addition, the use of linear calculations is based on the assumption of Gaussian error distributions for both observations and priors. In spite of the formal simplicity of the linear relation (5.5), it is not necessarily the most efficient way to evaluate maximum likelihood estimates. The minimisation relation (from which (5.5) is derived) is

$$\frac{1}{2}\nabla_{\mathbf{x}}\Theta = \mathbf{G}^{\mathsf{T}}\mathbf{X}[\mathbf{G}\,\mathbf{x} - \mathbf{z}] + \mathbf{W}[\mathbf{x} - \mathbf{x}_0] = \mathbf{0} \tag{5.7}$$

In using this, $\mathbf{G}\,\mathbf{x}$ represents an integration of the transport model, $\mathbf{z}$ is readily subtracted and multiplication by $\mathbf{X}$ is at worst a 'matrix $\times$ vector' operation. It is often much simpler because $\mathbf{X}$ is often block diagonal. What is required to complete the evaluation of the gradient is the ability to apply the operation represented by $\mathbf{G}^{\mathsf{T}}$ (the transpose of $\mathbf{G}$) without having to explicitly construct a high-dimension matrix representation of $\mathbf{G}$. This can be achieved by what are known as adjoint techniques. A model whose linearised response is given by $\mathbf{G}'$ can be transformed into a linear model whose response is given by $\mathbf{G}'^{\mathsf{T}}$. There are software tools that can automatically transform computer codes in this way.[69,70] Such adjoint techniques can get $\nabla_{\mathbf{x}}\Theta$ as single integration, to be used in iterative minimisation of Θ. In spite of the need for iteration, the resulting computational efficiencies have allowed trace gas inversions at high resolution.[71]

These gradient methods can be readily extended to non-linear cases with $\mathbf{G}(\mathbf{x}) = \mathbf{z} + \textit{noise}$ which have

$$\frac{1}{2}\nabla_{\mathbf{x}}\Theta = \mathbf{G}'^{\mathsf{T}}\mathbf{X}[\mathbf{G}(\mathbf{x}) - \mathbf{z}] + \mathbf{W}[\mathbf{x} - \mathbf{x}_0] \tag{5.8}$$

where $\mathbf{G}'$ denotes the linearisation of the relation $\mathbf{G}(\mathbf{x})$. The operator $\mathbf{G}'$ is the Green's function of what is known as the tangent linear model, and $\mathbf{G}'^{\mathsf{T}}$ is adjoint of this Green's function. Alternatively, $\mathbf{G}'^{\mathsf{T}}$ is Green's function of the adjoint model,' i.e. the adjoint of the tangent linear model. An implementation of the adjoint model is computationally desirable if the dimensionality of the space of $\mathbf{x}$ is large.

Gradient methods are particularly appropriate for process inversion which usually involves non-linear relations. Since iterative solutions are generally needed in such cases, the fact that gradient methods require iteration is not a specific disadvantage. The generic relation for applying consecutive processes, represented by a relation $\mathbf{T}(\mathbf{S}(\mathbf{x}))$ lead to the gradient relation:

$$\nabla_{\mathbf{x}}\mathbf{T}(\mathbf{S}(\mathbf{x})) = \nabla_{\mathbf{s}}\mathbf{T}\nabla_{\mathbf{x}}\mathbf{S} \tag{5.9}$$

but in the case of atmospheric transport, $\mathbf{T}$, relating fluxes and concentrations, linearity leads to

$$\nabla_{\mathbf{x}}\mathbf{T}\mathbf{S}(\mathbf{x}) = \mathbf{T}\nabla_{\mathbf{x}}\mathbf{S} \tag{5.10}$$

This has been applied to using concentration data to estimate parameters in a 'curve-fitting' model, SDBM (Simple Diagnostic Biosphere Model),[72] and the more mechanistic BETHY model.[73]

5.3.3. *Deducing emission targets*

The usual form of forward modelling determines concentrations from sources as:

Change in concentration
= source − loss from atmosphere

where the 'loss from atmosphere' term is calculated by a model. Given such a model, the calculation can be transformed to give:

Required source
= Change in target concentration
+ loss from atmosphere

Enting et al.[74] described a parameterisation that captured a range of options from early reductions (allowing a more gradual change) to later reductions that, for any given concentration target, need to be more abrupt once commenced. Figures 5.2 and 5.3 show such families of emission profiles for CO_2 concentration targets of 550 ppm and 450 ppm respectively.

For lower target concentrations, delay in beginning emission reductions greatly reduces the scope for a smooth transition to stabilisation. These calculations used models that did not include any explicit representation of climate-to-carbon feedbacks (see section 5.4.2) although some feedback process may be included implicitly. If feedbacks greater than those included

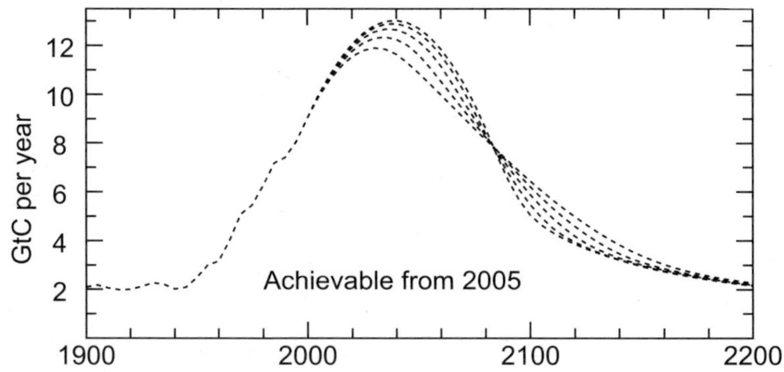

Fig. 5.2. Family of profiles, with smoothly changing emissions, for stabilisation at 550ppm, showing the trade-off between higher peak and rapid subsequent reductions. (From[23] using the parameterisation from[74]).

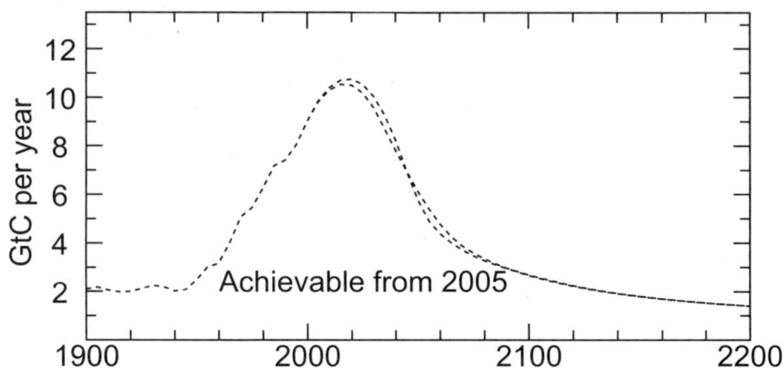

Fig. 5.3. Family of profiles, with smoothly changing emissions, for stabilisation at 450ppm, showing the trade-off between higher peak and rapid subsequent reductions. (From[23] using the parameterisation from[74]).

implicitly occur then additional emission reductions will be required in order to reach specific targets.[75,76]

5.3.4. *Future directions*

The synthesis inversion calculations carried out to date have all been linear estimates. This is appropriate if:

(i) the relation between sources and concentrations is linear; and

(ii) the measurement statistics (and the statistics of the Bayesian prior distributions) are Gaussian.

Condition (i) will hold for conserved tracers; condition (ii) is problematic, especially for trace gas records that represent a mixture of 'baseline' conditions and 'pollution events'.[52] However, even with Gaussian statistics, the experience of synthesis inversion has revealed a number of subtleties. In particular, measurement error is often used as a proxy for the combination of true measurement error and model error. In general this can be justifiable mathematically, but if such an approach is used it has to be realised that for repeat measurements the contributions from true measurement error *may* be independent, but the contributions from model error are likely to have correlations near to 1. Failure to take such correlations into account can give (and has given) severe bias in inversion calculations. Similarly, a neglect of correlation in space or time can lead to unjustifiably small estimates for the uncertainty in aggregated values.

As noted above, the analysis of data records containing both 'baseline' and 'pollution events' is more complicated than cases analysed to date. Some of the difficulties are due to the complex states arising from advective transport. Even if the advective field has simple time variation (e.g. strictly periodic) and only large-scale variations, tracer transport will still be chaotic — the trajectories of neighbouring particles will diverge at some point. Correspondingly, contours of concentration will be stretched into fractal shapes. Two contrasting approaches were presented by Enting[37] and Prinn[36] at the Heraklion conference on *Inverse Methods in Global Biogeochemical Cycles*. Prinn emphasised the advective nature while Enting emphasised current practice, which uses a combination of baseline data selection and time averaging to produce records that are effectively smoothed and that make the atmospheric transport appear diffusive. This removes the effects of complexity in the advective transport, at the expense of neglecting much useful information. Techniques to utilise more of the information in non-baseline data are currently being investigated.[77] Extensions of trace gas inversions to achieve a process-level description are described in the following section. A review of carbon flux estimation and the ways in which the inversions have been developed and extended is given by Wang et al.[78]

5.4. Earth System Modelling

5.4.1. *Carbon cycle process inversions*

In a series of recent developments, trace gas inversions have been generalised from calculations that estimate gas fluxes (as in section 5.3.2) to calculations that estimate parameters and properties of the processes responsible for these fluxes. One example is the study by Knorr and Heimann.[79] The theoretical basis is described by Rayner.[80] Enting[52] notes a number of inversion calculations that can be regarded as forerunners of this type of process inversion.

The requirements of earth system modelling suggest that there is a need to go one step further an estimate the 'role' of the processes in the earth system, whether as drivers, responses or feedbacks (positive or negative). Figure 5.4 shows an analysis of carbon cycle processed in terms of their functional role. Taken from Figure 14.2 of my book,[52] this classifies the roles as:

- equilibrium — the gross fluxes where both oceans and terrestrial ecosystems exchange large amounts of carbon with the atmosphere;
- forcing — the anthropogenic processes that are perturbing the equilibrium:
- carbon cycle response — fluxes induced by changes in atmospheric CO_2 concentration;
- variability — the effects of climate variability on the gross fluxes and response.

	Equilibrium	Forcing	Response	Variability
Fossil	–	Y	–	–
Land-use change	–	Y	–	–
Cycle	Y	climatic	–	climatic
CO_2 fertilisation	–	–	Y	–
Ocean	Y	climatic	Y	climatic

Fig. 5.4. Carbon fluxes: system component vs. functional role. This is based on the characterisation of CO_2 fluxes given in Figure 14.2 of my book[52] and considers the carbon cycle in isolation.

Figure 5.5, based on Box 3 of the IPCC Third Assessment Synthesis Report,[81] gives a schematic of how earth system modelling is evolving from

Process	1985	1990	2000	2005	Emerging
Atmosphere	AGCM	AGCM	AGCM	AGCM	AGCM
Land surface	Land	Land	Land	Land	Land
Ocean:	slab/swamp	OGCM	OCGM	OGCM	OCGM
Carbon	off-line	off-line	off-line	coupled?	coupled
Sulphate	—	—	—	aerosol	aerosol
Non-sulphate	—	—	—	aerosol	aerosol
Vegetation	—	—	—	—	dynamic
Chemistry	—	—	—	—	chemistry

Fig. 5.5. Schematic of development of climate models. Slightly modified from Box 3 of IPCC Third Assessment Synthesis Report.[81]

early climate modelling efforts. The current status is that only a small number of models have the carbon cycle included as an interactive component, rather than just being an off-line forcing. The models have given disparate predictions of the significance of carbon-climate feedback processes.[82–84]

The main process affecting terrestrial carbon are:

- photosynthesis;
- phenology;
- pool turnover;
- heterotrophic respiration.

Each of these is subject to modulation by changes in temperature and water availability, constrained by nutrients and subject to competition and disturbance. There is considerable heterogeneity in both the environments and these modulating factors. The land-surface components of climate models include both biological and physical processes in the exchange of energy, momentum, moisture and carbon because of the strong couplings between processes.

Within Australia, development of a new generation of climate model is taking place under a framework called ACCESS: Australian Community Climate and Earth System Simulator. This is a combined effort involving the Bureau of Meteorology, CSIRO and Australian universities.

The terrestrial carbon model in ACCESS is intended to be constructed from several existing components that cover rather different time scales:

CABLE This is the current (circa 2005) CSIRO land-surface scheme, somewhat upgraded.[85] It is based on the earlier CBM with processes described by Wang and Leuning.[86] CABLE runs at the time-step of the atmospheric model.

CASA This is a carbon pool model,[87] which is being extended to include the key nutrients: nitrogen and phosphorus.[88] It can be run at time steps of days or longer.

LPJ This is a dynamic vegetation model describing competition and succession.[89] It can be run at time steps of about a year.

Given the importance of statistics even in simple problems such as those of equation (5.1), any consideration of data assimilation on terrestrial carbon models will need to identify inverse problems and their statistical characteristics. One needs to quantify the statistical considerations[68] for particular cases. While a detailed case-by-case analysis is beyond the scope of this chapter, and indeed remains very much a work-in-progress, some generic cases can be identified.

calibration: Calibrations of components of the land surface model are the most immediate need for ACCESS. They are also the simplest since the factorisation (5.10) suggests that the tangent linear model can be used (rather than needing an adjoint model) without excessive loss of computational efficiency.

feedback detection: Earth system models can play a role in the detection of the onset of processes involving carbon-climate feedbacks.

carbon assimilation in natural resource management: This would be adding a predictive capability to the type of 'now-casting' that can be achieved with the data sets of the National Carbon Accounting System,[90] and applications such as satellite-based assessment of fire danger.[91]

calibrations involving assimilation: A hybrid case arises when calibrations, whose primary task is parameter estimation, use data assimilated into the model, in order to specify boundary conditions. An obvious example is the use of satellite-derived vegetation indices as part of a model calibration.

joint assimilations of carbon and water: The coupling between water and carbon on a range of scales may lead to carbon assimilation providing a constraint on water content and possibly relevant data for seasonal or synoptic forecasting. This is a somewhat speculative possibility that remains to be tested.

The other likely link between assimilation of carbon and soil moisture is methodological. The two problems have many similarities, each involving mainly local processes, each having a high degree of spatial heterogeneity and each involving a wide range of time-scales.

The general absence of long-range connections between components of terrestrial systems suggests that, depending on the data, there may be a hierarchy of levels of model calibration:

- Analysis of a single homogeneous site is possible if the data are purely local from a location with little spatial variability.
- Consideration of a mixed 'pixel' will be needed in most cases of analysis of satellite data and other medium-scale data such as boundary-layer measurements.
- Calibrations of biome-specific parameters can be performed separately for each single biome if the observations for that biome can be separated, e.g. using satellite data for a homogeneous region.
- Inversions is which the whole world needs to be analysed simultaneously are those involving global constraints. The most common case is that of atmospheric CO_2 concentrations, i.e. the extension of the flux inversion problem to process inversions. Other data that constrain terrestrial carbon processes are the radiocarbon data form the period after nuclear testing, when the seasonal modulation of the 'bomb-spike' gives a measure of biospheric turnover.[92] The minority stable carbon isotope, ^{13}C, gives less-direct information.[93]
- Analysis of global constraints will probably have to consider inhomogeneities within biomes in order to avoid the type of truncation error that arises in low-resolution synthesis inversions. This effect (and the efficacy of a statistical 'correction') has been analysed by Kaminski et al.[94] A schematic representation of the 'correction' formalism is given in Figure 8.3 of my book.[52]

When the multiple data streams are independent, the calibration has the potential to exploit the separation that appears in the sum of squares (5.4) or more generally by iterative application of Bayes relation:

$$\mathrm{Pr}_{\mathrm{posterior}}(\mathbf{x}|\mathbf{z}_1 \dots \mathbf{z}_N) \propto \mathrm{Pr}_{\mathrm{prior}}(\mathbf{x}) \prod_{n=1}^{N} \mathrm{Pr}(\mathbf{z}_n|\mathbf{x}) \tag{5.11}$$

This factorisation has the potential to greatly simplify the calibration. The approach used in flux estimation from Bayesian synthesis inversion had all the non-atmospheric constraints feeding into the atmospheric inversion as 'priors,' with specified uncertainty, and treated as independent in most cases.

Some of the criteria for achieving computational efficiency in more general cases are:

- maintain system separation for as long as possible;
- postpone any non-linear analysis (from non-linear processes and/or non-Gaussian distributions) to as late as possible in the inversion sequence;
- note that for parameter spaces of low dimension, use of the tangent linear model, rather than the adjoint, can incur a relatively low penalty in terms of computational cost, while allowing great simplification in the derivation.

5.4.2. *Feedbacks*

Recently, Pittock[95] has raised the concern that the projected climate changes are being underestimated. Note, however, that the existence of these various positive feedback processes does not imply any serious concern about 'run-away' climate feedback. The planet's natural response — radiating excess heat back into space — imposes a limit on the size of possible perturbations. In addition, there are a number of negative feedbacks. These include the partial saturation of the absorption bands for CO_2 and to a lesser extent CH_4.

The IPCC Fourth Assessment Report (AR4) noted two main caveats to the projections of 21st century climate change:

- The magnitude of the positive feedback between climate change and the carbon cycle is uncertain. (AR4: Technical Summary 5.5).
- Dynamical processes not included in current models but suggested by recent observations could increase the vulnerability of the ice sheets to warming, increasing future sea level rise. (AR4: Technical Summary 5.5).

The classification of carbon fluxes in Figure 5.4 above addresses the analyses of the carbon cycle. When considering the carbon cycle within the earth system, feedbacks connecting the carbon cycle and the physical climate system become important. Various cases, particularly those involving glacial-interglacial cycles have been noted by Falkowski et al.[6] and Rial et al.[96]

Feedbacks and non-linearities play separate roles, but:

- strong positive feedbacks can drive systems into non-linear regimes of behaviour;
- feedback processes can often act in a non-linear manner;
- the characterisation of a feedback, i.e. having a system's output affecting its input is dependent on the definition of the system boundary.

Chaos provides a further connection between feedbacks and non-linearities since non-linearity is essential for chaotic behaviour and delays, common in feedback processes, can allow chaotic behaviour even in single component systems such as the discrete logistic map.[97]

Abrupt climate change can occur as the earth system passes through some sort of threshold. Among the cases that have been identified of concern, the so-called tipping points,[98] are:

- weakening, or even collapse, of the thermohaline circulation due to freshening and warming of waters in the North Atlantic;
- transition of Amazonian rainforest to savannah in response to warming and drying of the region;
- accelerated climate change due to increased CH_4 emissions as tundra warms;
- release of CH_4 from clathrates on ocean shelves is a less certain, but potentially larger climate feedback;
- rapid Arctic warming due to albedo changes as the Arctic ocean becomes ice-free in summer;
- instability of ice-sheets as the lower levels begin to melt is another potential threshold effect.

5.4.3. *Analysing feedbacks*

For the 'data-assimilation' tasks of detection, estimation and attribution, feedbacks pose particular challenges. The context-dependence of the term feedback requires a definition of the system(s) without feedbacks. For the CO_2-climate feedbacks, the 'natural' description is to consider the carbon cycle and the physical climate system as two 'stand-alone' systems and define feedbacks as the processes that couple them.

The basic linear equations describing feedback can be written as

$$c = kf \tag{5.12}$$

where c is a response, f is the forcing and k is the sensitivity to the forcing. In a feedback case, the forcing is a combination of an 'external' component,

f_0, and a feedback component, αc, giving

$$c = k(f_0 + \alpha c) \tag{5.13}$$

whence

$$c = \frac{k}{1 - k\alpha} f_0 \tag{5.14}$$

where $k\alpha$ represents the 'gain' around the feedback loop and $1/(1 - k\alpha)$ represents the amplification due to the feedback. Negative feedbacks ($k\alpha <$ 0) reduce the response and positive feedbacks ($k\alpha > 0$ lead to amplification.

The relation (5.14) diverges at $k\alpha = 1$. Real-world systems will, of course, not be literally divergent and will rarely have a change of sign in the response. What the divergence in (5.14) is telling us is that the linearity assumption will break down at some point with $\alpha < 1/k$. Non-linear behaviour may be qualitatively similar to the linear case, but it can also lead to the various abrupt changes described by catastrophe theory. Such instabilities and threshold phenomena have been described in many human and natural systems such as power grids, ecosystems and human physiology.[7]

Modelling the changes as a system passes through such thresholds has proved rather difficult, and at the time of writing, much of the motivation for considering such possibilities as an important risk from climate change is that such abrupt changes have occurred in the past.

5.4.4. *Laplace transform analysis*

Laplace transforms have similar properties to Fourier transforms in analysing linear systems. Laplace transforms are more appropriate for one-sided causal relationships than Fourier transforms. The Laplace transform maps from time, t to an inverse time variable, p. Integro-differential relations map onto algebraic relations. In particular, convolution relations transform to products and integration multiplies a transform by $1/p$.

Using lower case to denote Laplace transforms:

$$f(p) = \int_0^\infty F(t)\, e^{-pt}\, dt \tag{5.15}$$

In terms of a CO_2 perturbation, $Q(t) = C(t) - C(t = 0)$, carbon response relations (5.1) transform to $q(p) = r(p)\, s(p)$ so that the inverse problems are represented by $r(p) = q(p)/s(p)$ (calibration) and $s(p) = q(p)/r(p)$ (deconvolution).

For a linearised model of the coupled carbon-climate system, warming, $W(t)$, is a response to CO_2 perturbation $Q(t)$ and other forcing $F(t)$

$$w(p) = u(p)[f(p) + \alpha q(p)] \tag{5.16}$$

A response function $H(t)$ describes additional CO_2 source from warming:

$$q(p) = r(p)[s(p) + h(p)w(p)] \tag{5.17}$$

whence

$$w(p) = \frac{u(p)f(p) + \alpha u(p)r(p)s(p)}{1 - \alpha u(p)r(p)h(p)} \tag{5.18}$$

$$q(p) = \frac{r(p)[s(p) + f(p)h(p)u(p)]}{1 - \alpha u(p)r(p)h(p)} \tag{5.19}$$

Thus forcing from radiative forcing, $f(p)$ or CO_2 emissions, $s(p)$ is amplified by feedback factor: $\kappa_{\mathrm{FB}} = 1/[1 - \alpha u(p)r(p)h(p)]$.
For multi-decadal time-scales, the C4MIP study[99] gives $\kappa = 1.18 \pm 0.11$ from 11 models, ranging from $\kappa = 1.04$ to 1.44. Another way of expressing the feedback factor uses more common quantities

$$\alpha\, u(p)\, r(p)\, h(p) = \frac{T_{2*\mathrm{CO2}}}{280 \ln 2} \times \frac{u(p)}{u(0)} [p\, r(p)]\, [h(p)/p] \tag{5.20}$$

where $1 - u(p)/u(0)$ is proportion of committed warming for time-scale $1/p$, $p\,r(p)$ is the CO_2 airborne fraction and $h(p)/p$ is feedback response as integrated flux.

In considering the carbon cycle 'calibration' problem for the 20th century, the form (5.19) raises the question of whether a calibration using $C(t)$ and $S(t)$ is estimating the model response $r(p)$ or $r(p)/[1 - \alpha u(p)r(p)h(p)]$? Considering multiple constraints[67] (even if only the inclusion of isotopic data) raises the question of whether models are being calibrated with 'incommensurate' data with some data reflecting $r(p)$ and some data reflecting $r(p)/[1 - \alpha u(p)r(p)h(p)]$.

To the extent that models are being calibrated to reproduce $r(p)/[1 - \alpha u(p)r(p)h(p)]$ then associated feedbacks are not something extra to add when considering 21st century.

5.5. Gaia

The concept of a combined science of the earth system — atmosphere, biosphere, hydrosphere and lithosphere — has evolved over recent decades with a recognition that life and the environment co-evolve. A somewhat different perspective is given by James Lovelock's Gaia hypothesis,[100,101] proposing that the earth system acts like a self-regulating lifelike entity with the biota maintaining the planet as a suitable habitat for life. At times Lovelock has sometimes used the term 'geophysiology' to distance the concept from the more mystical associations of the name Gaia.

In answer to the question *what type of complex system is the earth system?* the Gaia concept[100,101] proposes that the physico-chemical Earth System (the geosphere) and the biota act as a coupled system, with the biota acting to stabilise physical and chemical conditions in a way that preserves conditions on earth as favourable for life. An analysis of Gaia in specifically complex systems terms is given by Lenton and van Oijen.[102]

The Gaia hypothesis has been subject to many criticisms. Three important strands of criticism have been:

- The Gaia concept lacks scientific content. In particular the hypothesis is claimed to be untestable. The most extreme form of this is to ask: *Is the Gaia hypothesis science or religion?*
- Gaian self-regulation implies a goal-seeking behaviour that cannot occur without a sensory/nervous system and the existence of an underlying rationale;
- A Gaian state is unachievable by evolution through natural selection.

Dawkins[103] has argued that Gaian self-regulation is not occurring — such a system could only have been produced by evolution, rather than by chance, and that the essential requirements for evolution, reproduction with differential survival cannot apply to a single system. Specifically:[103] *"The fatal flaw in Lovelock's hypothesis would have instantly occurred to him if he had wondered about the level of natural selection process which would be required in order to produce the Earth's supposed adaptations."* In other words, the planetary biota as a whole is not a unit of selection. Recent work by Lenton[104] aims to address this criticism, but can at best be regarded as only a first step. However, seen from a complex systems perspective, the wording of Dawkins' criticism may contain the seed of an answer — the assumption of a *single level* for natural selection may turn out to be unjustified.

Dawkins' argument can be countered, in a rather unsatisfying way, by invoking an anthropic principle, supposing that Gaian self-organisation arises only by chance but is essential for long-term survival of life and so intelligent observers will only arise on the tiny fraction of planets that have achieved Gaian self-organisation. In contrast, using this argument, life failed to survive on planets (possibly including Mars) where life arose but failed to achieve Gaian homeostasis.

Analysis of the Gaia concept has been complicated by a degree of ambiguity in Lovelock's formulation — common in such exploratory work but needing clarification as a prerequisite for further progress. Lovelock,[105] in a non-standard usage of the words, distinguishes the original *Gaia hypothesis* (that the biota stabilise earth's physical and chemical environment for the benefit of life) from the *Gaia theory* (that self-regulation is achieved by the interaction of the biota and the environment).

Analysis by James Kirchner, originally at the 1989 AGU Chapman conference and then more recently,[106] distinguishes concepts such as 'strong Gaia' (with strong goal-seeking behaviour) vs 'weak Gaia' (essentially conventional earth system science).

A recent critical overview of the Gaia hypothesis by Crutzen[107] builds on this to categorise the variations as:

i. *Optimising Gaia* — Gaia shaped the planet to make it optimal for life – dismissed as untestable
ii. *Homeostatic Gaia* — life stabilises environment; and
iii. *CoEvolutionary Gaia* — a 'weak' form dismissed as not saying anything new.

Crutzen combines (i) and (ii) as *Healing Gaia* and asserts that is not possible. In particular he notes that evolutionary processes are just as likely to produce positive feedbacks as negative feedbacks. A weak point of this argument is that the combination of (i) and (ii) is used to imply that arguments against *Optimising Gaia* also preclude *Homeostatic Gaia*.

An interesting possibility where *Homeostatic Gaia* could arise 'automatically' (*Innate Gaia*) is proposed by Lenton.[108] He notes that *if* (i) physical system is stable, and (ii) biological system has self-increasing growth and (iii) there is a physical optimum for growth *then* the steady state will be whichever side of the optimum leads to negative feedbacks, thus enhancing the stability of the physical system.

A final question is what, if anything, the Gaia hypothesis would imply about anthropogenic global warming, assuming that the hypothesis is true. At a simple level, there are two diametrically opposed views:

- The Earth system will self-regulate to absorb the effect of human impacts; **OR**
- The continued existence of homo sapiens (or any other vertebrate) was never an essential part of Gaian self-regulation and extinction of our species may be part of the self-regulation process.

Something close to this latter view is proposed by Lovelock in his recent books *Revenge of Gaia*[109] and *The Vanishing Face of Gaia*.[105] However the term 'revenge' seems to be implying even more of the problematic goal-seeking behaviour than was implied by the original Gaia hypothesis.

However, even if the Earth has Gaian self-regulation, it is common for self-regulating systems to fail catastrophically when pushed beyond their limits. What Lovelock's recent books seem to be envisaging is that, in addition to the warm and cold quasi-stable climate states represented by the glacial-interglacial cycles, there is a third hotter stable state with greatly depleted life and little scope for return to present conditions if increased CO_2 emissions take us there. This is not 'revenge' — rather the envisioned Man-Gaia interaction is one of 'mutually-assured-destruction.'

5.6. Concluding Remarks

This chapter has attempted to scope some of the data assimilation problems involved in building a science of the integrated earth system in the face of complexity. The discussion of inverse problems in terrestrial carbon modelling has highlighted some of the strategies that are being implemented in on-going model development.

The themes of this chapter come from two of my earlier works, my book *Inverse problems in Atmospheric Constituent Transport*[52] and a CSIRO Technical Paper *Inverse Problems in Earth Systems Science: A Complex System Perspective*.[43] The main messages are:

- inverse problems need to be addressed as statistical estimation, implying a need to combine statistics and applied mathematics;[61]
- data assimilation in meteorology can be seen as a possible paradigm for studying more general complex systems.[43]

The complex systems perspective draws on my early research training in statistical physics which led to my involvement in the CSIRO Complex System Emerging Science Initiative. My thoughts on modelling expanded while writing *Twisted: The Distorted Mathematics of Greenhouse Denial*[23] which addressed the misrepresentations by so-called 'greenhouse-sceptics.' The misrepresentations are so blatant that the term 'sceptics' ceases to be appropriate and should probably be replaced by 'pseudo-sceptics' or more explicitly 'doubt-spreaders' reflecting the similarities between their activities and the "*doubt is our product*" approach used by the tobacco industry to delay regulation. Finally, as in noted in the preface to my earlier book,[52] working on the carbon cycle, as I have done since 1980, provides an excuse for engaging with almost any area of science — in the background of such biogeochemical studies has been the Gaia hypothesis of James Lovelock.[100,101]

Hulme[110] has remarked that "*the point of climate science is not to enable humanity to predict its future climate. It is to enable humanity to choose its future climate.*" In an absolute sense, neither prediction nor choice is possible, given the complexity of the earth system. An achievable objective needs to be characterised in terms of reduction of risks.[111] In discussing climate feedbacks, Rial et al.[96] have argued against 'prediction' and in favour of 'vulnerability analysis.' In *Uncertain Science ... Uncertain World* Pollack[112] describes how the uncertainties in climate modelling are already much smaller than in many cases of human decision making.

Making effective choices for the reduction of climate risk needs to take into account the role of human systems in implementing such choices. For a global problem such as climate change, linking global decisions to local and regional decisions remains the barrier to action. Inclusion of the human component potentially leads to what Kates et al.[113] have termed 'sustainability science.'

Acknowledgments

The ARC Centre of Excellence for Mathematics and Statistics of Complex Systems (MASCOS) is funded by the Australian Research Council. The author's fellowship at MASCOS is funded in part by CSIRO. The work on the global carbon cycle builds on long-term collaborations with many colleagues, particularly Drs. Etheridge, Francey, Lassey, Law, Pearman, Rayner, Trudinger and Wang. The discussion of climate-to-carbon feedbacks draws on on-going work with Nathan Clisby. The discussions of complexity draw on many discussions with participants in the CSIRO Complex Systems Emerging Science Activity. Particular thanks are due to Inez Fung and the participants at the 2006 Carbon Data Assimilation workshop at MSRI Berkeley for many stimulating discussions. Andrew Glickson's helpful comments on the manuscript are greatly appreciated.

Note added in proof:
The July 2009 issue of *European Physical Journal: Special Topics* is entitled 'Understanding the Earth as a Complex System' and contains papers from the First International Workshop on Data Analysis and Modelling in Earth Sciences (Potsdam, 2008), complementing the material in this chapter.

References

1. D. Rind, Complexity and climate, *Science.* **284**, 105–107, (1999).
2. P. D. Guthrie, The CH_4-CO-OH conundrum: a simple analytic approach, *Global Biogeochemical Cycles.* **3**, 287–298, (1989).
3. J. C. Farman, B. G. Gardiner, and J. D. Shanklin, Large losses of total ozone in Antarctica reveal ClO_x/NO_x interaction, *Nature.* **315**, 207–201, (1985).
4. M. Scheffer, S. Carpenter, J. Foley, C. Folke, and B. Walker, Catastrophic shifts in ecosystems, *Nature.* **413**, 591–596, (2001).
5. R. H. Bradbury, D. G. Green, and N. Snoad. Are ecosystems complex systems? In eds. T. R. J. Bossomaier and D. G. Green, *Complex Systems*, pp. 339–365. CUP, Cambridge, UK, (2000).
6. P. Falkowski, R. J. Scholes, E. Boyle, J. Canadell, D. Canfield, J. Elser, N. Gruber, K. Hibbard, P. Höberg, S. Linder, F. T. Mackenzie, B. Moore III, T. Pedersen, Y. Rosenthal, S. Seitzinger, V. Smetacek, and W. Steffan, The global carbon cycle: A test of our knowledge of the earth as a system, *Science.* **290**, 291–296, (2000).
7. A. Bunde, J. Kropp, and H. J. Schellnhuber, Eds., *The Science of Disasters: Climate Disruptions, Heart Attacks, and Market Crashes.* (Springer-Verlag, Berlin, 2002).
8. R. Gallagher and T. Appenzeller, Beyond reductionism, *Science.* **284**, 79, (1999).
9. J. Cohen and I. Stewart, *The Collapse of Chaos: Discovering Simplicity in a Complex World.* (Penguin, London, 1994).
10. G. F. R. Ellis, Physics, complexity and causality, *Nature.* **435**, 743, (2005).
11. A. J. McMichael, *Human Frontiers, Environments and Disease.* (CUP, Cambridge, UK, 2001).
12. E. O. Wilson, *Consilience: The Unity of Knowledge.* (Abacus, London, 1998).
13. R. Courant and D. Hilbert, *Methods of Mathematical Physics.* (Interscience (John Wiley), New York, 1953).
14. S. Wolfram, Ed., *Theory and Applications of Cellular Automata.* (World Scientific, Singapore, 1986).
15. W. J. Karplus, The spectrum of mathematical modelling and systems simulation, *Math. Comput. Simulation.* **19**, 3–10, (1977).
16. I. G. Enting, A modelling spectrum for carbon cycle studies, *Math. Comput. Simulation.* **29**, 75–85, (1987).
17. H. Lieth. Modeling the primary productivity of the world. In eds. H. Lieth and R. H. Whittaker, *Primary Productivity of the Biosphere*, pp. 237–264. Springer-Verlag, New York, (1975).
18. H. Oeschger, U. Siegenthaler, and M. Heimann. The carbon cycle and its perturbations by man. In eds. W. Bach, J. Pankrath, and J. Williams, *Interactions of Energy and Climate*, pp. 107–127. Reidel, Dordrecht, (1980).
19. M. V. Thompson and J. T. Randerson, Impulse response functions of terrestrial carbon cycle models: method and application, *Global Change Biology.*

5, 371–394, (1999).
20. P. Young, S. Parkinson, and M. Lees, Simplicity out of complexity in environmental modelling: Occam's razor revisited, *J. Appl. Statist.* **23**, 165–210, (1996).
21. P. Young and H. Garnier, Identification and estimation of continuous-time data-based mechanistic models for environmental systems, *Environmental Modelling and Software.* **21**, 1055–1072, (2006).
22. I. G. Enting, T. M. L. Wigley, and M. Heimann, *Future emissions and concentrations of carbon dioxide: Key/ocean/atmosphere/land analyses.* CSIRO Division of Atmospheric Research Technical Paper no. 31, (CSIRO, Australia, 1994). http://www.dar.csiro.au/publications/enting_2001a.htm.
23. I. G. Enting, *Twisted: The Distorted Mathematics of Greenhouse Denial.* (I. Enting/Australian Mathematical Sciences Institute, Melbourne, 2007).
24. M. Christie, *The Ozone Layer: A Philosophy of Science Perspective.* (CUP, Cambridge, U.K., 2000).
25. W. W. Kellogg. Effects of human activities on global climate. Technical Note 156, WMO, Geneva, (1977).
26. P. Gwynne, The cooling world, *Newsweek.* **28 April 1975**, 64, (1975).
27. I. G. Enting, Assessing the information content in environmental modelling: A carbon cycle perspective, *Entropy.* **10**, 556–575, doi: 10.3390/e10040556, (2008).
28. E. N. Lorenz. Predictability: Does the flap of a butterfly's wings in Brazil set off a tornado in Texas? [Lecture at AAAS meeting. Reprinted in E. Lorenz (1993) *The Essence of Chaos.* (UCL Press: London)], (1972).
29. E. N. Lorenz, *The Essence of Chaos.* (UCL Press, London, 1993).
30. P. Bak, C. Tang, and K. Wiesenfeld, Self-organized criticality: an explanation for 1/f noise, *Phys. Rev. Lett.* **59**, 381–384, (1987).
31. P. Bak, C. Tang, and K. Wiesenfeld, Self-organized criticality, *Phys. Rev. A.* **38**, 364–374, (1988).
32. P. Bak, *How Nature Works: The Science of Self-Organized Criticality.* (Springer-Verlag, New York, 1996).
33. M. Buchanan, *Ubiquity: The Science of History ... Or Why the World is Simpler than We Think.* (Weidenfeld and Nicholson, London, 2000).
34. S. J. Gould, *Wonderful Life: The Burgess Shale and the Nature of History.* (Penguin, London, 1989).
35. S. Conway Morris, *The Crucible of Creation: The Burgess Shale and the Rise of Animals.* (OUP, Oxford, 1998).
36. R. G. Prinn. Measurement equation for trace chemicals in fluids and solution of its inverse. In Ref. 114, pp. 3–18.
37. I. G. Enting. Green's function methods of tracer inversion. In Ref. 114, pp. 19–31.
38. R. J. P. Williams, A systems view of the evolution of life, *J. R. Soc. Interface.* **4**, 1049–1070, (2007).
39. P. Crutzen and E. F. Stoermer, The "Anthropocene", *Global Change Newsletter.* **41**, 17–18, (2000).
40. Royal Society. Ocean acidification due to increasing atmospheric carbon

dioxide. Policy document 12/05, (2005).

41. A. D. Moy, W. R. Howard, S. G. Bray, and T. W. Trull, Reduced calcification in modern Southern Ocean planktonic foraminifera, *Nature Geoscience.* **407**, 364, (2009).
42. E. N. Lorenz, Irregularity: A fundamental property of the atmosphere, *Tellus.* **36A**, 98–110, (1964).
43. I. G. Enting, *Inverse problems in earth system science: A complex systems perspective.* CSIRO Atmospheric Research Technical Paper no. 62, (CSIRO, Australia, 2002). http://www.dar.csiro.au/publications/enting_2002c.pdf.
44. K. Ziemelis, Nature insight: Complex systems, *Nature.* **410**, 241, (2001).
45. L. P. Kadanoff, Turbulent heat flow: Structures and scaling, *Physics Today.* **54(8)**, 34–39, (2001).
46. J. D. van der Waals. *Over de Continuiteit van der Gas- en Vloeistoftoestand.* Doctoral thesis, Leiden, (1873).
47. B. B. Mandelbrot, *Fractals: Form Chance and Dimension.* (W. H. Freeman, San Francisco, 1977).
48. J. Feder and A. Aharony, Eds., *Fractals in Physics: Essays in Honour of Benoit B. Mandelbrot.* (North Holland (Elsevier), Amsterdam, 1990).
49. S. Wolfram, *A New Kind of Science.* (Wolfram Press, 2002).
50. M. E. Fisher, The renormalization group in the theory of critical phenomena, *Rev. Mod. Phys.* **46**, 597–616, (1974).
51. S. Kauffman, *Investigations.* (OUP, Oxford, 2000).
52. I. G. Enting, *Inverse Problems in Atmospheric Constituent Transport.* (CUP, Cambridge, UK, 2002).
53. G. Backus and F. Gilbert, The resolving power of gross earth data., *Geophys. J. R. astr. Soc.* **13**, 247–276, (1968).
54. F. Press, Earth models obtained by Monte Carlo inversion, *J. Geophys. Res.* **73**, 5223–5234, (1968).
55. I. J. D. Craig and J. C. Brown, *Inverse Problems in Astronomy: A Guide to Inversion Strategies for Remotely Sensed Data.* (Adam Hilger, Bristol, 1986).
56. C. Wunsch, *The Ocean Circulation Inverse Problem.* (CUP, Cambridge, UK, 1996).
57. C. D. Rodgers, *Inverse Methods for Atmospheric Sounding: Theory and Practice.* (World Scientific, Singapore, 2000).
58. A. Tarantola, *Inverse Problem Theory: Methods for Data Fitting and Model Parameter Estimation.* (Elsevier, Amsterdam, 1987).
59. A. Tarantola, *Inverse Problem Theory: Methods for Data Fitting and Model Parameter Estimation.* (SIAM, Philadelphia, 2005).
60. E. Kalnay, *Atmospheric Modeling, Data Assimilation and Predictability.* (CUP, Cambridge, 2003).
61. S. N. Evans and P. B. Stark, Inverse problems as statistics, *Inverse Problems.* **18**, R55–R97, (2002).
62. R. Daley, *Atmospheric Data Analysis.* (CUP, Cambridge, UK, 1991).
63. A. Lorenc, Analysis methods for numerical weather prediction, *Q. J. Roy. Meteorol. Soc.* **112**, 1177–1194, (1986).

64. C. B. Field and M. R. Raupach, Eds., *The Carbon Cycle: Integrating Humans, Climate and the Natural World.* (Island Press, Washington, 2004).
65. E. T. Sundquist. Geological perspectives on carbon dioxide and the carbon cycle. In eds. E. T. Sundquist and W. S. Broecker, *The Carbon Cycle and Atmospheric CO_2: Natural Variations Archean to Present*, Geophysical Monograph 32, pp. 5–59. AGU, Washington, (1985).
66. J. P. Canadell, H. A. Mooney, D. D. Baldocchi, J. A. Berry, J. R. Ehleringer, C. B. Field, S. T. Gower, D. Y. Hollinger, J. E. Hunt, R. B. Jackson, S. W. Running, G. R. Shaver, W. Steffen, S. E. Trumbore, R. Valentini, and B. Y. Bond, Carbon metabolism of the terrestrial biosphere: a multitechnique approach for improved understanding, *Ecosystems.* **3**, 115–130, doi: 10:1007/s100210000014, (2000).
67. B. Kruijt, A. J. Dolman, J. Lloyd, J. Ehleringer, M. Raupach, and J. Finnigan, Assessing the regional carbon balance: Towards an integrated, multiple constraints approach, *Change.* **56 (March–April 2001)**, 9–12, (2001).
68. M. R. Raupach, P. J. Rayner, D. J. Barrett, R. S. DeFries, M. Heimann, D. S. Ojima, S. Quegan, and C. C. Schmullius, Model-data synthesis in terrestrial carbon observation: methods, data requirements and data uncertainty specifications, *Global Change Biology.* **11**, 378–397, doi: 10.1111/j.1365–2486.2005.00917.x, (2005).
69. A. Griewank, *Evaluating Derivatives: Principles and Techniques of Algorithmic Differentiation.* (SIAM, Philadelphia, 2000).
70. R. Giering. Tangent linear and adjoint biogeochemical models. In Ref. 114, pp. 33–48.
71. C. Rödenbeck, S. Houwerling, M. Gloor, and M. Heimann, CO_2 flux history 1982–2001 inferred from atmospheric data using a global inversion of atmospheric transport, *Atmospheric Chemistry and Physics.* **3**, 1919–1964, (2003).
72. T. Kaminski, W. Knorr, P. J. Rayner, and M. Heimann, Assimilating atmospheric data into a terrestrial biosphere model: A case study of the seasonal cycle, *Global Biogeochemical Cycles.* **16**, 1066, (2002). doi:10.1029/2001GB001463.
73. P. J. Rayner, M. Scholze, W. Knorr, T. Kaminski, R. Giering, and H. Widmann, Two decades of terrestrial carbon fluxes from a carbon cycle data assimilation system (CCDAS), *Global Biogeochemical Cycles.* **19**, GB2026, doi: 10.1029/2004GB002254, (2005).
74. I. G. Enting, D. M. Etheridge, and M. J. Fielding, A perturbation analysis of the climate benefit from geosequestration of carbopn dioxide, *Int. J. Greenhouse Gas Control.* **2/3**, 289–296, doi: 10.1016/jiggc.2008.02.005, (2008).
75. C. D. Jones, P. M. Cos, and C. Huntingford, Climate-carbon cycle feedbacks under stabilization: uncertainty and observational constraints, *Tellus.* **58B**, 603–613, (2006).
76. M. D. Matthews, Emission targets for CO_2 stabilization as modified by carbon cycle feedbacks, *Tellus.* **58B**, 591–602, (2006).
77. R. M. Law, P. J. Rayner, L. P. Steele, and I. G. Enting, Using high temporal

frequency data for CO_2 inversions, *Global Biogeochemical Cycles.* **16**, (2002). doi:10.1029/2001GB001593.

78. Y.-P. Wang, C. M. Trudinger, and I. G. Enting, A review of applications of model-data fusion to studies of terrestrial carbon fluxes at different scales, *Agricultural and Forest Meteorology.* (in press), (2009).
79. W. Knorr and M. Heimann, Impact of drought stress and other factors on seasonal land biosphere CO_2 exchange studied through an atmospheric tracer transport model, *Tellus.* **47B**, 471–489, (1995).
80. P. J. Rayner. Atmospheric perspectives on the ocean carbon cycle. In eds. E.-D. Schulze, M. Heimann, S. Harrison, E. Holland, J. Lloyd, I. C. Prentice, and D. Schimel, *Global Biogeochemical Cycles in the Climate System.* Academic, San Diego, (2001).
81. R. T. Watson and THE CORE WRITING TEAM, Eds., *Climate Change 2001: Synthesis Report.* (Published for the IPCC by CUP, Cambridge, UK, 2001).
82. P. M. Cox, R. A. Betts, and C. D. Jones, Acceleration of global warming due to carbon-cycle feedback in a coupled climate model, *Nature.* **408**, 184–187, (2000).
83. P. Friedlingstein, L. Bopp, P. Ciais, J.-L. Dufresne, L. Fairhead, H. le Treut, P. Monfray, and J. Orr, Positive feedback between future climate change and carbon cycle, *Geophys. Res. Lett.* **28**, 1543–1546, (2001).
84. P. Friedlingstein, J. L. Dufresne, P. M. Cox, and P. J. Rayner, How positive is the feedback between climate change and the carbon cycle?, *Tellus.* **55B**, 692–700, (2003).
85. E. A. Kowalczyk, Y.-P. Wang, R. M. Law, H. L. Davies, J. L. McGregor, and G. S. Abramowitz. The CSIRO Atmosphere Biosphere Land Exchange (CABLE) model for use in climate models and as an offline model. CMAR technical paper, 013, (2006).
86. Y.-P. Wang and R. Leuning, A two-leaf model for canopy conductance, photosynthesis and partitioning of available energy I: Model description and comparison with multi-layered model, *Agricultural and Forest Meteorol.* **91**, 89–111, (1998).
87. C. B. Field, J. T. Randerson, and C. M. Malmström, Ecosystem net primary production: combining ecology and remote sensing, *Remote Sensing and Environment.* **51**, 74–88, (1995).
88. Y.-P. Wang, B. Houlton, and C. B. Field, A model of biogeochemical cycles of carbon, nitrogen and phosphorus including nitrogen fixation and phosphotase production, *Global Biogeochemical Cycles.* **21**, GB1018, doi: 10.1029/2006GB002797, (2007).
89. S. Sitch, B.Smith, I. C. Prentice, A. Arneth, A. Bondeau, W. Cramer, J. O. Kaplans, S. Levis, W. Lucht, M. T. Sykes, and K. Thonicke, Evaluation of ecosystem dynamics, plant geography and terrestrial carbon cycling in the LPJ dynamic global vegetation model, *Global Change Biology.* **9**, 161–185, (2003).
90. W. Steffen, Ed., *Blueprint for Australian Carbon Cycle Research.* (Australian Greenhouse Office, Canberra, 2005).
91. A. C. Dilley, M. Edwards, D. M. O'Brien, and S. J. Millie, Satellite-based

monitoring of grassland curing in Victoria Australia, *International Geoscience and Remote Sensing Symposium.* pp. 597–599, (2001).
92. J. T. Randerson, I. G. Enting, E. Schuur, K. Caldeira, and I. Y. Fung, Seasonal and latitudinal variability of tropospheric $\Delta^{14}CO_2$: Post bomb contributions from fossil fuels, oceans and the stratosphere, and the terrestrial biosphere, *Global Biogeochemical Cycles.* **16**, 1112, doi:10:1029/2002GB001876, (2002).
93. I. Y. Fung, C. B. Field, J. A. Berry, M. V. Thompson, J. T. Randerson, C. M. Malmström, P. M. Vitousek, G. J. Collatz, P. J. Sellers, D. A. Randall, A. S. Denning, F. Badeck, and J. John, Carbon 13 exchanges between atmosphere and biosphere, *Global Biogeochemical Cycles.* **11**, 507–533, (1997).
94. T. Kaminski, P. J. Rayner, M. Heimann, and I. G. Enting, On aggregation errors in atmospheric transport inversions., *J. Geophys. Res.* **106**, 4703–4715, (2001).
95. A. B. Pittock, Are scientists undersestimating climate change?, *EOS (trans. AGU).* **87**, 340, (2006).
96. J. Rial, R. A. Pielke sr., M. Beniston, M. Claussen, J. Canadell, P. Cox, H. Held, N. de Noblet-Docourdé, R. Prinn, J. F. Reynolds, and J. D. Salas, Nonlinearities, feedbacks and critical thresholds within the earth's climate system, *Climatic Change.* **6**, 11–38, (2004).
97. R. May, Biological populations with nonoverlapping generations: Stable points, stable cycles, and chaos, *Science.* **186**, 645–647, (1974).
98. T. M. Lenton, H. Held, E. Kriegler, J. W. Hall, W. Lucht, S. Rahmstorf, and H. J. Schellnhuber, Tipping elements in the earth's climate system, *Proc. Nat. Acad. Sci.* **105**, 1786–1793, (2008).
99. P. Friedlingstein, P. Cox, R. Betts, L. Bopp, W. von Bloh, V. Brovkin, P. Cadule, S. Doney, M. Eby, I. Fung, G. Bala, J. John, C. Jones, F. Joos, T. Kato, M. Kawamiya, W. Knorr, K. Lindsay, H. D. Matthews, T. Raddatz, P. Rayner, C. Reick, E. Roeckner, K.-G. Schnitzler, R. Schnur, K. Strassmann, A. J. Weaver, C. Yoshikawa, and N. Zeng, Climate-carbon cycle feedback analysis: Results from the C4MIP model intercomparison, *Journal of Climate.* **19**, 3337–3353, (2006).
100. J. E. Lovelock and L. Margulis, Atmospheric homeostasis by and for the biosphere: the Gaia hypothesis, *Tellus.* **26**, 2–10, (1974).
101. J. E. Lovelock, *Gaia: A New Look at Life on Earth.* (OUP, Oxford, 1979).
102. T. Lenton and M. van Oijen, Gaia as a complex adaptive system, *Phil. Trans. R. Soc. Lond.* **357**, 683–695, (2002).
103. R. Dawkins, *The Extended Phenotype.* (OUP, Oxford, 1982).
104. T. M. Lenton, Gaia and natural selection, *Nature.* **394**, 439–447, (1998).
105. J. Lovelock, *The Vanishing Face of Gaia: A Final Warning.* (Allen Lane (Penguin), London, 2009).
106. J. W. Kirchner, The Gaia hypothesis: Fact theory and wishful thinking, *Climatic Change.* **52**, 391–408, (2002).
107. P. J. Crutzen, A critical analysis of the Gaia hypothesis as a model for climate/biosphere interactions, *GAIA.* **2**, 96–103, (2002).
108. T. Lenton, Testing Gaia: the effect of life on earth's habitability, *Climatic*

Change. **52**, 409–422, (2002).

109. J. Lovelock, *The Revenge of Gaia: Why the Earth is Fighting Back — And How We Can Still Save Humanity.* (Allen Lane (Penguin), London, 2006).
110. M. Hulme, Choice is all, *New Scientist. 4 November, 2000.* pp. 56–57, (2000).
111. M. Webster, C. Forest, J. Reilly, M. Babiker, D. Kicklighter, M. Mayer, R. Prinn, M. Sarofim, A. Sokolov, P. Stone, and C. Wang, Uncertainty analysis of climate change and policy response, *Climatic Change.* **61**, 295–320, (2003).
112. H. N. Pollack, *Uncertain Science... Uncertain World.* (CUP, Cambridge, UK, 2003).
113. R. W. Kates, W. C. Clark, R. Corell, J. M. Hall, C. C. Jaeger, I. Lowe, J. McCarthy, H. J. Schellnhuber, B. Bolin, N. M. Dickson, S. Faucheux, G. C. Gallopin, A. Grübler, B. Huntley, J. Jäger, N. S. Jodha, R. E. Kasperson, A. Mabogunjie, P. Matson, H. Mooney, B. Moore, T. O'Riordan, and U. Svedin, Sustainability science, *Science.* **292**, 641–642, (2001).
114. P. Kasibhatla, M. Heimann, P. Rayner, N. Mahowald, R. G. Prinn, and D. E. Hartley, Eds., *Inverse Methods in Global Biogeochemical Cycles. (Geophysical Monograph no. 114).* (American Geophysical Union, Washington D.C., 2000).

Chapter 6

Applied Fluid Chaos: Designing Advection with Periodically Reoriented Flows for Micro to Geophysical Mixing and Transport Enhancement

Guy Metcalfe

Commonwealth Scientific & Industrial Research Organisation (CSIRO)
Division of Materials Science and Engineering
Box 56, Highett Vic 3190 Australia
guy.metcalfe@csiro.au

Beginning with motivating examples of chaotic fluid advection applied to control mixing and scalar transport in devices, in biophysical flows, and in environmental and geophysical flows, I discuss basic features and organizing ideas of chaotic advection and transport in planar flows. Due to the combination of kinematics and continuity, steady planar flows must mix poorly; the flow designer must add some time-dependence that causes streamlines to cross at successive times (a simple though effective heuristic). Unstable manifolds of hyperbolic points in flows are the "highways" along which transport takes place. These structures emerge from the way the fluid is stirred and change as stirring parameters vary, leading to fractal distributions of transport rates in the control parameter space of flows. This has important implications for designing applications because it means that small variations in parameters can lead to large changes in performance. Optimum designs require knowledge of the complete parametric variation of transport properties for a given flow. As an example, I explore mixing and scalar advection-diffusion transport with the class of periodically reoriented flows, which is often simple to implement for applications, yet imposes, via its symmetries, a rich structure onto chaotic transport that includes generating sub-harmonic resonance "tongues" of the mixing and transport rates in the flows' control parameter space. My objective in this chapter is to impart to the reader the ability to recognize, design, and use chaotic advection to profitably shape and constrain his or her own investigations.

Contents

6.1. Introduction and Motivation

6.1.1. *What is chaotic advection?*

What is chaotic advection, and what can it be used for? The classical Greeks thought chaos was the void before time in which all things existed in a confused and amorphous shape. In his Metamorphoses (Book I) the poet Ovid wrote that

> what we call chaos: a raw confused mass, nothing but inert matter, badly combined discordant atoms of things, confused in the one place. ...Nothing retained its shape, one thing obstructed another, because in the one body, cold fought with heat, moist with dry, soft with hard, and weight with weightless things.

The common view is that chaos is equivalent to disorder or randomness. But it's not true! chaotic advection produces structures in stirred fluids that are highly ordered—just not regularly ordered into simple geometries. But the Greeks were correct in their idea that chaos in stirred fluids is pervasive.[a]

This chapter describes some basic aspects of mixing and transport in fluids that are stirred in simple, even symmetric, fashions. Some questions I'll try to answer include: What are chaotic flows? Am I likely to encounter

[a]As pervasive means to become spread throughout all parts of something, it seems an apt word to describe mixing.

them? What can you do with chaos? How can I design a good mixing flow? Other good questions, such as, what's the difference between chaotic advection and turbulence and what do you mean by well-mixed, I won't be able to find room for in this chapter. Before revealing some of the order in stirred flows, let's see an overview of situations where chaotic advection controls important processes in devices, in biology, and in environmental and geophysical flows.

6.1.2. *Examples of chaotic advection in practice*

Figure 6.1 places a selected set of subjects, for which chaotic advection is important to understanding mixing and transport properties, into a Reynolds number versus length scale graph. The scales are representative, not exact, and the Reynolds number

$$Re = \frac{UL}{\nu} \tag{6.1}$$

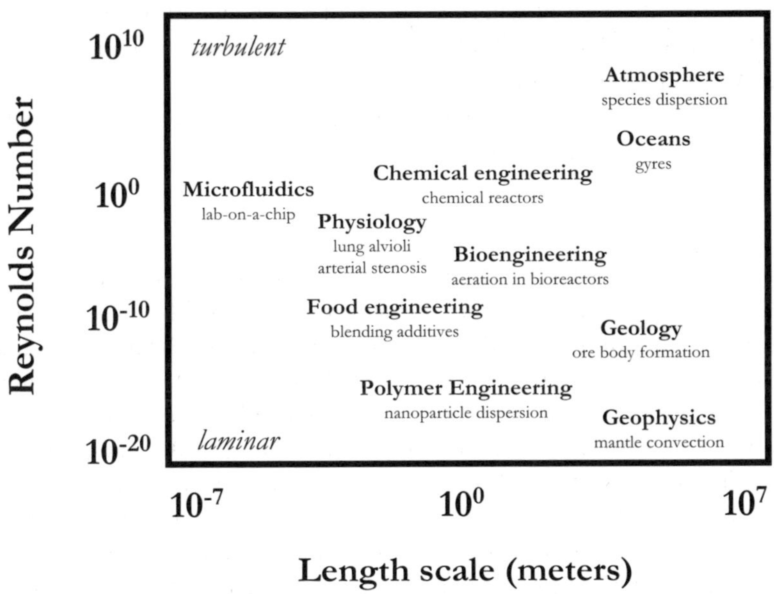

Fig. 6.1. A selected set of subjects for which chaotic advection is important to mixing and transport properties. The scales are representative, not exact. A modified and updated version of a figure by Ottino.[1]

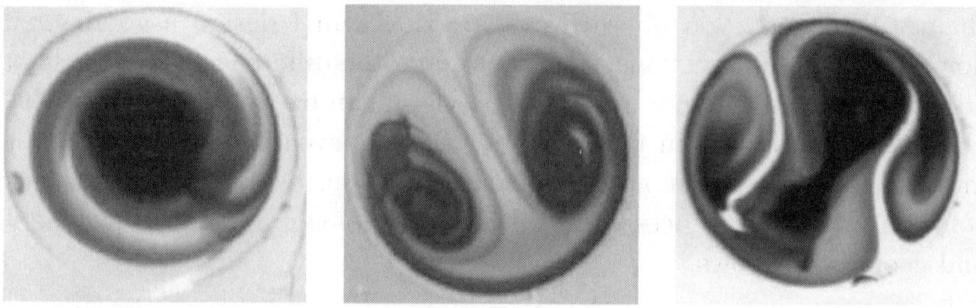

(a) (b) (c)

Fig. 6.2. Chaotic acoustic micromixing. (a) Dye moves with a one vortex flow pattern at one acoustic frequency. (b) Dye moves with a two vortex pattern at a different acoustic frequency. (c) Dye chaotically advects when periodically switching between frequencies. Tho et al.[4] describe the device, and the data is from Karolina Petkovic-Duran (2008, personal communication).

is the ratio of inertial to viscous forces in a fluid, where U is a characteristic velocity scale, L a characteristic length, and ν the kinematic viscosity of the fluid. Even without considering quantum[2] or astrophysical flows,[3] examples of chaotic advection span 14 orders of magnitude in length and almost 30 in Re. As to the question of whether one is likely to encounter chaotic flows, figure 6.1 suggests that, if a fluid in motion is involved in the question, then the answer is quite possibly yes.

Now let's see several examples of chaotic advection across the small, medium, and large length scales.

6.1.2.1. *Small scale*

When grabbing mRNA from single neurons in order to show the gene being expressed at a particular instant, samples must be quickly mixed with certain enzymes to stop mRNA degradation; however, mixing nanoliter quantities of liquid quickly enough is difficult. Chaotic acoustic microstreaming mixes the molecules nearly two orders of magnitude faster than otherwise. Figure 6.2 shows acoustic micromixing at the micron (10^{-6} meters) length scale.[4] The figure shows a droplet in a well with an even smaller droplet of dyed fluid inside. For chaotic acoustic microstreaming an externally applied acoustic wave of frequency f_1 causes a flow pattern inside the droplet. The dye deforms with the moving fluid to reveal the flow is a nearly 2-dimensional single vortex (figure 6.2a). A different frequency f_2 causes a different flow pattern, in this case a two vortex flow (figure 6.2b). I'm ignoring the mildly three-dimensional nature of the flows in the droplet and consider them predominantly to occur in the two-dimensional plane shown. Likewise, why acoustic waves produce these flows is beyond the scope of this chapter, but, perhaps surprisingly, none of the results on mixing and transport depend on precisely what physical mechanism is used to stir fluid; it only matters that the fluid can be stirred by some means.[b] Returning to the micromixing example, if one switches between frequencies f_1 and f_2, turning on flow 1 for a specific duration then switching from flow 1 to flow 2 for a (possibly different) duration, then repeating this sequence many times, it only requires a few iterations of this switching sequence to arrive at the dye pattern in figure 6.2c.

Note two things. One, the dye isn't following the flow any more. The velocity fields/streamlines of the flows are always as in figures 6.2a and b. If the dye lines stayed with the streamlines, figure 6.2c would look like figure 6.2a or b, depending only on which frequency we stopped with. But it doesn't. Something different happens when we compose flows into sequences. As I'll discuss in §6.2, the something different is that *fluid particles*, small bits of fluids, no longer follow the streamlines of the flow; instead, they follow what are called pathlines, in which the totality of the fluid particle paths makes what is called the *advection field*. When the advection field differs from the velocity field and the velocity field is simple, that's a first order definition of chaotic advection. The second thing to note is that figure 6.2c has several characteristics of what our intuition might tell us corresponds to being on the way to being well-mixed, while figures 6.2a and b don't. For instance, the distance between individual dye lines has decreased and the lines themselves have thinned and the dye appears to cover more of the drop volume than it did originally. In other words, the distinction between dyed and un-dyed fluid is decreasing.

Now consider breathing. Deep in the lung, gas exchange takes place in alveoli, which are closed-end sacs in the pulmonary ancinus. Alveoli expand and contract with breathing rhythm, and alveolar diameters are a few hundred microns. Air flows in the alveoli may be modeled approximately as in figures 6.3c and d, which show calculated flow streamlines due to expansion and contraction of the alveolar sac (c), plus streamlines due to air going backwards and forwards across the opening (d). During breathing the vortex in figure 6.3d switches orientation, and the flow looks like a mirror image of itself—in the biological object the symmetry is only approximate—during inhalation and expiration; likewise with figure 6.3c

[b]Many other micromixers work by chaotic advection. The special journal issue edited by Ottino & Wiggins[5] contains many examples.

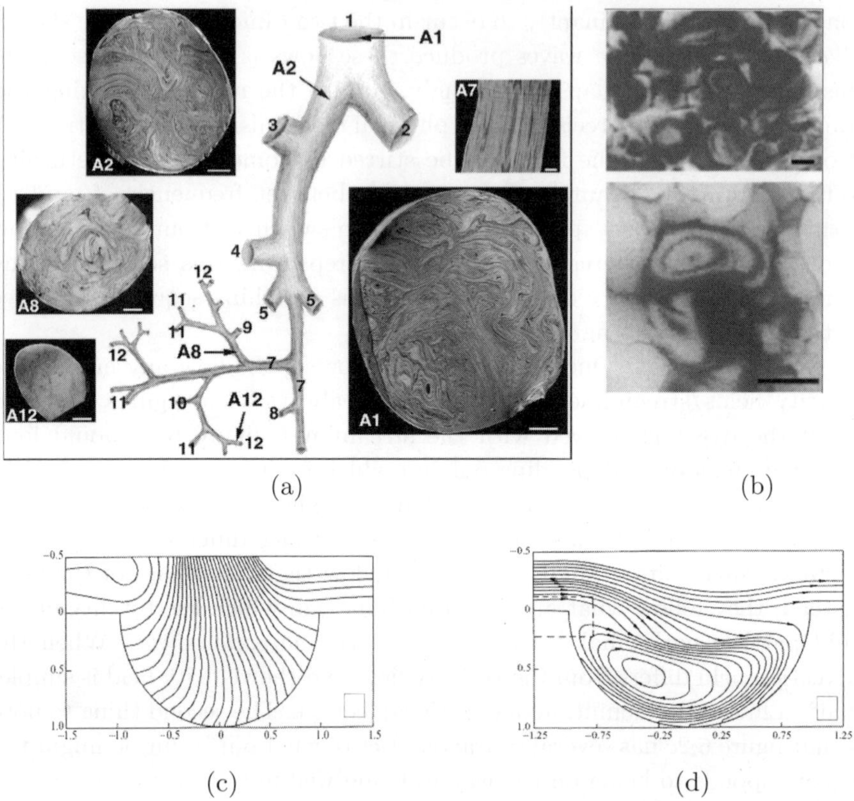

Fig. 6.3. Chaotic advection deep in the lung. (a) Typical flow patterns observed on airway cross sections at different locations in the tracheobronchial tree after one ventilatory cycle. Images A1 (bar = 500 μm), A2 (bar = 500 μm), A8(bar = 200 μm), and A12 (bar = 100 μm) show patterns on the transverse section of the trachea, the right main stem bronchus, a medium-size airway, and a small airway, respectively. Image A7 (bar = 100 μm) shows a longitudinal section through a medium-size airway. Images shown are representative of five rat lungs analyzed. (b) Alveolar recirculation. In many alveoli, circular (or ellipsoidal) blue/white color patterns were observed. (Upper) A field of multiple alveoli, each containing a recirculating pattern. (Lower) Enlarged view of an alveolar recirculation flow. (Bar = 100 μm.) (c) and (d) Streamline maps for the combined shear (c) and expansion (d) flows in an alveolar sac deep in the lung during inspiration and exhalation. (a) and (b) adapted from Tsuda et al.;[6] (c) and (d) adapted from Haber et al.[7]

for the expansion and contraction flow. Note a similarity with the micromixing example above: periodically the geometry of the flow changes between one type and another. Based on finding chaotic advection in the previous example, we might expect chaotic advection to appear deep in the lung too. Figures 6.3a and b show results of experiments with rat lungs from Tsuda et al.[6] Figure 6.3a shows the interaction after one breathing cycle of inhaled air and residual alveolar gas being exhaled. The rat lungs were filled with white fluid to represent alveolar gas, then blue fluid was slowly ventilated into the lungs. After some number of breathing cycles at the end of expiration, ventilation was stopped and the fluids allowed to harden. After becoming solid, the lungs were sectioned and the observed stretch and fold patterns emerged. Figure 6.3b shows vortical recirculation patterns, similar to the instantaneous circulations in figure 6.3d, in the alveolar sacs themselves. As stretching and folding is the essence of mixing, that and other quantitative measures are strong evidence that chaotic advection occurs deep in the lung.

But is chaotic advection important to the lung's natural function, which is gas exchange? To suggest an answer to this question, I'll introduce another dimensionless number, the Péclet number

$$Pe = \frac{UL}{\mathcal{D}}, \tag{6.2}$$

where $\mathcal{D}$ is a diffusivity of either heat (thermal diffusivity) or species (a concentration diffusivity). Pe quantifies the relative size of the advection rate to the diffusion rate: $Pe \to 0$ says diffusion is rapid and the flow is slow, while $Pe \to \infty$ means diffusion is slow and the flow is rapid.

The Péclet number for gas molecules in the alveoli is $Pe \sim 10^{-1}$, indicating gas exchange in the alveoli is determined by diffusion, i.e. pure advection is not important for the alveoli's normal function of gas exchange. However, for 50 nanometer particles in the alveoli $Pe \sim 10^3$, so the fate of nanoparticles deep in the lung is determined by chaotic advection. (Though, in fact, diffusion is not negligible, and alveolar nanoparticle transport is a problem in chaotic advection-diffusion, which I discuss in §6.4.1.) So chaotic advection will be important for both inhaled drug delivery[8,9] and for ascertaining potential environmental toxicity effects of nanoparticles.[10] Nonetheless, to my knowledge experiments to quantify chaotic advection-diffusion effects on nanoparticle deposition deep in the lung have not yet been done.

6.1.2.2. *Medium scale*

In 1995 Ottino, Metcalfe & Jana[11] could give a fairly complete list of experimental reports of laboratory-scale chaotic advection. Now there are too many examples to list them all. Commercial applications of chaotic advection, all of which currently exist on this medium length scale of order tens to hundreds of centimeters, are concentrated on flows of highly viscous fluids, for example polymers[12–17] and foods; although, extension to delicate fluids, such as viscous fermentations[18] or cell bioreactors[19] would be straightforward. Here I give two examples of industrial device-scale chaotic flows.

The first is the Rotated Arc Mixer (RAM) flow, the streamlines of which are shown at the top of figure 6.4f. Fluid is contained in a circular domain, and the boundary conditions prescribe that one arc—or possibly several arcs—of the boundary of angular extent Δ has a constant tangential velocity, while the rest of the circular boundary has zero velocity.[21] One moving arc produces a single vortex inside the circular domain whose center is closer to the moving boundary arc. At the end of a time interval τ of flow, the moving boundary arc instantaneously rotates its position by an angle Θ, which can be positive or negative, and flow resumes. The rotated boundary arc produces a *periodically reoriented flow*, and fluid in the domain is stirred at successive intervals by reoriented copies of the single basic flow. The bottom of figure 6.4f shows the superposed streamlines of three successive $\Theta = 2\pi/3$ reorientations.

Figure 6.4a shows schematically how a RAM flow device can be constructed.[20] It consists of two tightly wrapped coaxial tubes that rotate differentially. The wall of the inner tube has rectangular windows cut through it. Fluid flows through the inner tube in the axial direction. At each window viscous stress from the differential rotation generates the RAM flow (figure 6.4f) in the tube's cross-section transverse to the axial flow. Each successive window is offset from its upstream neighbor by the reorientation angle Θ. As fluid flows along the tube, it is stirred by reoriented copies of the basic RAM flow. The initial condition of a dye line across the tube diameter becomes stretched and folded, as in figure 6.4g where a laser sheet illuminates a cross-section perpendicular to the axis after the fluid has passed nine windows.

Figures 6.4b–e show experimental results for a clear Newtonian fluid injected with two colored dye streams (red and green) of the same fluid. Flow is from top to bottom. Dye diffusion was negligible, and the flow was steady and laminar ($Re = 4$). Figures 6.4b and c show the full length of a RAM tube of 15 windows. In figure 6.4c the two dye lines stretch and fold and mix well. Figure 6.4d shows a close-up of 6.4c about two-thirds down the tube. However, figure 6.4b, which differs from 6.4c only in the sign of Θ, is very different It shows a feature that was not apparent in previous

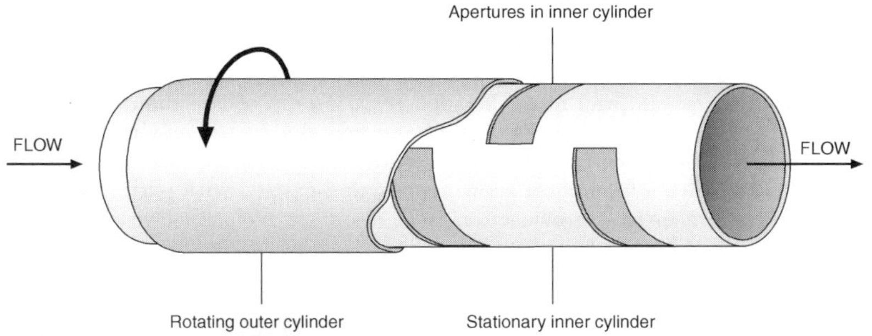

(a)

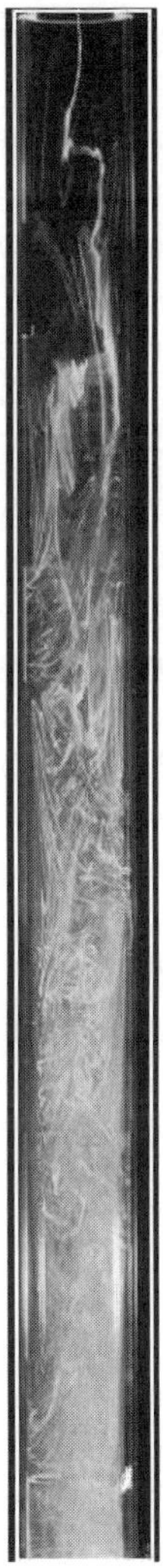

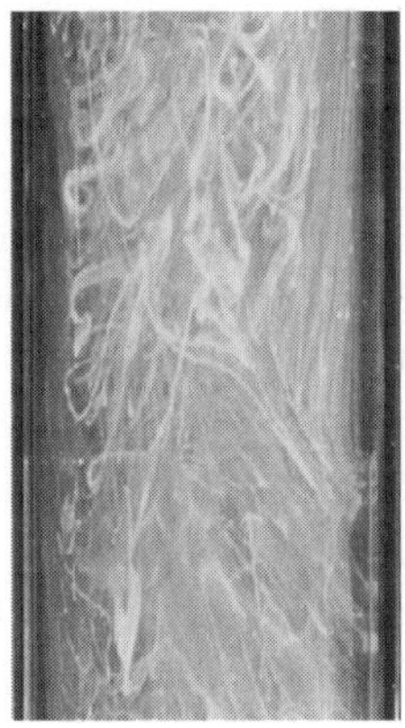

(d)

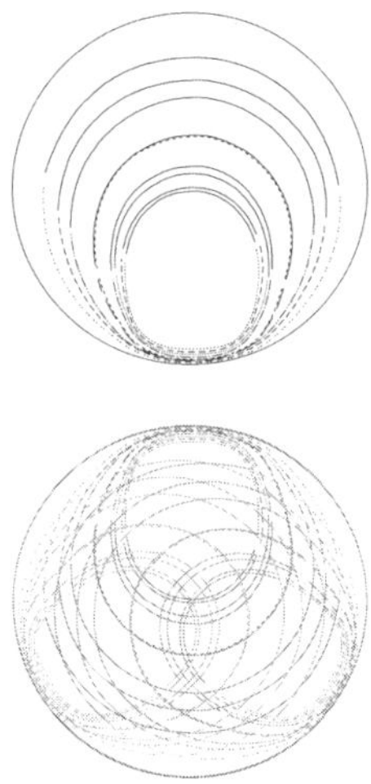

(f)

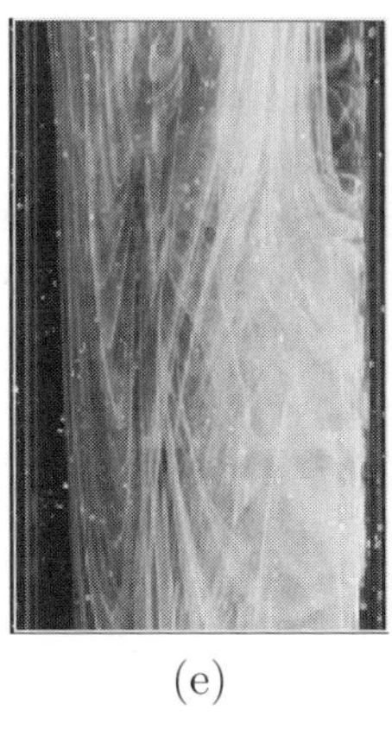

(e)

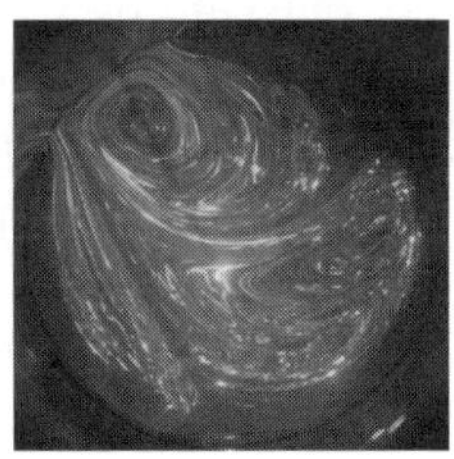

(g)

(b) (c)

Rotated Arc Mixer (RAM). See next page for caption.

Fig. 6.4. *(preceding page)* Rotated Arc Mixer (RAM) flow and experiments. (a) Schematic of a RAM device. (b,c) Red and green dye streams are injected into clear fluid flowing from top to bottom; in (b) an island forms; in (c) mixing is global. (d) Close-up of (c) about two-thirds of the way from the top. (e) Close-up of (b) about two-thirds of the way from the top. (f) Streamlines of the basic flow due to the moving boundary arc, and superposed streamlines from three reorientations. (g) Cross-section near the RAM exit illuminated with a laser sheet showing a mixing pattern with partially refined striations. The aperture in this cross-section is at about 11 o'clock. Dye diffusion is negligible, and $Re = 4$. Adapted from Metcalfe et al.[20]

examples. While the green line stretches, folds, and mixes through part of the tube, the red line now passes through the RAM virtually unaffected, and there is a large tubular volume containing the red line from which the green dye is excluded. Figure 6.4e shows a close-up of 6.4b about two-thirds down the tube. The unmixed region is called a regular region, or an *island.* In general islands and chaotic regions coexist.

In the class of periodically reoriented flows there are two intrinsic *control parameters*—the reorientation angle Θ and the time between reorientations τ. There may also be secondary parameters particular to the fluid and how reorientation is accomplished, but (τ, Θ) are the primary parameters for the class of periodically reoriented flows.

The generic picture of chaotic advection tiles the plane with both chaotic and regular regions. The job of someone who wants to harness chaotic advection for an engineered purpose is two-fold. One is to determine the ratio of chaotic and regular regions as a function of the control parameters for a particular flow for a particular function. For instance, a mixer that produces large island regions will be poorly suited to its purpose. On the other had, one can, by varying parameters, use islands as time-release capsules,[22] or to collect and separate particles.[23–25] The other job is to choose flows and ways to modulate them so that at least some accessible part of their control parameter space will produce the desired behavior.[20,26–30]

A household example of chaotic advection is the cake mixer of figure 6.5. Here the object is to only partially mix the white batter (white) and raspberry filling (red) to leave visually interesting chaotic patterns. Cake batter and filling lie in the annular space between two eccentric cylinders. Each cylinder can rotate about its own axis. This cake mixer is similar in spirit to more considered experiments with two- and three-dimensional eccentric cylinder flows.[31,32] Figures 6.5d and 6.5e show two-dimensional Stokes flow computations from Ottino[33] of the flow streamlines obtained respectively for the outer cylinder rotating and the inner cylinder rotating. Alternately

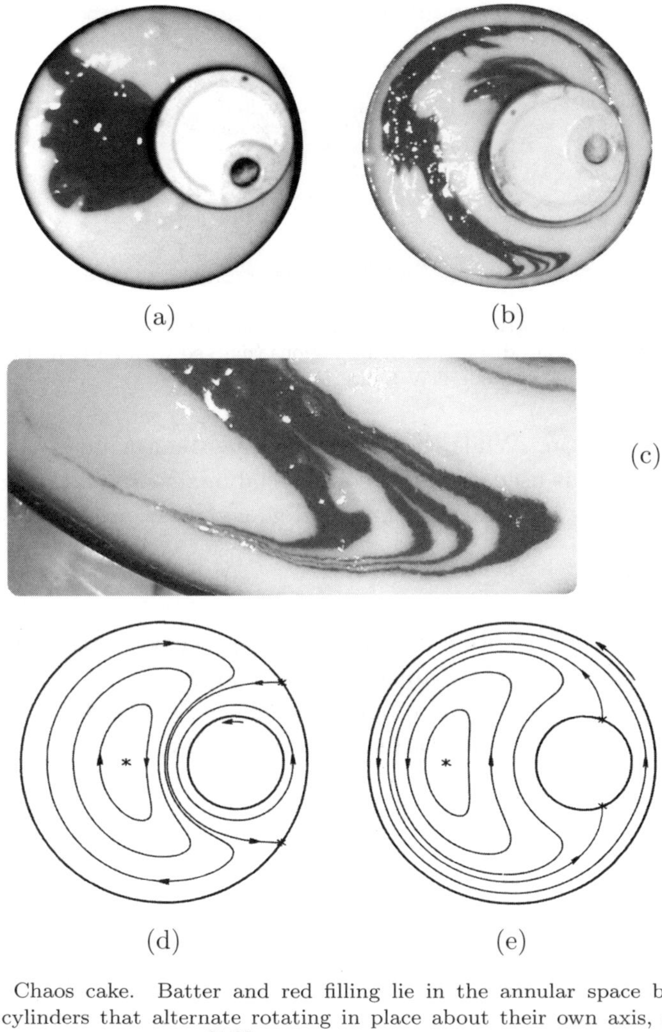

Fig. 6.5. Chaos cake. Batter and red filling lie in the annular space between two eccentric cylinders that alternate rotating in place about their own axis, similarly to more considered experiments.[31,32] (a) Initial condition. (b) After 2 cycles of 180° rotation of each cylinder the filling is partially stretched and folded. (c) Close-up of the folds at the bottom of (b). Stretching and folding is the essence of mixing. Flow patterns for inner cylinder rotation (d) and outer cylinder rotation (e) from Ottino.[33]

turning on the inner and outer cylinders' rotations generates chaotic advection, similarly to the alternating flows in the micromixer example. Figure 6.5a shows the initial condition and figure 6.5b shows the filling distribution after two cycles of 180° rotation of each cylinder. The close-up of figure 6.5c highlights the stretching and folding seen once more as a hallmark

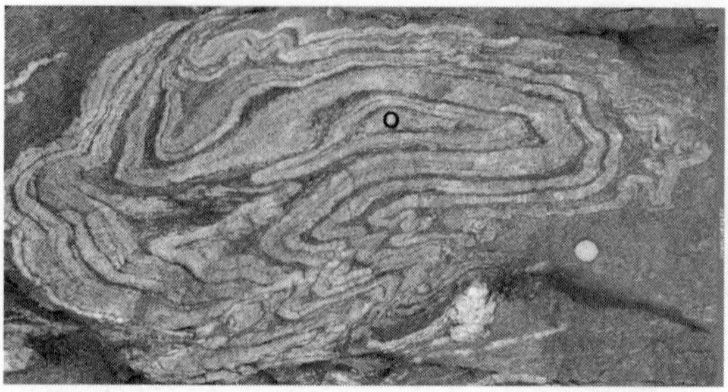

Fig. 6.6. Flow-induced structures in rock due to (postulated) chaotic advection. Deformed leucogranite dykes displaying island regions (marked with an **O** and folds in the Karakoram Shear Zone, Ladakh, NW India. From Weinberg et al.[57]

of chaotic advection. While the chaos cake may be a whimsical example, chaotic advection is used to design commercial mixers and heat exchangers for low energy use handling of "thick and gooey" food products.[34,56]

6.1.2.3. *Large scale*

Chaotic advection also occurs in geophysical processes. Figure 6.6 shows a rock face exposed in the Karakoram Shear Zone of Ladakh, NW India. Of course, I can only speculate on the sequence of rock flows that could bring about this particular pattern, and, aside from this flow probably being three-dimensional, rock also has chemical reactions and porous matrix evolution closely coupled with flow. However, as islands (marked with an **O**) and ramified folds are clearly visible and bear a striking resemblance to other patterns known to be generated by chaotic advection, they are plausibly caused by chaotic advection in the earth. Moreover, it is known that observed chemical and age heterogeneities from rock at upwelling zones can be explained by assuming chaotic advection in mantle convection,[35,36] though mantle computations may need to be very high-resolution to detect the signature characteristics of chaotic advection [37, see figure A2].

Figure 6.7 shows the South Polar vortex at about 18 km altitude during a rare event where the vortex has split into two parts. The initial wind patterns and chemical conditions are derived from satellite data while color in the figure shows the evolution of chemical reactant distribution (shown is the HCl mixing ratio) simulated by Grooß, Konopka & Müller.[38] Note the filamentary structures on which reaction takes place in the stirred fluid around and in the vortices, and the island zones into which little or no reactant transport takes place. Keep in mind that these are static pictures.

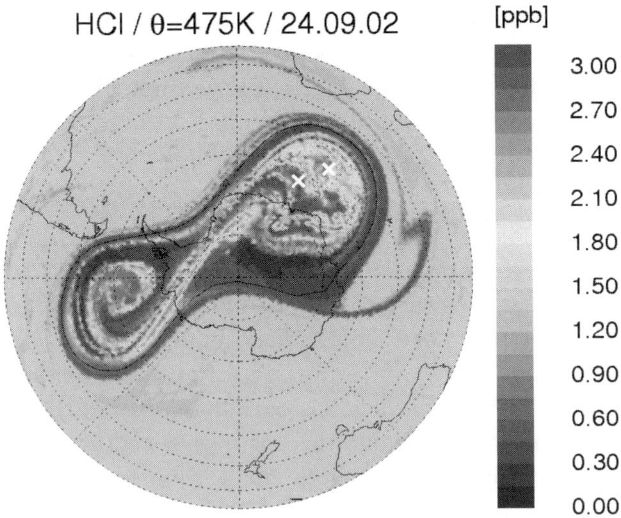

Fig. 6.7. Mixing and chemical reaction in the polar vortex. Hydrogen chloride (HCl) distribution (in parts per billion (ppb)) above the South Pole at about 18 km altitude on September 24, 2002. Reactions leading to ozone depletion are simulated over 23 days in a flow given by meteorological wind data during a rare event where the polar vortex splits into two parts. From Grooß et al.[38]

In reality the vortices move and change with time. Islands in this approximately planar atmospheric flow are seen to be barriers to transport. One might also note a similarity in appearance of the concentration distributions of the chaotic acoustic micromixer of figure 6.2c and the polar vortex of figure 6.7, despite some 11 orders of magnitude difference in characteristic length scales and completely different physical mechanisms driving the flows.

The polar vortex is the only high Reynolds number example I will show ($Re \sim (10^6 \text{ m})(10 \text{ m/s})/(10^{-5} \text{ m}^2/\text{s}) = 10^{12}$). $Re \sim 10^{12}$ indicates a highly turbulent flow. Why do I say that transport is due to chaotic advection instead of to turbulent transport mechanisms? Because, even though turbulent, the flow of interest is that of reactants into and out of the polar vortex, which is a *coherent structure* within a turbulent background. Coherent structures are observed frequently in planetary-scale oceanic and atmospheric flows,[39–41] and further examples include Jupiter's red spot, meddies, and gyres. Coherent structures within larger bodies of fluid generate their own local dynamical systems that govern collections of fluid particle trajectories over regions of fluid much larger than the coherent structure itself. And it's through the dynamical system associated with a coherent structure that chaotic advection determines most facets of mixing and transport in and around the coherent structure; turbulence provides a noise background that can often be negligible.

6.1.3. *Objectives and major points*

From the breadth of examples above chaotic transport may seem to be of daunting complexity. It is easy to get lost in the complexities of individual cases and to forget that there is also underlying, unifying structure. My objectives with these notes are to introduce the basics of the structure of chaotic advection, a subject which has been developed in the mathematics, physics, and engineering communities for over a century, and to show a simple, but useful, way to build chaotic flows from steady planar flows and then analyze them. At the end of this course you should be able to:

Recognize, given a physical flow picture, whether or not chaotic transport can or is likely to arise;
Design flows that can mix and enhance scalar transport;
Use some basic tools to begin qualitative or quantitative analysis.

I aim to strip away all but the essentials and to build intuition of what is possible or likely, elucidate fundamental physical mechanisms, and leave you with senses that can profitably shape and constrain more elaborate investigations.

Chaotic advection can be used both to design flows for technological ends and to understand transport occurring in natural flows. Engineered flows have control parameters that we can change and exploit, but we must also accept the challenge of understanding how transport behavior changes over the entire accessible control parameter space. Natural flows must be to some extent taken "as is", with parameters changing only slowly or periodically through some natural range. As my own work falls mostly on the applied science side, the examples in this chapter will be biased toward engineered flows. This neither slights natural flows as fascinating subjects of study, nor reduces the utility of the general results of chaotic advection and dynamical systems for application to natural flows.

To start, here are some of the general points about chaotic advection, illustrated by the examples above, that I'll elaborate and, to some extent, systematize through the rest of the chapter.

The general picture of chaotic advection has both island and chaotic regions. Islands are barriers to mixing and transport, while in chaotic regions mixing is rapid with every blob of fluid in the chaotic region pervading (eventually) the entire chaotic region. This makes highly stretched and folded structures a hallmark of chaos. While I will discuss only planar flows, all features carry over to three-dimensional flows, but fully three-dimensional flows may add additional features.

Kinematics is the term for describing the motion of an object without reference to the cause of the motion. None of the important features of chaotic advection depend on how precisely the fluid is stirred, and they occur over an enormous variety of length and time scales. The kinematics of chaos apply equally well to the polar vortex as to the micromixer. However, additional physics can come into play by admitting additional modes of transport, e.g. diffusion (as in the lung or viscous heat exchangers), or reaction (as in the polar vortex), or by coupling other fields to the flow, e.g. advection, reaction, and matrix deformation in the solid earth, and also inertia, electric, magnetic, etc. forces.

Crossed Streamlines. Notice that in all the examples above chaotic transport occurs when the streamlines of the flow at one instant of time cross the streamlines of the flow at a subsequent instant of time. This may be the simplest heuristic of how to build a chaotic flow from a steady planar flow and guarantee some chaotic transport.[c] In some of the example flows streamline crossing was obtained by composing sequences of different flows (acoustic micromixer and cake mixer). In others there was a basic flow followed by sequences of symmetries of the basic flow (a reflection for alveoli and a reorientation for the RAM). In others, such as the polar vortex, the flow itself was unsteady.

Symmetry. Taking a steady planar flow and composing it with reoriented copies of itself is one of the simplest ways to generate chaotic transport. I'll go into detail about the rich transport structure imposed on reoriented flows where the time-reversed flow coincides with a geometric symmetry.

Parametric variation. I have just called streamline crossing a heuristic for mixing, but it is really only half that. Getting streamlines to cross at successive times solves the problem of how to generate chaotic advection and, combined with symmetry, gives insight into how to do that simply for any steady flow. But, blind choice of flows and how to compose them

[c] A heuristic is an informal method that helps solve a problem and that can rapidly come to a solution that is reasonably close to optimal.

will not in general come close to an optimal solution for mixing or scalar transport. Control parameters of the flow feed back into qualitative and quantitative transport properties. Understanding how the transport varies over the *entire* accessible control parameter space is the only way to guarantee designs that are robustly close to optimum.

The rest of the chapter is organized as follows. Section 6.2 illustrates basic concepts of transport in planar flows, progressing from why steady flows must mix poorly to examples of chaotic tangles of manifolds of homo- and hetero-clinic connections. Section 6.3 presents a basic selection of nonlinear dynamics tools for analyzing flows: Poincaré sections, hyperbolic and elliptic periodic points, dye advection plots, and stretching distributions. Section 6.4 shows mixing and scalar transport properties for a periodically reoriented chaotic flow and finds mixing and transport optima in a fractal control parameter space. Section 6.5 uses symmetry to analyze the flows of §6.4 and to explain the physical origin of some its generic transport properties. I summarize in §6.6.

6.2. Illustrating Basic Concepts of Transport in Planar Flows

I will discuss mixing and transport in two-dimensional (2D) flows. All results extend naturally to three-dimensional (3D) flows; however, 3D flows may have additional transport mechanisms. The approach is to examine the geometry created by all the paths that can be taken by fluid particles. These paths are governed by the kinematic equation (6.3) under the constraint of mass conservation (continuity) and form a conservative dynamical system with a restricted set of possible flow topologies and special manifold lines that partition the fluid domain and govern transport. One of the most important consequences of continuity is that steady 2D flows must mix poorly. However, time-dependent, 2D flows can mix well, as I'll show through the important heuristic of designing flows with streamlines that cross at successive instants of time. Time-dependence in the flow induces transport by a particular geometric folding mechanism of the manifold lines called turnstile lobes, and results in a highly interwoven geometry of manifolds leading to a chaotic tangle and thereby chaotic fluid motion along the tangled manifolds. But inducing a chaotic tangle does not guarantee chaotic motion everywhere in the fluid domain. The generic outcome is coexisting chaotic and non-chaotic, regular, regions. These results lead naturally into questions of how to solve the parametric variation problem of where in the control parameter space of a flow is an optimum or globally well-mixed state to be found.

6.2.1. *Motion, kinematics, continuity*

There are two ways to look at fluid motion. In one you put a fixed frame of reference onto the flow domain and measure what happens at any point in the fixed frame. For instance, the fluid flowing through any point flows with a particular speed and direction at a given instant, defining a velocity vector at that point and instant of time. The collection of velocity vectors at every point in space defines a vector field that can vary in time. This is called the *Eulerian* point of view. In the other way you follow particular fluid particles. The totality of all the pathlines of all the fluid particles is the advection field. This is called the *Lagrangian* point of view.

A fluid particle may seem a strange idea, so I digress briefly to explain. A fluid particle is not one molecule of the fluid. It is a coherent region of the fluid that can be marked—even if only conceptually—in some way. Within the necessary continuum assumption of scale separation, fluid particles can be whatever size is convenient, but their volume is usually taken to be much smaller than the length scale over which the geometry of the flow or domain changes appreciably. An example is a blob of dyed fluid placed within otherwise identical undyed fluid. Fluid particles can be deformed and are not required to keep their original shape.

The kinematic equation

$$\frac{d\mathbf{X}}{dt} = \mathbf{V}(\mathbf{X}, t) \tag{6.3}$$

is the equation of motion for a point-like fluid particle whose location $\mathbf{X}$ is being moved by a velocity field $\mathbf{V}$ that may vary in space and time. Particle positions are advected by the velocity, and the particle paths are well-defined whenever $\mathbf{V}$ is, in particular when $\mathbf{V}$ is deterministic. Other common terminology is to call the pathlines of fluid particles *orbits* of the dynamical system given by equation 6.3. If the advected fluid particles don't change the velocity field and match their own velocity to changes in $\mathbf{V}$ instantaneously, then the particles are called *passive* or Lagrangian particles, and they truly just mark the fluid and trace the motion of a fluid within itself. To equation 6.3 one might add terms to account for diffusion, inertia, reactions, surface tension, magnetic or electric fields, etc. In this chapter I consider mostly passive particle transport, but take up advection-diffusion in §6.4.1.

Equation 6.3 gives a dynamical system for the fluid particle orbits in whatever flow $\mathbf{V}$ is specified. However, fluid flows are also constrained by mass conservation, often called continuity. The continuity equation for the fluid density ρ is

$$\frac{\partial \rho}{\partial t} + \nabla \cdot (\rho \mathbf{V}) = 0. \tag{6.4}$$

Using Gauss's law it is easy to see that continuity requires that if any control volume of fluid is squeezed, fluid must leave the control volume instead of the fluid density increasing (unless there's a source or sink of fluid). For a constant density fluid equation 6.4 becomes

$$\nabla \cdot \mathbf{V} = 0, \tag{6.5}$$

which is called the incompressibility condition. In dynamical systems terminology, for an incompressible fluid the phase-space of the dynamical system is conserved[d].

For the special, but important, case of a constant density, 2D flow, the incompressibility condition is

$$\frac{\partial V_x}{\partial x} + \frac{\partial V_y}{\partial y} = 0, \tag{6.6}$$

which can be satisfied by any scalar function ψ whose derivatives are

$$\frac{\partial \psi}{\partial y} = V_x, \quad \frac{\partial \psi}{\partial x} = -V_y. \tag{6.7}$$

ψ is called a stream function. Aside from affording the convenience of specifying a vector field by a scalar function, what is the stream function? Consider the level sets of ψ. Level sets are lines where $\psi =$ constant. These lines are called *streamlines*. Along lines of constant ψ, the differential of ψ is zero:

$$d\psi = \frac{\partial \psi}{\partial x} dx + \frac{\partial \psi}{\partial y} dy = 0. \tag{6.8}$$

Combining the last two equations gives

$$\frac{dx}{V_x} = \frac{dy}{V_y} \tag{6.9}$$

[d]Some readers may note the deep connection between conservative dynamical systems and Hamiltonian mechanics. A planar flow's streamfunction is the Hamiltonian of the system, and the conjugate variables are the physical space of the fluid flow. However, I will not pursue the connection here.

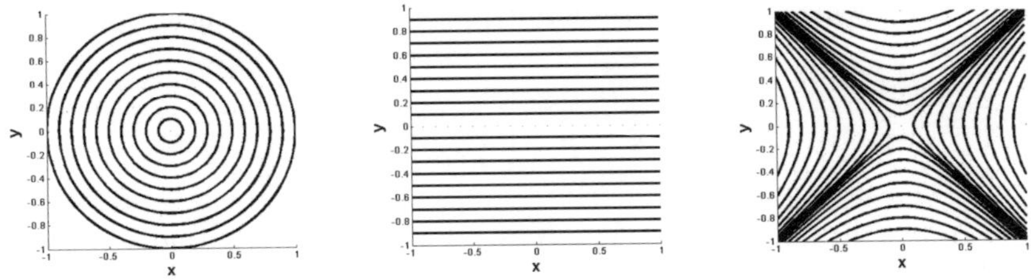

Fig. 6.8. Streamlines of the general linear flow for (a) circular streamlines, (b) pure shear, and (c) hyperbolic flow, about the origin.

which can be further expanded by parameterizing the length along the streamline by a parameter s and writing in vector notation

$$\frac{d\mathbf{x}}{ds} = \mathbf{V}. \tag{6.10}$$

This says that the tangent along a streamline is the vector velocity at that point, and the level sets of ψ are lines tangent to the velocity at any fixed instant of time. This can also be shown by verifying that $\mathbf{V} \cdot \nabla\psi = 0$.

There are several far-reaching consequences of continuity for 2D steady flows:

(1) Except at points with $\mathbf{V} = 0$, streamlines cannot cross. If they did, it would mean that the same fluid particle could have two different velocities at the crossing point.
(2) As $\mathbf{V}$ is everywhere tangent to streamlines, kinematics ($d\mathbf{X}/dt = \mathbf{V}$) says that fluid particles follow streamlines.
(3) Consequences 1 and 2 imply that a fluid particle on a streamline must stay on the same streamline.

Taken together these say that when fluid particles follow streamlines, the Eulerian and Lagrangian viewpoints are identical, and that the most important consequence of continuity is that *2-dimensional steady flows must mix poorly.*

6.2.2. *Steady flow local topologies*

That streamlines cannot cross except at points with $\mathbf{V} = 0$ also restricts planar flow topology to combinations of the elliptic, shear, or hyperbolic flows shown in figure 6.8. Shown are streamlines of the General Linear

Flow (GLF), which encompasses elliptic, hyperbolic, and shear flows in one matrix

$$\nabla \mathbf{V} = \dot{\gamma} \begin{pmatrix} 0 & 1 \\ K & 0 \end{pmatrix}, \tag{6.11}$$

where $\dot{\gamma}$ is a shear rate and $-1 \leq K \leq +1$ is a parameter that determines the character of the flow. The GLF is the most general representation of a 2D, volume-preserving flow. Elliptic flows are regions of closed recirculation. Hyperbolic flows are regions of exponential stretching, which because of volume-preservation, must be accompanied by an equal amount of compression. Shear flows are a degenerate case in-between, with limited stretching, compression, and rotation. Figure 6.9 illustrates a hypothetical planar flow made up of several elliptic and hyperbolic regions. Arrowheads indicate the direction of flow. Red dots are the centers of elliptic recirculating regions, called *elliptic points*. The other dots are *hyperbolic points*. Each hyperbolic point has a local expanding and contracting axis. Compare the hyperbolic points of figure 6.9 and figure 6.8c. The lines terminating exactly on the hyperbolic point have special properties and are called *manifolds*. Fluid approaches the hyperbolic point along the *stable manifold*, but exactly on the manifold takes an infinite amount of time to arrive at the point. Fluid moves away from the hyperbolic point along the *unstable manifold*. The unstable manifold is defined exactly as the stable manifold, but with time reversed. Time-reversal reverses the direction of all the arrows. The blue dots are hyperbolic points whose stable/unstable manifolds coincide with the unstable/stable manifolds of a different hyperbolic point. These are called *heteroclinic points*, and they can chain together in heteroclinic cycles. The green dot is a hyperbolic point whose stable manifold coincides with its own unstable manifold. These are called *homoclinic points*. It is important to keep clear that stable and unstable manifolds are different objects that are forced to coincide by the steady flow. (The fun begins when you make them diverge.) As fluid cannot cross streamlines, note how manifolds partition the plane, dividing stretching/compressing regions from circulating regions.

6.2.3. *Chaotic tangles*

As steady flows mix poorly, consider introducing periodic time-dependence to the steady flow. Except in special cases, time-dependence will cause streamlines at subsequent times to cross when superposed. I'll illustrate

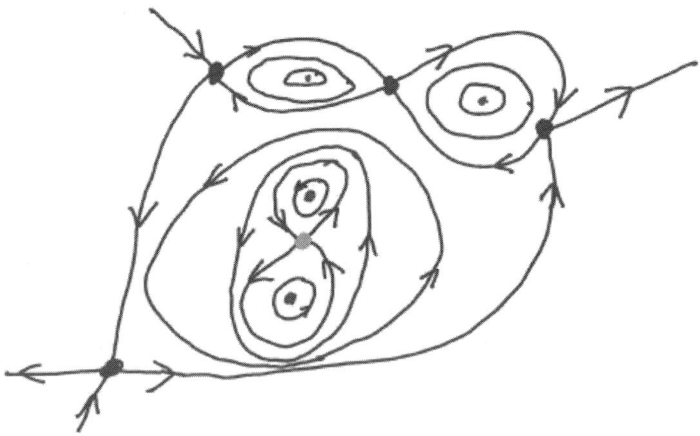

Fig. 6.9. A hypothetical stream function partitioning the plane. Arrows indicate direction of flow. Blue dots are heteroclinic points. The green dot is a homoclinic point. Red dots are elliptic points.

first using a homoclinic connection with time-dependence made by alternating symmetric flows and then using a heteroclinic connection with the time-dependence made by applying small periodic perturbations.

6.2.3.1. *Homoclinic connection*

Material cannot cross manifold lines: they are invariant curves. Unstable manifolds cannot intersect themselves or other unstable manifolds; the same applies to stable manifolds.[e] However, stable and unstable manifolds can intersect each other, and if there is one intersection, then there must be an infinite number of intersections.[42] In steady flow this makes the stable and unstable manifolds of a homoclinic point coincide. In an unsteady flow, the generic situation of the manifolds of the homoclinic point is illustrated in figure 6.10. The unstable manifold is the red line, and the stable manifold is the blue line. Flow time-dependence causes the flow's instantaneous streamlines to cross at subsequent instances of time, and the stable and unstable manifolds no longer coincide everywhere, but they still intersect at an infinite number of points. The intersection point p is advected forward in time to the red points and backward in time to the blue points. In a fluid experiment a blob of dye placed initially on the homoclinic point will

[e]If (un)stable manifolds intersected themselves then there would be an ambiguity as to where the intersection point goes as $t \to (-) + \infty$. Implicitly I've assumed the map is invertible.

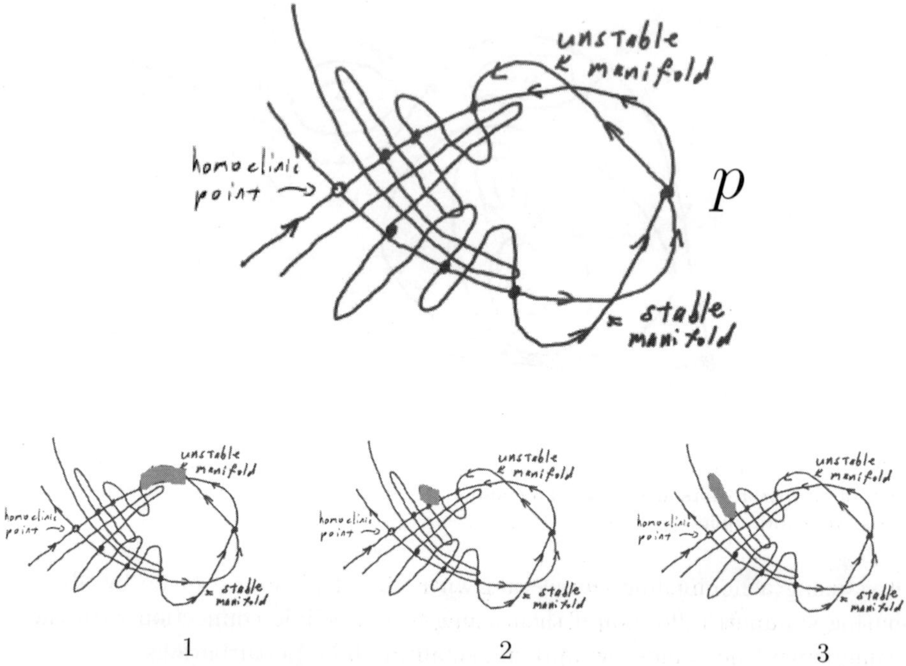

Fig. 6.10. Homoclinic tangle and lobe transport. (top) Red lines are the unstable manifold and blue lines are the stable manifold of the homoclinic point under a suitable perturbation of the steady flow. Dots are intersection points of the stable and unstable manifolds that are mapped to each other under forward and backward iteration of the map. (bottom) Turnstile lobe mechanism of chaotic transport. The green area denotes one lobe. At successive iterations from 1 to 3 the green material is mapped to a succeeding lobe. This moves material starting inside the separatrix to outside, and material starting outside the separatrix to inside.

trace out the unstable manifold. There are an infinite number of lobe segments, but each segment must contain the same area to conserve mass. (Continuity at work again.) This means that the lobes become thinner and longer as they approach the hyperbolic point along the stable manifold—even to being infinitely long in the $t \to \infty$ limit. For a confined flow domain this means that the filaments must fold and become highly ramified in order to fill space with finer and finer striations. This is what is meant by well-mixed. However, the actual fraction of the domain that becomes well-mixed varies with exactly how the flow deviates from its steady state.

Note how the tangle transports fluid across steady-flow manifold lines. In the steady case the homoclinic loop completely separates fluid inside the loop from fluid outside the loop. With time-dependence, lobes of fluid cross the steady-state invariant manifold. This is illustrated at the bottom of figure 6.10 where the green material starts inside the homoclinic loop, but at periods 1,2,3, etc. moves into the exterior domain.

Figure 6.11 illustrates an experimental homoclinic tangle. On the left are the streamlines for steady flow in a rectangular box with the two short sides sliding in opposite directions at the same speed. The flow possesses a homoclinic connection via the hyperbolic point in the center of the box. The flow is made periodic by varying the ratio of speeds of the sliding walls: periodically one wall moves faster than the other—sliding direction is unchanged. The hyperbolic point will move toward the faster wall. Topologically the flow is still a homoclinic connection, but the location of the hyperbolic point moves back and forth over a line segment of the axis of symmetry of the flow, in this case the y-axis. What happens to the stable and unstable manifolds of the hyperbolic point? On the right of figure 6.11 is an experiment—video in the lectures—where each wall of the box of fluid moves sequentially with equal and opposite displacement.[43] A small blob of dye was placed near the hyperbolic point of the time-averaged flow. Evident is the development of the chaotic tangle from the repeated stretching

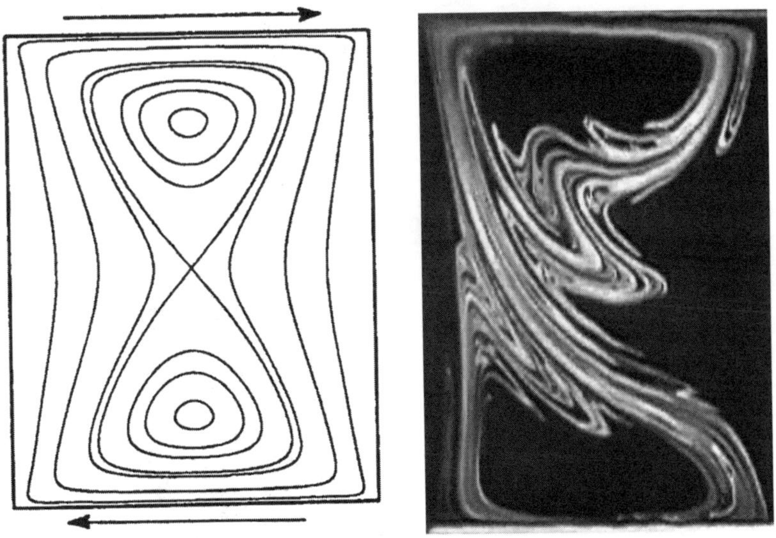

Fig. 6.11. Experimental homoclinic tangle in a box of fluid with periodically moving walls. (left) Steady flow streamlines. (right) Time-dependent flow—video in the lectures. Adapted from Jana et al.[43]

and folding of the unstable manifold.

6.2.3.2. *Heteroclinic*

An example of lobe transport and the chaotic tangle for a heteroclinic connection with a periodic perturbation is shown in figures 6.12–6.14. The scheme of the experiment (top of figure 6.12) is a channel flow over a cavity. The top wall of the channel slides parallel to the flow direction to drive the flow across the mouth of the cavity, generating circulation in the cavity. In steady state a separating streamline at the cavity's mouth is anchored to up- and down stream points on the walls, and there is no transport across this separatrix Ψ_s between the cavity and the channel. Perturbations are introduced by making the moving wall wavy. The bottom of figure 6.12 shows several choices of perturbation waveform; see Horner et al.[44] for

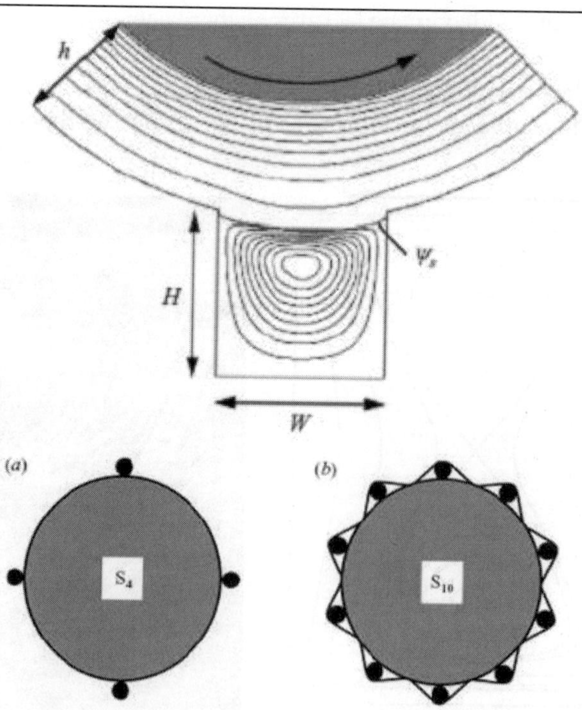

Fig. 6.12. Streamlines of channel flow driven by a moving wall across a cavity (top), and several ways to modify the moving wall to perturb the flow (bottom). Adapted from Horner et al.[44]

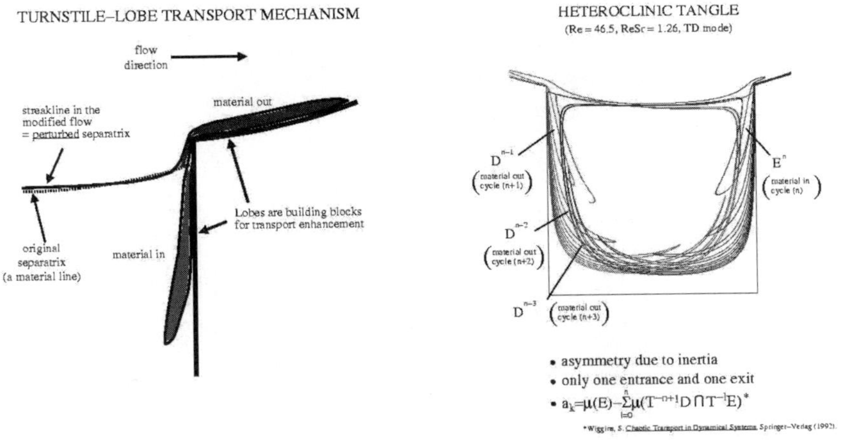

Fig. 6.13. Lobe formation at a heteroclinic point. (left) Near downstream wall attachment point a lobe of red fluid enters the cavity, while an equal amount of blue fluid exits. (right) Stable and unstable manifolds. Adapted from Horner et al.[44]

other choices. As a wave passes over the cavity mouth, the perturbed separatrix forms a lobe pair that propagates to the vicinity of the downstream attachment point, where one lobe carries material across the unperturbed separatrix from the cavity to the channel and the other lobe brings channel material into the cavity (figure 6.13 left); figure 6.13 (right) shows the stable and unstable manifolds computed for this heteroclinic connection.

Figure 6.14 shows chaotic transport of fluid into a cavity across the heteroclinic separatrix at the cavity's mouth. All dyed fluid is initially outside the cavity. Dye enters via the lobe seen in the upper right corner to be stretched, folded, and perhaps leave the cavity again after intersecting the stable manifold of the upstream heteroclinic point. Note the three islands persevering along the centerline of the cavity.

Figures 6.11 and 6.14 experimentally illustrate the key point that in general a chaotically advecting flow has both chaotic regions (the repeatedly stretched and folded parts) and regular regions (the islands).

6.2.4. *Where are applications?*

The kinematic equation $dX/dt = \mathbf{V}(x,t)$ applies without restriction on $\mathbf{V}$. Low *Re* flows are called laminar and high *Re* flows are called turbulent.

Fig. 6.14. Chaotic transport of fluid into a cavity. Video in the lectures.

A turbulent **V** varies rapidly in time and space. If the dispersion of passive tracers is dominated by the time and space statistics of the changing velocity field, then mixing and transport properties must be approached by methods of turbulence statistics.[45] It's not that the dynamical systems results become invalid. On the contrary, being basic kinematics they must still apply. But, if the dynamical system itself, given by **V**, changes unpredictably in time and space, then the dynamical systems approach can no longer give useful answers about long-term transport properties. So applications of chaotic advection tend to lie with flows that modulate reasonably predictably, and these flows are typically low *Re* flows or flows with coherent structures. Looking at the definition of Reynolds number (equation 6.1) allows a listing of some likely, though by no means exhaustive, low *Re* application areas for fluid chaos:

- Large viscosity—(bio)polymers, foods, viscous fermentations and bio-products.
- Small lengthscale—microfluidics, transport within cells, tiny swimming organisms, on-chip cooling.
- Small velocities—geology, giant ore body formation, sub-surface and geoengineered flows, porous media.

6.3. Tools for Chaotic Advection Studies

This section briefly outlines some of the standard nonlinear dynamics tools for characterizing chaotic advection. Chaotic advection and nonlinear dynamics are much more fully canvassed in many books and review articles. Among them Ottino[33] writes from a fluid mechanics perspective. Wiggins[42] and Sturman, Ottino & Wiggins[46] cover the mathematical structure of chaotic advection and transport. In particular, Sturman et al. cover what proofs there are to justify the heuristic of streamline crossing. And Aref[47] reviews nonlinear dynamics theory and its role in the development of chaotic advection in fluids.

6.3.1. *Poincaré sections*

The asymptotic-in-time topology of flows can be examined by means of a *Poincaré section.* Poincaré sections plot the locations of a small number ($\mathcal{O}(10)$) of fluid particles after every period of the flow. Sections are constructed by following these trajectories in the flow. As these trajectories pierce designated planes, a dot is recorded on the plane. Different choices of planes can give different, though topologically equivalent, pictures. Trajectories are followed for hundreds or thousands of periods to obtain the asymptotic time advection behavior. Examples showing both regular and chaotic regions are presented in figure 6.15 for the RAM flow defined in figure 6.4 and at the top of figure 6.16 for the driven-lid cavity flow. Chaotic regions of the flow have a seemingly featureless jumble of points spread evenly over a region. Regions of the section into which no points enter or that have regular structure indicate transport barriers and islands. Poincaré sections graphically show the location and size of islands. As Poincaré sections are computationally inexpensive, once they are computed for any selected dense grid of parameters, any areas of interest can be further examined with computationally more expensive methods.

Duct flows are 3D flows such that the velocity field consists of a uniaxial flow, which marks out a special axial flow direction, and a 2D flow transverse to the axial direction. Making Poincaré sections for duct flows, such as the flow in figure 6.4 or other duct flows[48] requires further consideration because the plane that orbits pass through every period is displaced in space. The locations of points are taken at the end of each aperture and translated to a common plane with the necessary proviso that they are suitably rotated to maintain a common orientation. Poincaré sections are built up over

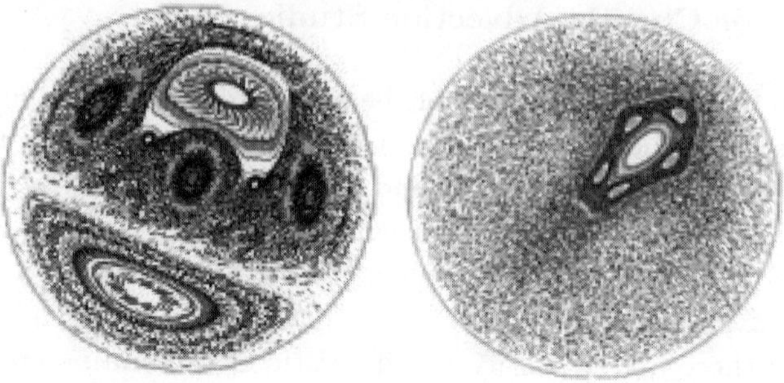

Fig. 6.15. Poincaré sections for the RAM flow with non-Newtonian fluid with window opening size Δ of (left) $\Delta = 135°$ and (right) $\Delta = 90°$. Adapted from Metcalfe et al.[20]

several thousand apertures and provide a clear picture of asymptotic-in-time advection behavior, revealing the geometry of the interlaced regions of chaos and regular islands.

6.3.2. *Periodic points*

The basic building blocks of planar chaotic flows are the flows around elliptic and hyperbolic points and the separatrices between them. In Poincaré sections the regular regions have elliptic points at their heart, and the chaotic seas have hyperbolic points at unknown locations. It is not obvious from static pictures, such as figures 6.15 and 6.16, but some of the islands are connected in island chains whereby a fluid particle's orbit moves it from island to island in a periodic cycle.

In order to find and analyze periodic points, flows may by usefully represented by a map. A map is a matrix that operates on a fluid particle to move it from its current position to its new position at a subsequent time. For instance, to go from the nth to the $(n+1)$th time the map $\mathbf{M}$ would be

$$\mathbf{X}_{n+1} = \mathbf{X}_n + \mathbf{V}\tau = \mathbf{M} \cdot \mathbf{X}_n, \tag{6.12}$$

where τ is the time interval. Maps are particularly useful when the velocity is periodic: $\mathbf{V}(x, t+\tau) = \mathbf{V}(x, t)$. Periodic flows can occur when a steady flow experiences a periodic perturbation, when a finite number of different flows are composed together in a repeating sequence, or when a steady flow is composed together with symmetric copies of itself in a repeating

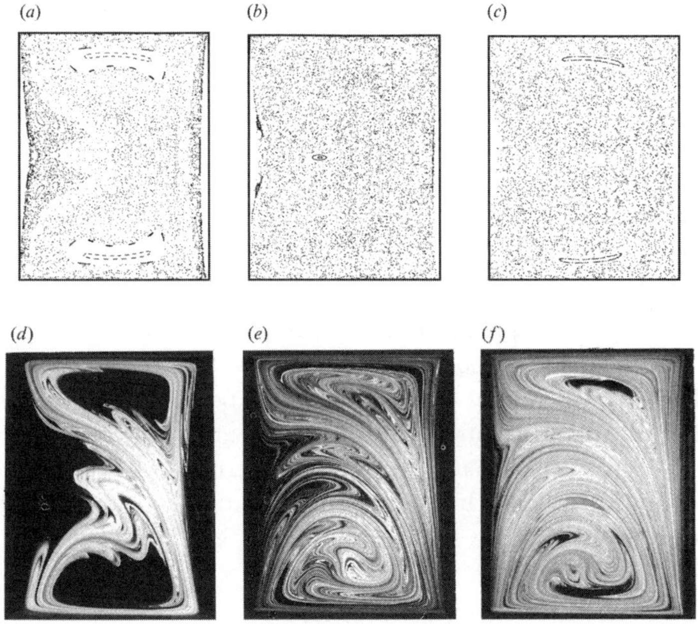

Fig. 6.16. Driven-lid cavity flow. Streamlines of time-averaged steady flow are in figure 6.11. (a–c) Poincaré sections. (d–f) Dye advection experiments. The movie shown in the lecture corresponds to (d). Adapted from Jana et al.[43]

sequence.[f]

Some points advected by $\mathbf{M}$ return to their starting position. These are *periodic points*, and they are closed orbits of the flow. Clearly every point on the orbit is also periodic, and the orbit is a periodic orbit. Periodic points satisfy $\mathbf{X}(p\tau) = \mathbf{X}(t = 0)$. Period-1 points ($p = 1$) end up in the same location after each iterate of $\mathbf{M}$: $\mathbf{X}_n = \mathbf{M} \cdot \mathbf{X}_n$. Period-2 points end up in the same location after 2 iterates: $\mathbf{X}_n = \mathbf{M}^2 \cdot \mathbf{X}_n$. In general a period-$p$ point ends up in the same location after p iterates:

$$\mathbf{X}_n = \mathbf{M}^p \cdot \mathbf{X}_n. \tag{6.13}$$

To find periodic points, convert equation 6.13 to the eigen-problem

$$(\mathbf{M}^p - \mathbf{1}) \cdot \mathbf{X}_n = 0, \tag{6.14}$$

which, by standard results from linear algebra, only has non-trivial solutions when $\det(\mathbf{M}^p - \mathbf{1}) = 0$.

[f]More complicated ways to force time-dependence, e.g. quasi-periodic, is of course possible and leads to chaotic motion, but this is beyond the scope of this chapter.

Even for an analytic $\mathbf{M}$ the algebra to find the location of periodic points above $p > 2$ usually becomes quite involved. Luckily most often the advection properties are dominated by lower order periodic points ($p < 4$). If the location of higher order points is desired or $\mathbf{M}$ is only available from a numerical computation, it may be easier to search for periodic points by placing N points in the domain of interest of area A and integrating through p periods. Then all points such that

$$|\mathbf{X}_{n+p} - \mathbf{X}_n| < \epsilon_n = \sqrt{A}/N \tag{6.15}$$

are candidate periodic points. Put $N_1 < N$ points into the ϵ-balls centered on the few candidate locations and integrate again. Convergence is usually rapid. However, the initial coverage needs to be sufficiently dense to discover all the periodic points. And there is no guarantee of discovering all periodic points. Further manipulation can determine whether found points are elliptic or hyperbolic. In this way one can build a skeleton of the flow and transport topology.

6.3.3. *Dye advection and stretching*

While the Poincaré sections show the asymptotic structure for a given set of parameters, they provide no information on the rate at which this structure arises. Physically this is the rate at which fluid striations thin. For instance, it can happen that well-mixed states predicted by Poincaré sections remain poorly mixed for hundreds of periods, which may be an impractically long time. Short-time and rate information is crucial for application designs. To get some short-time and rate information one can use dye traces.

Numerical dye traces mimic experiments where small amounts of dyed additive is either placed initially as a blob or continuously injected (bottom of figure 6.16). The numerical dye lines are an initially dense collection of points (tens of thousands or more) that are advected and followed. As the points separate exponentially quickly over most of the line, it may be desirable to interpolate new points to maintain the integrity of the numerical dye. This exponential separation also means that it is impractical to follow the numerical dye interfaces for a large number of periods.

Another style of numerical dye advection is shown in figure 6.17. Here $\mathcal{O}(10^4)$ fluid particles are randomly placed in the flow domain. One half of the particles are initially marked black and the other half red, and all particles are advected for a small ($\mathcal{O}(10)$) number of periods. The result is a qualitative picture of the striation distribution and whether good mixing

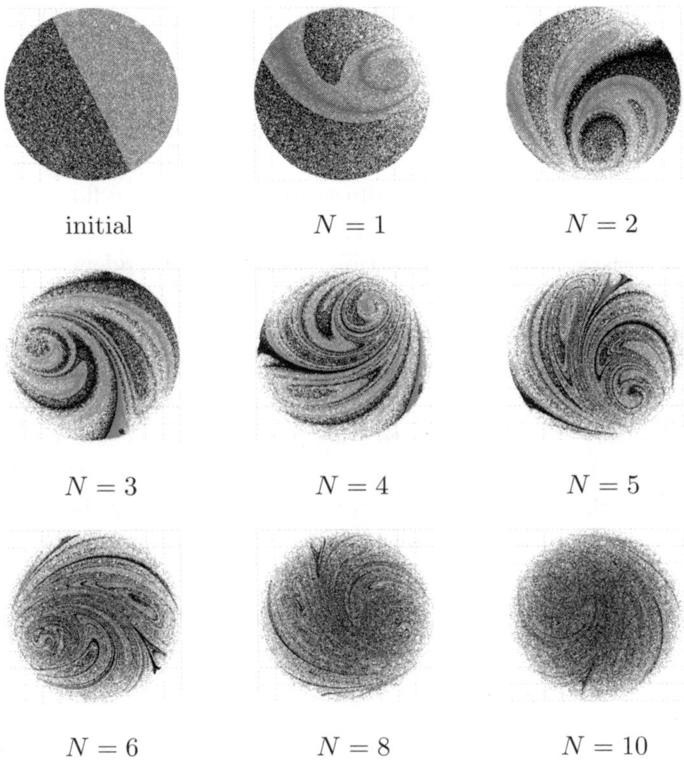

Fig. 6.17. Dye advection plots for a Newtonian fluid in the RAM flow of figure 6.4 after N windows. Adapted from Metcalfe et al.[20]

is achieved in a reasonable time. Dye advection plots are also predictions of how mixing will proceed, and can be easily compared with experiments.

Stretching distributions can be used to further refine parameter searches for well-mixed states. They can indicate local rates of mixing and how uniformly rates are distributed over the mixing region. On the other hand, interpretation of stretching distributions can be problematic, and they do not always discriminate very well between states. Moreover, it is extremely difficult to measure a stretching distribution, particularly for 3D flows, in order to validate computations. Finally, the computational cost of dynamical systems tools for chaotic advection ranks in order from least to most costly: Poincaré sections, dye traces, and stretching distributions.[43] Stretching distributions are more than 10^3 times more costly than the other tools, while a dye trace is about a factor of ten more costly than a Poincaré section. For these reasons stretching distributions are used sparingly.

6.4. Periodically Reoriented Flows: Mixing, Scalar Transport, and Parametric Variation

The examples in the introduction illustrated that making flow streamlines cross at successive times generates the stretching and folding of material lines characteristic of chaotic advection. This time-dependence, or *stirring*, of the fluids gets around the fact that steady flows must mix poorly. One of the simplest ways to engineer time-dependence into devices is with a periodically reoriented flow, of which an example is the RAM flow of figure 6.4. A Periodically Reoriented Flow (PRF) has one non-rotationally-symmetric base flow that at every time interval τ is rotated instantaneously through an angle Θ. τ and Θ are the two primary intrinsic control parameters of a PRF, but there may also be secondary parameters particular to the fluid or how reorientation is accomplished. (Elaborations on the basic PRF described here are, of course, possible.) Figure 6.18 shows the control parameters and steady flow for the RAM cross-section.

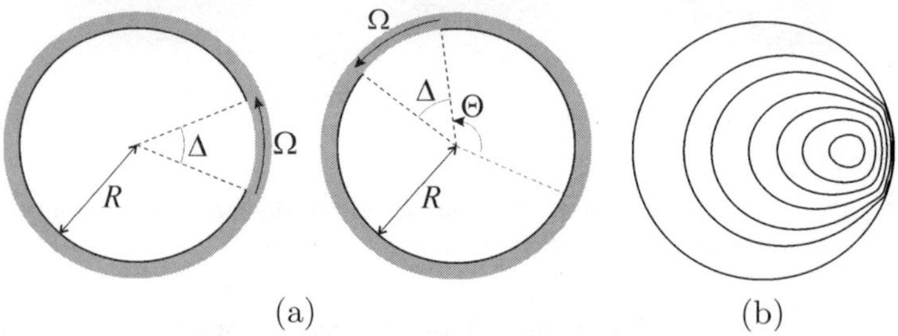

Fig. 6.18. A periodically reoriented flow, the Rotated Arc Mixer (RAM) flow (a) geometry and parameters, (b) streamlines for steady Stokes flow from Hwu et al.[21]

The RAM is a 3D duct flow, so it may not be obvious how streamlines can cross. Duct flows are 3D flows such that the velocity field consists of a uniaxial flow, which marks out a special axial flow direction, and a 2D flow transverse to the axial direction. What's important is that with duct flows the transverse flows are the planar flows in which the dynamical system of equation 6.3 operates, and the axial space direction acts like a time axis.[g] For a duct flow capable of chaos, the requirement is that, when superposed,

[g]Strictly true only for plug flow, but any other uniaxial flow profile gives the same result with a change of coordinate.

the in-plane streamlines at one axial location cross the streamlines from a downstream cross-section. Figure 6.19 illustrates the successive streamline crossings for the RAM flow. The bottom row shows photographs of an experiment viewed from the inlet end of the mixing section of the duct. Flow is into the page, and "time" increases left to right. The two horizontal shadows at 3 and 9 o'clock are dye injection ports. The top row shows streamlines of the analytical steady Stokes solution of Hwu, Young & Chen.[21] In (a) the streak undergoes transverse rotation at the first window of boundary motion and should be compared with the analytical velocity solution on the right. At a later time (b) the orientation of the streak's transverse motion has rotated; the orientation of the analytic solution has been rotated and superposed onto that of the original orientation. At a later time again (c) there is a further reorientation of the transverse velocity field. After this reorientation at later time (d) the dye streak is losing its regular structure. With flow through further reorientations the dye streak is stretched and folded into a globally chaotic advection field with striation refinement down to the diffusion limit for this particular set of experimental parameters.

For the RAM flow the time between reorientations is $\tau = \Omega R/\bar{U}$, the ratio of rotational to axial velocities, or the amount of transverse circulation or linear stretching within the interval between flow reorientations. Metcalfe et al.[20] and Singh et al.[28] have computed and experimentally verified values of (τ, Θ) that optimize mixing for the RAM flow.

6.4.1. *Scalar transport*

To move beyond straight advection of passive fluid particles, the addition of diffusion brings up the important problem of scalar transport. Dispersion of a passive scalar ϕ, which can be heat, a concentration of chemical or species, and many other things, is important in a variety of phenomena from epidemiology to geophysics and across length scales from the molecular to the celestial.[49] Transport of ϕ proceeds simultaneously through organized motion (the flow $\mathbf{v}(\tau, \Theta)$) and through disorganized motion (molecular diffusion)—vastly different modes and scales of transport. In this section I will illustrate the kind of scalar transport obtained with PRFs, and touch on some of the fundamental differences between purely advective transport and advective-diffusive transport.

Passive scalar transport is described by the advection-diffusion equation

(a)
(b)
(c)
(d)

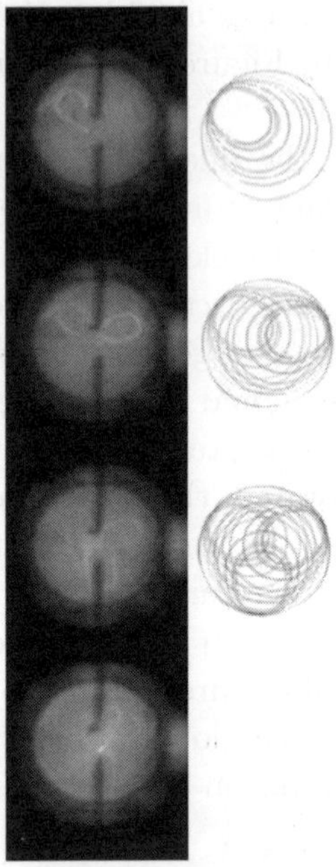

Fig. 6.19. Superposed crossing of successive streamlines. (bottom) Experiments injecting colored dye. The view is from the beginning of the mixing section; vertical shadows are dye injection needles; and flow is into the page. (a–d) are taken at successive times from the start of dye injection. (top) Streamlines computed from the analytical crossflow solution,[21] rotated and plotted superposed. Emerging jumble of chaotic orbits s apparent in (d).

(ADE)

$$\frac{\partial \phi}{\partial t} + \mathbf{v}(\tau, \Theta) \cdot \nabla \phi = \frac{1}{Pe} \nabla^2 \phi. \tag{6.16}$$

The Péclet number Pe balances diffusive to advective rates. Recent work has firmly established that "strange eigenmodes"—sets of naturally persistent spatial patterns $\varphi_k(\mathbf{x}, t)$ with decaying amplitudes—are fundamental solutions of the ADE.[50–52] This means that the scalar field ϕ is a finite sum of these patterns:

$$\phi(\mathbf{x}, t) = \sum_{k=0}^{K} \alpha_k \, \varphi_k(\mathbf{x}, t) \, e^{\lambda_k t} \to \alpha_0 \, \varphi_0(\mathbf{x}, t) \, e^{\lambda_0 t}, \tag{6.17}$$

where the sum is ordered by the magnitude of the real parts of the decay rates λ_k, with initial weights α_k. As the λ_k have to have negative real parts, at long times only the most slowly decaying term persists: the asymptotic transport rate is given by $Re(\lambda_0)$ and the asymptotic scalar distribution is given by $\varphi_0(\mathbf{x}, t)$.

Note that the dissipation introduced by diffusion means the dynamical system is no longer conservative of phase space volume and that in equation 6.17 the scalar field ϕ is attracted to φ_0. The addition of diffusion to advection converts a conservative to a dissipative dynamical system.

Figure 6.20 (from the video shown during the lectures) illustrates the cooperation between advection and diffusion as the pattern evolves through one reorientation interval. For illustration the effect of the exponentially decaying amplitude has been removed from figure 6.20. At 0 the boundary arc velocity indicated by the arrow reorients. Region (a) has been stretched and folded during the previous interval, and during the subsequent interval diffusion "heals" the pattern along this fold. Simultaneously advection stretches and makes new folds around the region (b). The pattern at τ resumes its original shape, rotated by Θ. Diffusion heals the pattern wherever folds bring parts of the patterns close together: stretching and folding still play a crucial role in scalar transport in chaotically stirred fluids.

6.4.2. *Parametric variation*

To use any dynamical system with parameters an important and natural requirement is characterizing the major qualitative changes as parameters vary; moreover, having complete parametric solutions is crucial for transport optimization or for the inverse problem of estimating parameter values from observations.[53] Generally it is difficult to compute the parametric variation of solutions over the full parameter space, because of the large cost of evolving a solution for each point on a fine grid throughout parameter space. But this is crucial for optimization, parameter estimation, and elucidating the global structure of scalar transport.

For scalar transport in a PRF the asymptotic complete parametric solution is knowing φ_0 and λ_0 as a function of τ and Θ. By exploiting the symmetry of PRFs Lester et al.[27] created a composite spectral method that is three to four orders of magnitude faster than other methods of similar accuracy at obtaining φ_0 and λ_0 over a dense grid in (τ, Θ) space. Figure 6.21 is a contour plot of $Re(\lambda_0)$ calculated at $\mathcal{O}(10^5)$ points over the τ–Θ plane for $Pe = 10^3$ and homogeneous Dirichlet boundary conditions. λ_0 is scaled by the diffusion rate (the most slowly decaying eigenvalue of the diffusion operator in the disc), i.e. figure 6.21 maps *transport enhancement* relative

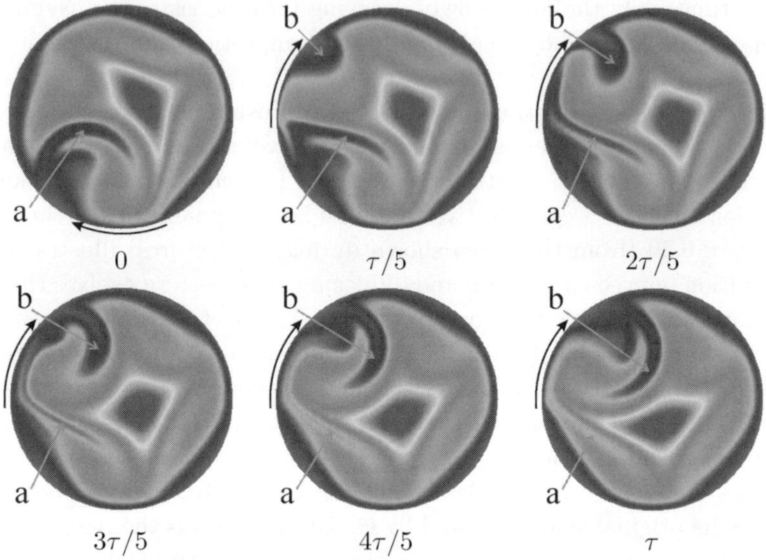

Fig. 6.20. How a pattern (point (i) in figure 6.21) evolves though one reorientation interval. At 0 the moving boundary indicated by the arrow reorients. Region (a) has been stretched and folded; diffusion "heals" the pattern along folds. Simultaneously advection stretches and makes new folds around the region (b). The pattern at τ resumes its original shape, rotated by Θ. Video in the lectures.

to diffusion alone. Figure 6.21 shows the first complete parametric solution for scalar transport in a physically realizable chaotic flow. Several features deserve comment.

At low values of τ, the enhancement distribution is fractal with many localized maxima, which can only be discerned with highly resolved solutions. The spiky regions originate from rational values of Θ/π at $\tau = 0$ and grow in width with increasing τ. Inside each region the scalar spatial distribution is locked into a symmetric pattern whose azimuthal wave number m is a rational multiple of the forcing wavenumber $k = 2\pi/\Theta$; these regions are symmetry-locked "tongues" similar to frequency-locked Arnold tongues.[54] Figure 6.22 shows asymptotic spatial distributions of the scalar field (φ_0) at the correspondingly labeled points of figure 6.21. Figures 6.22(a–d) show symmetry-locking for $m \in [3, 6]$. Figure 6.20 is at point (i) in figure 6.21. Section 6.5 will identify the physical origins of these tongues for a PRF.

As τ becomes $\mathcal{O}(1)$ the spreading tongues interact to produce an order-disorder transition, where the symmetric spatial patterns smoothly lose symmetry. Figures 6.22(b, e–h) show this transition along the $m = 4$

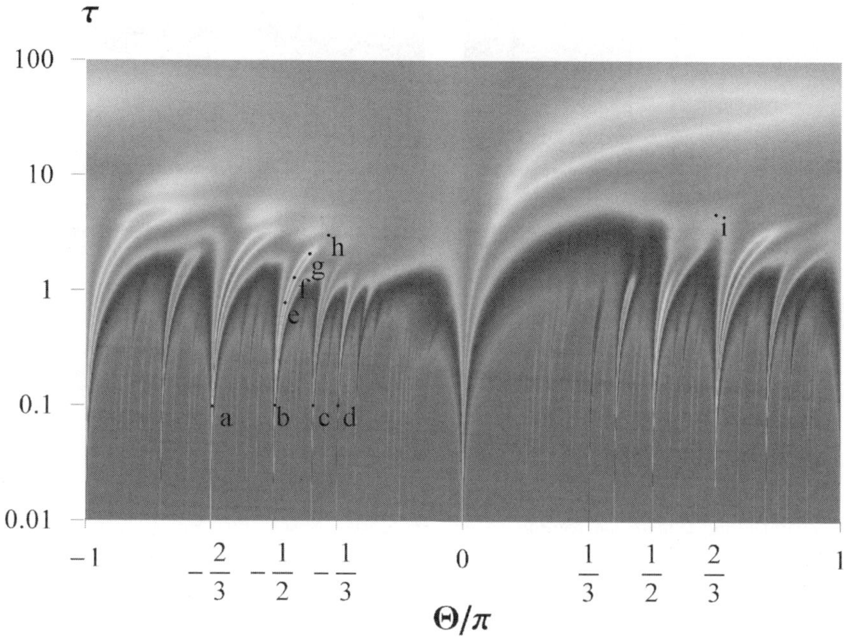

Fig. 6.21. Map of the asymptotic decay rate $Re(\lambda_0)$ scaled by the diffusion rate for Dirichlet boundary conditions, $\Delta = \pi/4$, and $Pe = 10^3$. Note logarithmic scaling of τ axis and λ_0 contours.

tongue at the points correspondingly labeled in figure 6.21. At large values of τ ($> 10^3$) the enhancement ratio everywhere takes on the same value as that on the $\Theta = 0$ line, around three.

For $Pe = 10^3$ the maximum transport enhancement ratio of six occurs in the $m = 3$ tongue at low values of τ. This seems counter-intuitive because the no-diffusion mixing optimum occurs[20,26,28] at $(\Theta, \tau) \approx (\pi/5, 15)$, and at low τ chaotic advection is not global over the domain. However, for higher Pe (not shown) the maximum transport enhancement increases, and its location moves toward the mixing optimum. Can one use this computational method to compute mixing optima, i.e. the $Pe \to \infty$ limit of figure 6.21? Yes, in principle; however the computational expense increases as $Pe^{1/2}$,[27] so for very large Pe ($> \mathcal{O}(10^7)$) complete parametric solutions are not currently feasible to compute.

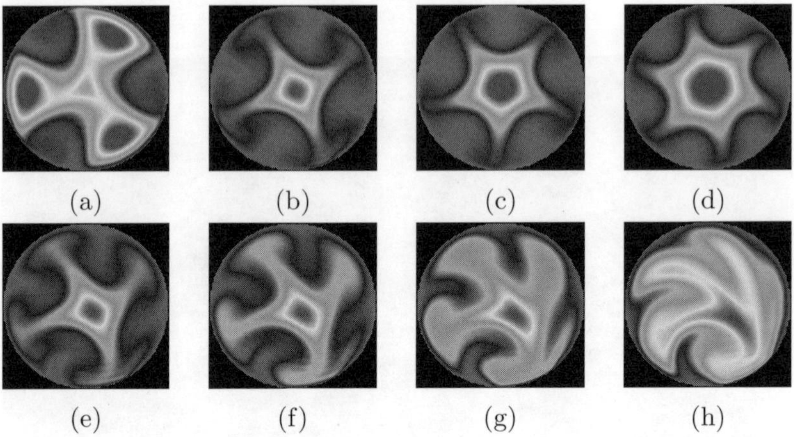

Fig. 6.22. Spatial patterns of the scalar distribution (φ_0) at the correspondingly labeled points in figure 6.21. (a-d) symmetry-locking at $m \in [3, 6]$. (b, e-h) order-disorder transition along the $m = 4$ tongue.

6.5. Symmetries of Periodic Reorientation and Their Effects

If the base flow of a PRF has a symmetry, then the symmetry manifests itself in non-obvious ways in both the advection fields and the scalar fields resulting from the PRF, affecting both mixing and diffusive transport properties. So, by designing in flow symmetry, it can be exploited to tailor the available range of transport properties and can greatly reduce the computational effort to establish transport optima. In this section I first briefly introduce notation for reflections and rotations in the plane, then show how a symmetry in the base flow along with periodic reorientation leads to doubly-periodic symmetry in the advection field and helps create the resonance tongues of decay rates in the scalar field.

6.5.1. *Reflection and rotation in the plane*

By way of background, the operations of reflection $\mathbf{S}$ and rotation $\mathbf{R}$ in the plane are illustrated in figure 6.23. I will use these to introduce symmetry operators in preparation for using reorientation of a base flow to design flows with chaotic advection fields that can mix and transport heat efficiently. $\mathbf{S}_x$ is a reflection across the x-axis, $\mathbf{S}_y$ is a reflection across the y-axis, and $\mathbf{S}_x\mathbf{S}_y$ is a reflection across both the x- and y-axes, which is equivalent to a 180° rotation. $\mathbf{R}(\theta)$ rotates a point through an angle θ, which can be positive or negative, with positive a counter clockwise rotation and negative a clockwise

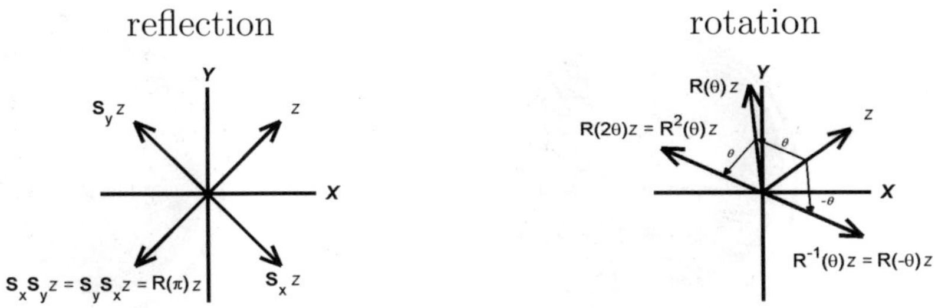

Fig. 6.23. Reflection and rotation operations in the plane and notation conventions.

rotation. Various notations for multiple rotations are given in the figure. The matrix for rotation in the plane is

$$\mathbf{R}(\theta) = \begin{pmatrix} \cos\theta & -\sin\theta \\ \sin\theta & \cos\theta \end{pmatrix}, \tag{6.18}$$

and the matrix for a general reflection about a line inclined at an angle α to the x-axis is

$$\mathbf{S}_\alpha = \begin{pmatrix} \cos 2\alpha & \sin 2\alpha \\ \sin 2\alpha & -\cos 2\alpha \end{pmatrix}. \tag{6.19}$$

For reflection about the x- and y-axes the matrices are

$$\mathbf{S}_x = \begin{pmatrix} 1 & 0 \\ 0 & -1 \end{pmatrix} \quad \mathbf{S}_y = \begin{pmatrix} -1 & 0 \\ 0 & 1 \end{pmatrix}. \tag{6.20}$$

Finally a double reflection about perpendicular axes is

$$\mathbf{S}_\alpha \mathbf{S}_{\alpha+\pi/2} = \mathbf{R}(\pm\pi) = \begin{pmatrix} -1 & 0 \\ 0 & -1 \end{pmatrix}. \tag{6.21}$$

As reflections are their own inverses, $\mathbf{S}_\alpha = \mathbf{S}_\alpha^{-1}$ and $\mathbf{S}_\alpha^2 = \mathbf{1}$. Applications of multiple operations can be composed by matrix multiplication, taking care to keep the correct order when applying matrix transformation to a point.

The determinant of the reflection matrices is -1, which means reflection reverses orientation. The determinant of the rotation matrix is $+1$, which means rotation preserves orientation. The determinant of products of matrices equals the product of the determinant of the individual matrices. The inverse of products of matrices is given by the product of the matrix inverses taken in reverse order: $(\mathbf{AB})^{-1} = \mathbf{B}^{-1}\mathbf{A}^{-1}$.

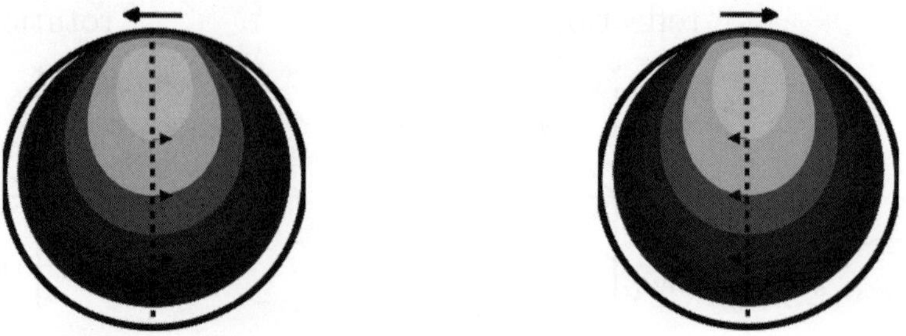

(a) (b)

Fig. 6.24. RAM base flow streamlines (a) forward-in-time flow and (b) time-reversed flow are related by a reflection symmetry about the dashed line.

6.5.2. *Time-reversal symmetry*

Two flows $\mathbf{A}$ and $\mathbf{B}$ are symmetric if $\mathbf{A} = \mathbf{SBS}^{-1}$ for some operator $\mathbf{S}$. What this says in the context of fluid flow is that if a point X advected by $\mathbf{A}$ goes from X_n to $X_{n+1} = \mathbf{A}X_n$, then the same point advected by $\mathbf{SBS}^{-1}X_n$ goes to the same point X_{n+1}.

The particular flow symmetry I want to focus on is one where the time-reversed flow coincides with a geometry symmetry of the flow-domain combination. For instance, figure 6.24a shows the streamlines of the RAM flow; the arrows indicate the time-forward direction of the flow. If time runs in reverse, the time-reversed picture of the RAM streamlines (figure 6.24b) looks the same except the sense of the arrows reverses.[h]

I want to specialize to a class of reoriented flow that has the property that the time-reversed flow $\Phi(-\tau)$ is symmetric with the time-forward flow:

$$\Phi(-\tau) = \Phi^{-1}(\tau) = \mathbf{S}\Phi(\tau)\mathbf{S}^{-1}, \tag{6.22}$$

where $\mathbf{S}$ is any geometric symmetry operation. It's fairly easy to see that the RAM Stokes flow has this property because the time-reversed base flow is a reflection of the time-forward flow. While the assumed symmetry restricts the generality of the results that follow, in practice it occurs frequently. Most chaotic flows composed from disjoint flow sequences have this

[h]For the RAM flow, time-reversal symmetry depends on it being a Stokes flow. Inertia or viscoelasticity (but not shear rate dependent or yield-stress non-Newtonian rheologies) tend to rotate the flow axis of symmetry in the downstream direction of the moving boundary. As time-reversal is equivalent to reversing the direction of boundary rotation, the flow axis of symmetry no longer coincides with the domain's axis of symmetry.

symmetry. On the other hand, it is easy to think of flows that aren't time-reversal symmetric, and most natural flows generating chaotic transport don't have this symmetry, so it is not applicable in all cases. Nonetheless, as I'll show, a time-reversal symmetry endows a rich structure on the mixing and transport properties of the entire class of reoriented time-symmetric flows. And as the class is particularly useful in engineering flows for particular transport jobs, it can be useful to know about.

At the kth period, the flow of a periodically reoriented flow is given by

$$\Phi_k = \mathbf{R}^{k-1}(\Theta)\Phi_0\mathbf{R}^{1-k}(\Theta). \tag{6.23}$$

The right-most rotation matrix for $k > 1$ *un*winds the point from all previous reorientations with the fixed reorientation angle Θ, then the base flow Φ_0, which in the math description is fixed in space, advects the point, and finally the advected point is wound forward to its current location. This is the basic operation of a PRF with one base flow. Generalization to multiple base flows is straightforward but can be cumbersome.[27]

Now assume that the forward and backward flows are symmetric:

$$\Phi_0 = \mathbf{S}\Phi_0^{-1}\mathbf{S}^{-1}, \tag{6.24}$$

where $\mathbf{S}$ is a geometric symmetry of the flow and boundary conditions. N products of equation 6.23 give the complete advection map Φ after N periods as

$$\begin{aligned}\Phi &= \prod_{1}^{k=N} \Phi_k \qquad\qquad (6.25)\\ &= \mathbf{R}^{N-1}\Phi_0\mathbf{R}^{-(N-1)}\mathbf{R}^{N-2}\Phi_0\mathbf{R}^{-(N-2)}\cdots\mathbf{R}^{1}\Phi_0\mathbf{R}^{-1}\Phi_0\\ &= \mathbf{R}^{N}\left(\mathbf{R}^{-1}\Phi_0\right)^{N}.\end{aligned}$$

The factor $\mathbf{R}^N$ is purely a phase that for convenience I will absorb into Φ and henceforward drop. Substitution of equation 6.24 into Φ gives

$$\Phi = \mathbf{R}^{-1}\mathbf{S}\Phi_0^{-1}\mathbf{S}^{-1}\mathbf{R}^{-1}\mathbf{S}\Phi_0^{-1}\mathbf{S}^{-1}\cdots\mathbf{R}^{-1}\mathbf{S}\Phi_0^{-1}\mathbf{S}^{-1}. \tag{6.26}$$

Collecting terms, defining $\mathbf{S}_\gamma = \mathbf{R}^{-1}\mathbf{S}$, and judiciously inserting $\mathbf{1}$ in the form of $\mathbf{1} = \mathbf{S}_\gamma\mathbf{S}_\gamma^{-1}$, gives

$$\Phi = \mathbf{S}_\gamma\left(\Phi_0^{-1}\mathbf{S}^{-1}\mathbf{R}^{-1}\mathbf{S}\right)^N\mathbf{S}_\gamma^{-1}. \tag{6.27}$$

To proceed further, I need to specify the symmetry $\mathbf{S}$. With reflections the choices are the single reflection, $\mathbf{S} = \mathbf{S}_\alpha$, or the double reflection, $\mathbf{S} =$

$\mathbf{R}(\pm\pi)$. I'll begin with the single reflection. Using the matrix definition, $\mathbf{S}_\alpha^{-1}\mathbf{R}^{-1}\mathbf{S}_\alpha = \mathbf{R}$, and from equation 6.25 note that

$$\Phi^{-1} = \left(\mathbf{R}^{-1}\Phi_0\right)^{-N} = \left[\left(\mathbf{R}^{-1}\Phi_0\right)^{-1}\right]^N = \left(\Phi_0^{-1}\mathbf{R}\right)^N . \tag{6.28}$$

Together this means that

$$\Phi = \mathbf{S}_\gamma \Phi^{-1} \mathbf{S}_\gamma^{-1}, \tag{6.29}$$

which is exactly the condition in which the forward flow and the time-reversed flow are symmetric under the operation of $\mathbf{S}_\gamma$. $\mathbf{S}_\gamma = \mathbf{R}^{-1}\mathbf{S}_\alpha$ is called a *reversal-reflection symmetry*, and

$$\mathbf{S}_\gamma = \begin{pmatrix} -\cos(2\alpha+\Theta) & \sin(2\alpha+\Theta) \\ \sin(2\alpha+\Theta) & \cos(2\alpha+\Theta) \end{pmatrix}. \tag{6.30}$$

With reference to equation 6.19, this is a reflection about an axis of symmetry inclined from the x-axis by $\theta_s = \alpha + \Theta/2$. If we take that the reflection symmetry of Φ_0 is the x-axis, then $\alpha = 0$ and

$$2\theta_s = \Theta. \tag{6.31}$$

This says that a reoriented time-symmetric flow with a reflection symmetry for the base flow has an axis of symmetry *in the advection field* that is doubly-periodic in the reorientation angle.

For the double reflection, $\mathbf{S} = \mathbf{R}(\pm\pi)$, and

$$\Phi = \mathbf{S}_\gamma \left(\Phi_0^{-1}\mathbf{R}^{-1}\right)^N \mathbf{S}_\gamma^{-1}. \tag{6.32}$$

Everything is the same as for the single reflection, except that because Φ^{-1} has to equal $\Phi^{-1} = \left(\Phi_0^{-1}\mathbf{R}^{-1}\right)^N$ and from equation 6.28 Φ^{-1} also has to equal $\Phi^{-1} = \left(\Phi_0^{-1}\mathbf{R}\right)^N$, this requires $\mathbf{R}^{-1} = \mathbf{R}$. For the double reflection there is a further symmetry such that advection for $+\Theta$ is symmetric to advection for $-\Theta$. The practical effect of having a doubly reflective flow is that the parameter space that must be searched for transport optima is cut in half.

A final note concerns the relation between geometric symmetries and flow symmetries. The fact that a flow symmetry coincides with a geometric symmetry is a result of the overall phases of the forcing protocols. Other phases may turn flow symmetries into curves that do not coincide with any geometric symmetries. If the phase is chosen poorly, then flow symmetries are hidden and, from a practical point of view, cannot be exploited.

6.5.3. *Origin of tongues*

Finally, the tongue structures and the transition from spatially regular to spatially chaotic solutions of figure 6.21 come from phase locking of Lagrangian orbits with the forcing of the reorientation. Let's consider how this occurs generically for PRFs by examining how the angular position θ of an orbit changes with time. I'll use a shorthand notation of $\dot{\theta}$ for $d\theta/dt$ and ask, what happens to $\dot{\theta}$?

First, $\dot{\theta}$ is the instantaneous angular rotation rate at any point in the flow, so $\dot{\theta} = \partial\psi/\partial r$, where ψ is the stream function of the base flow and r is the radial coordinate. Recall that the base flow is the unreoriented flow. Then, form the time average of $\dot{\theta}$ and expand it in a Fourier series in the angular coordinate θ. For simplicity assume a circular domain, which makes it obvious that ψ is 2π periodic in θ. The time average of $\dot{\theta}$ is

$$\dot{\theta} = N^{-1} \sum_{n=0}^{N-1} R^n \partial\psi/\partial r, \tag{6.33}$$

where R is the rotation operator and N is the number of periods of flow and reorientation. Expanding $\partial\psi/\partial r$ in a Fourier series gives

$$\dot{\theta} = N^{-1} \sum_{n=0}^{N-1} R^n \left[a_0 + \sum_{k=1}^{\infty} [a_k \sin(k\theta) + b_k \cos(k\theta)] \right], \tag{6.34}$$

where the expansion coefficients a_0, a_k, and b_k are functions of r, and a_0 is the average angular rotation rate at a particular value of r. Note that the expansion coefficients are obtained only from the base flow, which could aid in designing flows with tailored transport properties.

Normally reorientation would take a function $h(\theta)$ to $R^n h(\theta) = h(\theta + n\Theta)$, but for the class of reoriented flows with time-reversal symmetry, as discussed above, one effect of this symmetry (c.f. equation 6.31) is to make the action of the map doubly periodic in the reorientation angle Θ. This means the effect of the rotation matrix is $R^n h(\theta) \to h(\theta + n(2\Theta))$. With that effect of symmetry in mind, swapping the order of summation, and collecting terms in the coefficients gives

$$\begin{aligned}\dot{\theta} &= N^{-1}\sum_{n=0}^{N-1} a_0 + \\ &\sum_{k=1}^{\infty}\left[a_k\left(N^{-1}\sum_{n=0}^{N-1}\sin\left[k(\theta+2n\Theta)\right]\right)+b_k\left(N^{-1}\sum_{n=0}^{N-1}\cos\left[k(\theta+2n\Theta)\right]\right)\right] \\ &= a_0 + \sum_{k=1}^{\infty}\left[a_k\, f(\theta;k,\Theta)+b_k\, g(\theta;k,\Theta)\right]. \qquad (6.35)\end{aligned}$$

The functions f and g are sums of phase-shifted Fourier modes that become the *de facto* modes of the reoriented flow, while the coefficients a_k and b_k are obtained by Fourier transform of the unreoriented flow. The phase-summed modes exhibit two types of behavior depending on whether Θ is an irrational or a rational number. As f and g behave identically, I need only consider one of them, say f.

Recall that a rational number can be exactly expressed as the ratio of two integers p and q, such that $\Theta/\pi = p/q$. Examples of rational reorientation angles are $\Theta/\pi = 2/5$ or $1/4$. All values of Θ/π that are not expressible as rational number are called irrational numbers. There are many more irrational numbers than rationals.

For irrational Θ/π, reorientations never repeat exactly: $N \to \infty$ in the time average, and the sum becomes an integral of f over all possible phases:

$$\int_0^{2\pi}\sin(k(x+\phi))d\phi = -k^{-1}\cos(kx+\phi)\,|_0^{2\pi} = 0. \qquad (6.36)$$

This means the functions multiplying all the coefficients except a_0 are zero, so $\dot{\theta} = a_0 \equiv \omega$, and orbits move with a constant angular frequency ω (at fixed r). As $\partial\psi/\partial r = \omega$, this also says that the time-averaged stream function for irrational Θ/π is a set of nested circles whose stream values are given by the average rotation rate obtained from the base stream function. This is illustrated in figure 6.25a. This carries over to scalar transport because, as a consequence of the circular symmetry of the orbits, the orbits of the time-averaged flow offer no radial transport. Radial transport can only arise from diffusion, and that's why the scalar transport rates for the RAM flow of figure 6.21 are identical to those in the diffusion-only case for the large parts of the parameter space where Θ/π is irrational (but before the chaotic transition with increasing τ).

For rational Θ/π, reorientations are periodic, and there are a finite number of terms in f. For there to be constructive interference of the phase-

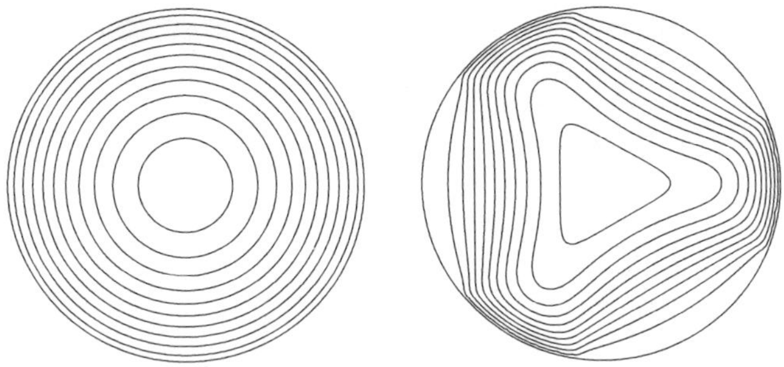

(a) (b)

Fig. 6.25. Irrational and rational orbits for the RAM flow calculated directly from equation 6.33 with $N = 30$. (a) $\Theta/\pi = 1/\sqrt{10}$, irrational. (b) $\Theta/\pi = 1/3$, rational and with a 3-fold angular symmetry.

shifted modes and a *resonance*, the resonance condition must be fulfilled. The resonance condition is that $k(\theta + 2n\Theta) = 2\pi n$: for rational $\Theta/\pi = p/q$ the modes of the time-averaged reoriented flow are non-zero only for angular wavenumbers satisfying $k = \pi/\Theta = q/p$. At resonance, equation 6.35 simplifies to

$$\dot{\theta} = \omega + a_k \sin\left(\frac{\pi\theta}{\Theta}\right) + b_k \cos\left(\frac{\pi\theta}{\Theta}\right). \tag{6.37}$$

This is the origin of the resonance tongues for reoriented flows. Since the flow symmetry halves the angular wavenumber, this is a *sub-harmonic resonance.*

Figure 6.26 illustrates sub-harmonic resonance by plotting the maximum value of f as a function of wavenumber k and reorientation angle Θ; $\max(f)$ is colored according to the bar at the top of the figure. Blue is zero and, as most values of Θ/π are irrational, most of the parameter space is zero; red is near one. Visible in figure 6.26a are resonance lines of $k = \pi/\Theta$ (and higher harmonics). Figure 6.26b shows a close-up around $\Theta/\pi = 1/3$. Resonance "lines" now resolve into narrow, isolated resonance peaks at rational values of Θ/π. The dashed lines show the 1/3 resonance lining up with $k = 3$, sub-harmonic as predicted. Other peaks occur at all other rational values of Θ/π, but finite numerical resolution doesn't allow every rational value to be displayed. For instance, the resonances resolved on either side of 1/3 are close to 53/160 and 81/240. In practice, higher order resonances can't be discerned either computationally or experimentally in the mixing or transport behavior of chaotically advected flows.

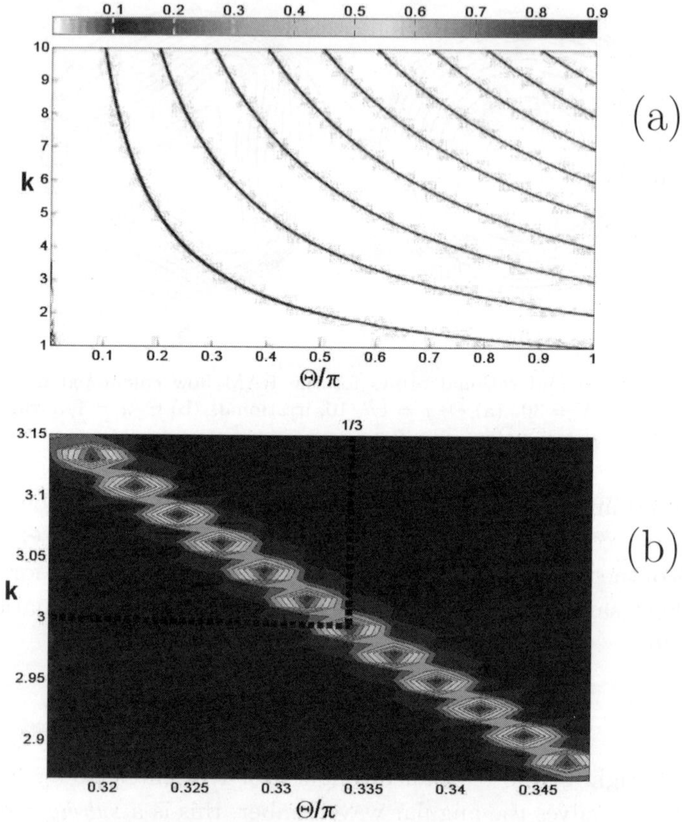

Fig. 6.26. Sub-harmonic resonances of time symmetric reoriented flows. (a) Fourier wavenumber k versus reorientation angle Θ. Color indicates the maximum value of the phase shifted sum in equation 6.35 at that point in k-Θ space, according to the colorbar at the top, which applies to both figures. Visible are resonance lines of $k = \pi/(n\Theta)$. (b) Close-up of (a) around $\Theta/\pi = 1/3$. The resonance "lines" resolve into resonance peaks at rational values (to numerical resolution) of Θ/π. The resonances resolved either side of 1/3 are close to 53/160 and 81/240.

Notice how the symmetry causes sub-harmonic resonance. Without the symmetry $k = 2q/p$ would be the resonance condition, e.g. for $\Theta/\pi = 1/3$, k would be $k = 6$, and $f(\theta)$ would give a six-fold spatial symmetry. But the flow symmetry selects the $k/2$ mode, giving instead a three-fold spatial symmetry (figure 6.25b).

But rational resonances don't complete the picture of mode-locked tongue regions of expanding widths with increasing τ. The next step to tongues is to notice that equation 6.37 simplifies to

$$\dot{\theta} = \omega + A_k \sin\left(\frac{\pi\theta}{\Theta}\right), \tag{6.38}$$

where $A_k = \sqrt{a_k^2 + b_k^2}$ and for simplicity I've absorbed an added phase of b_k/A_k in the sine function. The difference between Θ, the forcing, and θ the response is the phase difference $\phi = \Theta - \theta$, and its time derivative is $\dot{\phi} = \dot{\Theta} - \dot{\theta}$. If ϕ varies irregularly with time, then the forcing and response move independently; however, if ϕ is a constant, then the forcing and response bear a fixed relationship. Rearranging terms to give $\theta/\Theta = 1 - \phi/\Theta$, using $k = \pi/\Theta$, and recalling that $-\sin(x - \pi) = \sin(x)$, leads to an equation for $\dot{\phi}$ in the form of a nonlinear oscillator equation

$$\dot{\phi} = \dot{\Theta} - \omega - A_k \sin(k\phi). \tag{6.39}$$

Dividing both sides by A_k to nondimensionalize, rescales time in the phase time derivative to give

$$\phi' = \mu - \sin(k\phi) \tag{6.40}$$

with

$$\mu = \frac{\dot{\Theta} - \omega}{A_k}. \tag{6.41}$$

The parameter μ measures the ratio of the frequency difference between the reorientation forcing and the bare flow to the strength of the bare flow. These equations come about at resonances where $k = \pi/\Theta = q/p$: each rational resonance has the same form of governing equation, but k varies from resonance to resonance.

The behavior of equation 6.40 can easily be discerned from its graph, and in fact equation 6.40 is a standard form of nonlinear oscillator equation (e.g. Strogatz[55]). Figure 6.27 illustrates the graph of ϕ' versus ϕ for several values of μ. When $\mu = 0$, there is one stable fixed point at $\phi = 0$ and one unstable fixed point at $\phi = \pi/k$. How do I know they're stable and unstable? When ϕ goes negative, its speed ϕ' increases and an orbit moves more rapidly back toward zero, and vice versa for positive ϕ. Points intersecting their derivative at a negative slope are stable. Points intersecting their derivative at a positive slope are unstable. In figure 6.27 I've labeled stable fixed points as filled circles and unstable fixed points as open circles. The attracting solution at $\mu = 0$ is $\phi = 0$. There is no phase difference between Θ and θ: they are locked together. When $0 < \mu < 1$, the attracting solution shifts to $\phi > 0$. There is a solution, but it has a fixed phase difference between the forcing and response. The forcing and response are still locked, just now they do not happen at the same moment. When $\mu > 1$, there is no solution, and the forcing and response unlock. Similar results are obtained for $\mu < 0$. Phase-locked solutions exist for μ in the range $-1 < \mu < 1$ or

$$-1 < \frac{\dot{\Theta} - \omega}{A_k} < 1, \tag{6.42}$$

which, with $\dot{\Theta} = \Theta/\tau$, becomes

$$\tau(\omega - A_k) < \Theta < (\omega + A_k)\tau. \tag{6.43}$$

In other words, this says the width of the phase-locked region of Θ expands linearly with increasing τ. This is why resonance tongues exist as they do.

6.6. Summary

Laminar mixing and transport is essential in many applications, and chaotic advection along with tools from nonlinear dynamics is often the only way to enhance or analyze laminar flow transport. Examples and suggestions of application areas for fluid chaos are given throughout this chapter.

I have discussed basic features and organizing ideas of chaotic advection and transport in planar flows. Due to the combination of kinematics and continuity, steady planar flows must mix poorly; the flow domain is partitioned by the stable and unstable manifolds of hyperbolic and elliptic points, over which material cannot travel. Transport comes about via the time-dependence of the flow such that streamlines cross at successive times. This generates a self-kneading and self-refining motion around unstable manifolds of hyperbolic points, which are the "highways" along which transport takes place. The manifold tangles around homo- and heteroclinic connections generate transport, but the transport may be limited in extent. The generic picture of a chaotic advection field has interwoven regions of regular and chaotic motions. The regular regions are elliptic islands that are barriers to transport, while the chaotic regions are well-mixed by the tangles of unstable hyperbolic manifolds. The extent of mixing varies as control parameters of the flow change the stirring.

As an example, I explored in some detail mixing and scalar advection-diffusion transport with the class of PRFs, which is often simple to implement for applications, yet imposes, via its symmetries, a rich structure onto chaotic transport that includes generating sub-harmonic resonance "tongues" in the flows' control parameter space, spatial symmetry locking,

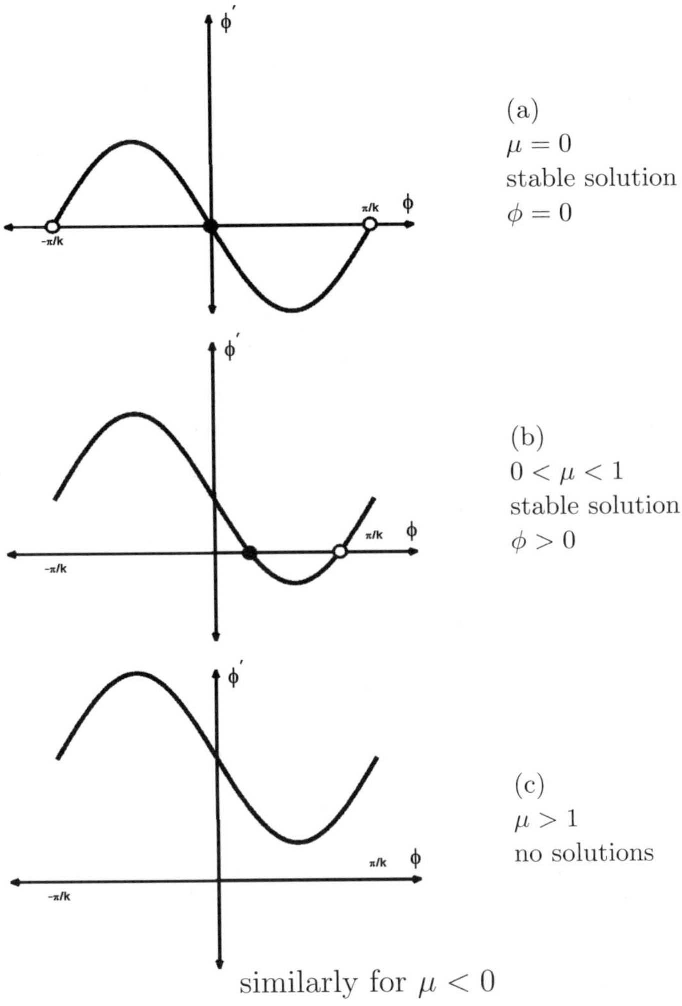

Fig. 6.27. Graphical solutions for the nonlinear oscillator equation 6.40 for several values of the bifurcation parameter μ. Filled (open) circles are stable (unstable) fixed points. Similar behavior is obtained for μ negative.

and order-disorder transitions. These structures led to fractal distributions of transport rates in the control parameter space of PRFs. This has important implications for designing applications because it means that small variations in parameters can lead to large changes in performance. Control parameters of the flow feed back into qualitative and quantitative transport properties. Understanding how the transport varies over the *entire* accessible control parameter space is the only way to guarantee designs that are close to optimum.

Acknowledgments

My research activities are supported within CSIRO by the Advanced Engineered Components Theme, the Minerals Down Under Flagship, and the Complex Systems Science initiative. I have benefited over the years from close interaction with many colleagues at CSIRO, particularly Dr Murray Rudman, Dr Michel Speetjens (now at the University of Eindhoven) who also gave a critical reading to a draft of the manuscript, Dr Daniel Lester, who also provided figures 6.20–6.22 and 6.25, Dr Richard Manasseh, Dr Yonggang Zhu, Dr Karolina Petkovic-Duran, Dr Kurt Liffman, and Dr Lachlan Graham. Ross Hamilton, Dean Harris, Tony Kilpatrick, Tony Swallow, and Robert Stewart have provided technical support for our group's experiments. Professor Julio Ottino of Northwestern University provided improving comments on a draft of these notes. Thank you to all.

References

1. J. M. Ottino, Mixing, chaotic advection, and turbulence, *Annual Review of Fluid Mechanics.* **22**, 207–253, (1990).
2. R. Klages, *Microscopic Chaos, Fractals and Transport in Nonequilibrium Statistical Mechanics.* vol. 24, *Advanced Series in Nonlinear Dynamics*, (World Scientific, Singapore, 2007).
3. A. Fridman, M. Marov, and R. Miller, Eds., *Observational Manifestation of Chaos in Astrophysical Objects.* (Springer, 2003).
4. P. Tho, R. Manasseh, Y. Zhu, and G. Metcalfe. Method for microfluidic mixing and mixing device. Patent Numbers WO2006105616-A1; AU2006230821-A1, (2006).
5. J. M. Ottino and S. Wiggins, Eds., *Transport and Mixing at the Microscale*, vol. A362, pp. 923–1129. Philosophical Transactions of the Royal Society of London, (2004).
6. A. Tsuda, R. A. Rogers, P. E. Hydon, and J. P. Butler, Chaotic mixing deep in the lung, *Proceedings of the National Academy of Science.* **99**(15), 10173–10178, (2002). www.pnas.org/cgi/doi/10.1073/pnas.102318299.
7. S. Haber, J. Butler, H. Brenner, I. Emanuel, and A. Tsuda, Shear flow over a self-similar expanding pulmonary alveolus during rhythmical breathing, *Journal of Fluid Mechanics.* **405**, 243–268, (2000).
8. P. Dames et al., Targeted delivery of magnetic aerosol droplets to the lung, *Nature Nanotechnology.* **2**, 495–499, (2007).
9. C. Plank, Nanomagnetosols: Magnetism opens up new perspectives for targeted aerosol delivery to the lung, *Trends in Biotechnology.* **26**(2), 59–63, (2008).

10. R. Behra and H. Krug, Nanoecotoxicology: Nanoparticles at large, *Nature.* **3**, 253–254, (2008).
11. J. M. Ottino, G. Metcalfe, and S. Jana. Experimental studies of chaotic mixing. In eds. W. Ditto, L. Pecora, M. Shlesinger, M. Spano, and S. Vohra, *Proceedings of the 2nd Experimental Chaos Conference*, pp. 3–20. World Scientific, (1995).
12. S. Kim and T. Kwon, Enhancement of mixing performance of single-screw extrusion processes via chaotic flows, *Advances in Polymer Technology.* **15**, 41–54, (1996).
13. S. Jana, E. Scott, and U. Sundararaj. Single extruder screw for efficient blending of miscible and immiscible polymeric materials. Patent number US 6132076, (1998).
14. M. Tjahjadi and F. William. Single screw extruder capable of generating chaotic mixing. Patent Number EP 0 644 034 B1, (1999).
15. S. C. Jana and M. Sau, Effects of viscosity ratio and composition on the development of morphology in chaotic mixing of polymers, *Polymer.* **45**, 1665–1678, (2004).
16. D. A. Zumbrunnen and O. Kwon. Chaotic mixing method and structured materials formed therefrom. United States Patent 6 770 340 B2, (2004).
17. D. A. Zumbrunnen, R. Subrahmanian, B. Kulshreshtha, and C. Mahesha, Smart blending technology enabled by chaotic advection, *Advances in Polymer Technology.* **25**(3), 152–169, (2006).
18. B. McNeil and L. Harvey, Viscous fermentation products, *Critical Reviews in Biotechnology.* **13**, 275–304, (1993).
19. C. Elias and J. Joshi. Role of hydrodynamic shear on activity and structure of proteins. In ed. T. Scheper, *Advances in Biochemical Engineering/Biotechnology*, vol. 59, pp. 47–71. Springer, (1998).
20. G. Metcalfe, M. Rudman, A. Brydon, L. Graham, and R. Hamilton, Composing chaos: An experimental and numerical study of an open duct mixing flow, *AIChE Journal.* **52**(1), 9–28, (2006). doi:10.1002/aic.10640.
21. T.-H. Hwu, D.-L. Young, and Y.-Y. Chen, Chaotic advections for Stokes flows in a circular cavity, *Journal of Engineering Mechanics.* pp. 774–782 (August, 1997).
22. L. Bresler, T. Shinbrot, G. Metcalfe, and J. M. Ottino, Isolated mixing regions: Origin, robustness and control, *Chemical Engineering Science.* **52**, 1623–1636, (1997).
23. A. Omurtag, V. Stickel, and R. Chevray. Chaotic advection in a bioengineering system. In ed. Y. Lin, *Proceedings of the 11th Engineering Mechanics Conference.* American Society of Civil Engineers, (1996).
24. T. Shinbrot, M. Alvarez, J. Zalc, and F. Muzzio, Attraction of minute particles to invariant regions of volume preserving flows by transients, *Physical Review Letters.* **86**(7), 1207–1210, (2001).
25. J. H. E. Cartwright, M. O. Magnasco, O. Piro, and I. Tuval, Bailout embeddings and neutrally buoyant particles in three-dimensional flows, *Physical Review Letters.* **89**(26), 264501 (Dec, 2002). doi: 10.1103/PhysRevLett.89.264501.

26. M. Speetjens, G. Metcalfe, and M. Rudman, Topological mixing study of non-Newtonian duct flows, *Physics of Fluids.* **18**(103103), 1–11, (2006).
27. D. Lester, M. Rudman, G. Metcalfe, and H. Blackburn, Global parametric solutions of scalar transport, *Journal of Computational Physics.* **227**, 3032–3057, (2007). doi:10.1016/j.jcp.2007.10.015.
28. M. Singh, P. Anderson, M. Speetjens, and H. Meijer, Optimizing the Rotated Arc Mixer (RAM), *AIChE Journal.* **54**(11), 2809–2822, (2008).
29. D. Lester, M. Rudman, and G. Metcalfe, Low Reynolds number scalar transport enhancement in viscous and non-Newtonian fluids, *International Journal of Heat and Mass Transfer.* **52**, 655–664, (2009). doi:10.1016/j.ijheatmasstransfer.2008.06.039.
30. M. Singh. *Design, Analysis, and Optimization of Distributive Mixing with Applications to Micro and Industrial Flow Devices.* PhD thesis, Technische Universiteit Eindhoven, The Netherlands, (2008).
31. P. D. Swanson and J. M. Ottino, A comparative computational and experimental study of chaotic mixing of viscous fluids, *Journal of Fluid Mechanics.* **213**, 227–249, (1990).
32. M. F. M. Speetjens, H. J. H. Clercx, and G. J. F. van Heijst, A numerical and experimental study of advection in three-dimensional Stokes flows, *Journal of Fluid Mechanics.* **514**, 77–105, (2004).
33. J. M. Ottino, *The Kinematics of Mixing: Stretching, Chaos, and Transport.* (Cambridge University Press, 1989).
34. G. Metcalfe and D. Lester. Mixing and heat transfer of highly viscous foods with the rotated arc mixer (RAM): Low energy use, low work input and a high degree of control. In ed. B. Ottoway, *International Review of Food Science and Technology*, International Union of Food Scientists, pp. 57–64. GIT Verlag, (2008).
35. D. L. Turcotte, *Fractals and Chaos in Geology and Geophysics.* (Cambridge University Press, 1997), 2nd edition.
36. G. Metcalfe, C. Bina, and J. M. Ottino, Kinematic considerations for mantle mixing, *Geophysical Research Letters.* **22**, 743–746, (1995).
37. W. Gorczyk, T. V. Gerya, J. A. D. Connolly, D. A. Yuen, and M. Rudolph, Large-scale rigid-body rotation in the mantle wedge and its implications for seismic tomography, *Geochemistry, Geophysics, Geosystems.* **7**(5), Q05018, (2006). doi:10.1029/2005GC001075.
38. J. Grooß, P. Konopka, and R. Müller, Ozone chemistry during the 2002 Antarctic vortex split, *Journal of the Atmospheric Sciences.* **62**, 860–870, (2005).
39. K. Ngan and T. G. Shepherd, A closer look at chaotic advection in the stratosphere. part I: Geometric structure. part II: Statistical diagnostics, *Journal of the Atmospheric Sciences.* **56**, 4134–4166, (1999).
40. S. Wiggins, The dynamical systems approach to Lagrangian transport in oceanic flows, *Annual Review of Fluid Mechanics.* **37**, 295–328, (2005). doi:10.1146/annurev.fluid.37.061903.175815.
41. K. V. Koshel and S. V. Prants, Chaotic advection in the ocean, *Physics—Uspekhi Fizicheskikh Nauk.* **176**(11), 1177–1206, (2006).

42. S. Wiggins, *Chaotic Transport in Dynamical Systems.* (Springer-Verlag, 1992).
43. S. Jana, G. Metcalfe, and J. M. Ottino, Experimental and computational studies of mixing in complex Stokes flows: The vortex mixing flow and multicellular cavity flows, *Journal of Fluid Mechanics.* **269**, 199–246, (1994).
44. M. Horner, G. Metcalfe, S. Wiggins, and J. M. Ottino, Transport mechanisms in open cavities: Effects of transient and periodic boundary flows, *Journal of Fluid Mechanics.* **452**, 199–229, (2002).
45. G. Falkovich. Introduction to developed turbulence. In eds. M. Shats and H. Punzmann, *Turbulence and Coherent Structures in Fluids, Plasmas and Nonlinear Media*, vol. 4, *Lecture Notes in Complex Systems*, pp. 1–20. World Scientific, (2006).
46. R. Sturman, J. M. Ottino, and S. Wiggins, *Mathematical Foundations of Mixing.* (Cambridge University Press, 2006).
47. H. Aref, The development of chaotic advection, *Physics of Fluids.* **14**, 1315–1325, (2002).
48. H. A. Kusch and J. M. Ottino, Experiments on mixing in continuous chaotic flows, *Journal of Fluid Mechanics.* **236**, 319–348, (1992).
49. T. Tél, A. de Moura, C. Grebogi, and G. Károlyi, Chemical and biological activity in open flows: A dynamical system approach, *Physics Reports.* **413**, 91–196, (2005).
50. R. T. Pierrehumbert, Tracer microstructure in the large-eddy dominated regime, *Chaos, Solitons & Fractals.* **4**(6), 1091–1110, (1994).
51. D. Rothstein, E. Henry, and J. P. Gollub, Persistent patterns in transient chaotic fluid mixing, *Nature.* **401**, 770–772, (1999).
52. W. Liu and G. Haller, Strange eigenmodes and decay of variance in the mixing of diffusive tracers, *Physica.* **D188**, 1–39, (2004).
53. J. Gollub (ed.). Research in fluid dynamics: Meeting national needs. Technical report, U.S. National Committee on Theoretical and Applied Mechanics, www.usnctam.org, (2006).
54. J. A. Glazier and A. Libchaber, Quasi-periodicity and dynamical systems: An experimentalist's view, *IEEE Transactions on Circuits and Systems.* **35** (7), 790–809, (1988).
55. S. H. Strogatz, *Nonlinear Dynamics and Chaos.* (Westview Press, 1994).
56. G. Metcalfe and D. Lester, Mixing and heat transfer of highly viscous food products with a continuous chaotic duct flow, *Journal of Food Engineering.* **95**, 21–29, (2009). doi:10.1016/j.jfoodeng.2009.04.032.
57. R. F. Weinberg, G. Mark, and H. Reichardt, Magma ponding in the Karakoram shear zone, Ladakh, NW India, *Geological Society of America Bulletin.* **121**(1–2), 278–285 (January, 2009). doi:10.1130/B26358.1, and the photo is also available at www.earth.monash.edu.au/~weinberg.

Chapter 7

Approaches to Modelling the Dynamical Activity of Brain Function Based on the Electroencephalogram

David T. J. Liley* and Federico Frascoli†

Brain Sciences Institute,
Swinburne University of Technology, Hawthorn, Victoria 3122, Australia

The brain is arguably the *quintessential* complex system as indicated by the patterns of behaviour it produces. Despite many decades of concentrated research efforts, we remain largely ignorant regarding the essential processes that regulate and define its function. While advances in functional neuroimaging have provided welcome windows into the coarse organisation of the neuronal networks that underlie a range of cognitive functions, they have largely ignored the fact that behaviour, and by inference brain function, unfolds dynamically. Modelling the brain's dynamics is therefore a critical step towards understanding the underlying mechanisms of its functioning. To date, models have concentrated on describing the sequential organisation of either abstract mental states (functionalism, hard AI) or the objectively measurable manifestations of the brain's ongoing activity (rCBF, EEG, MEG). While the former types of modelling approach may seem to better characterise brain function, they do so at the expense of not making a definite connection with the actual physical brain. Of the latter, only models of the EEG (or MEG) offer a temporal resolution well matched to the anticipated temporal scales of brain (mental processes) function. This chapter will outline the most pertinent of these modelling approaches, and illustrate, using the electrocortical model of Liley *et al*, how the detailed application of the methods of nonlinear dynamics and bifurcation theory is central to exploring and characterising their various dynamical features. The rich repertoire of dynamics revealed by such dynamical systems approaches arguably represents a critical step towards an understanding of

*dliley@swin.edu.au
†ffrascoli@swin.edu.au

the complexity of brain function.

Contents

7.1. Introduction

The brain is arguably the *quintessential* complex system as indicated by the patterns of behaviour it produces. The whole of human culture is a testament to such complexity. Yet despite many decades of concentrated research efforts we remain largely ignorant regarding the essential processes that regulate and define its function. While advances in functional neuroimaging (such as functional magnetic resonance imaging and positron emission tomography) have provided welcome windows into the gross organisation of the neuronal networks that underlie a range of cognitive functions, they have, arguably, largely ignored the fact that patterns of behaviour and brain activity are notable principally for how they unfold spatio-temporally. Recorded activity in the central nervous system (CNS) spans more than seven orders of magnitude in space and time (see Figure 7.1) from the behaviour of individual channels and receptors up to the bulk electrical and haemodynamic activity of the whole brain. In general the dynamical phenomena recorded from the CNS can be conceptualised as occurring at three distinct spatial scales:

(1) *Macroscopic*: Activity at this scale is typically recorded with centimetric resolution and with varying temporal precision. For example, the bulk electrical activity of cerebral cortex, recorded using electroencephalography (EEG) or magnetoencephalography (MEG) typically has a spatial resolution of the order of a cubic centimetre and a temporal resolution of milliseconds. In contrast, measurements of the changes

in cerebral blood flow using fMRI and PET that attend cognition, while having similar (if somewhat better) spatial resolution, have vastly inferior temporal resolution of the order of seconds. EEG, MEG and the various haemodynamic and metabolic measures reflect the activity of large contiguous neuronal populations.

(2) *Mesoscopic*: By using invasive recording strategies CNS population neuronal activity can be recorded with spatial resolutions of the order of a cubic millimetre. Because of the different recording configurations, this scale reveals a range of dynamical phenomena not seen in the corresponding non-invasive macroscopic recordings. For instance, electrical recordings made with electrodes placed within (local field potential, LFP) or on the surface (electrocorticogram, ECoG) of the brain are capable of revealing a range of dynamical phenomena that includes the summed spiking activity of many tens of neurons (multi-unit activity, MUA).

(3) *Microscopic*: This scale typically corresponds to the recorded activity of single neurons and individual ion channels. Phenomena measured include axonal and dendritic action potentials, single channel ligand and voltage gating, postsynaptic potentials and their activity modulated long term potentiation (LTP) and long term depression (LTD).

Can coherent theories of brain activity be developed that bridge these multiple temporal and spatial scales? Logic would seem to dictate that activity at all these scales would be pertinent to understanding brain function and thus any theory purporting to describe brain function should incorporate and reveal activity at these various levels. Nonetheless this provides an ambiguous basis for the development of a specific theory regarding brain function.

A decision has to be made regarding what features of brain function can be meaningfully modelled. Ideally these features should be i) quantifiable and non-invasively measurable ii) have a time scale commensurate with that of cognition and behaviour and iii) have been empirically shown to be sensitively correlated with behaviour (normal and abnormal). Currently the only objectively measurable feature of brain activity that fulfils these criteria is the EEG, and possibly the MEG—though the biophysical generators of the latter are somewhat more uncertain.

The electrical potentials recorded from the scalp, the EEG, or the cortical surface, the ECoG, are generated by the current flowing in response

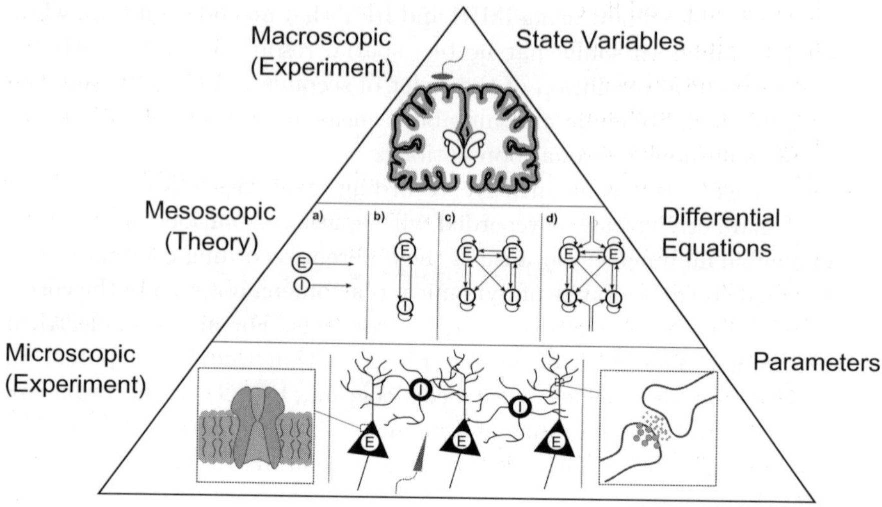

Fig. 7.1. Dynamic phenomena in the brain can be thought of as occurring on three distinct, but interrelated, spatial scales. Bulk or macroscopic neural activity is typically defined over a linear spatial scale of the order of centimetres. Exemplary activity recorded at this scales includes bulk electrophysiological measures such as the EEG or MEG, and haemodynamic measures such as BOLD (blood oxygen level dependent) MRI (functional MRI, fMRI). Activity at the microscopic level corresponds to activity at the cellular and subcellular/molecular scales. Activity recorded at this level of spatial resolution typically arises as a consequence of the flow of transmembrane ionic current. The corresponding phenomena include the nerve action potential and the postsynaptic potentials induced by neurotransmitter action. The typical linear scale of these phenomena are of the order of micrometres. An intermediate scale can also be defined which typically corresponds to the activity of populations of neurons. Activity recording at this scale reveals dynamical phenomena not seen in macroscopic or microscopic neural recordings. Theories of brain function are often constructed at this mesoscopic level (though of course they can be constituted at any level) as they can act as a bridge between the physiological properties of single neurons and the macroscopic manifestations of behaviour. Figure reproduced from Ref. 1.

to the combined activity of many thousands of cortical neurons. When recorded from the scalp they typically have an amplitude of the order of 10–50 μV RMS and possess a range of spectral components that span frequencies from about 0.1 Hz up to 40–60 Hz—though typically most spectral power ($> 90\%$) resides at below 20 Hz in an awake and restful state. While the EEG can be sampled with arbitrary precision, experience indicates that events having a temporal extent of the order of milliseconds can be successfully resolved, which appears to be more than sufficient to resolve any corresponding cognitive processes. For many decades, until the advent of a range of functional neuroimaging modalities, the EEG was a mainstay method of cognitive neuroscience because of its sensitive, and often specific, correlation with a range of normal and abnormal/pathological neurological states. For example the EEG shows a clear progression of states during sleep (that forms the basis of sleep stages and staging) that are in part similar to those observed during the action of a range of anaesthetic agents. The most significant dynamical feature of the EEG, and arguably of brain activity, is undoubtedly the well known, but poorly understood, alpha rhythm. The alpha rhythm appears as near sinusoidal fluctuations in the waking EEG/ECoG within the frequency range of 8-13 Hz and is regarded as the archetypal cerebral rhythm as it is cortically widespread and regionally attenuated by a diverse range of specific and non-specific stimuli and behaviours. For example, states of wakeful restfulness in which the eyes are closed enhance it, whereas the opening of the eyes as well as a range of cognitive and motor activity can topographically selectively attenuate it.

For these reasons, and the fact that the biophysical generators of the EEG are known with some certainty, many attempts to develop general theories of brain function have started with approaches aimed at modelling the dynamics of the EEG. Given that the EEG arises as a consequence of the flow of synaptic currents in the dendrites of neurons (see Figure 7.2) it would seem that the most logical approach by which to model the dynamics of EEG would be to consider the behaviour of large networks of interconnected model neurons. Given our extensive knowledge of the electrical properties of neuronal membranes, particularly in terms of the mechanisms underlying the generation of the action potential, this would seem to be the most veracious, and thereby fruitful, path to follow. However action potentials are highly nonlinear phenomena. Indeed decades of detailed empirical research has taught us that virtually all important aspects of neural function are inherently nonlinear. Thus any attempts to model networks of physiologically realistic neurons at a scale sufficient to represent the EEG would result in coupled nonlinear systems of extremely high dimension. Such systems would be impossible to mathematically analyse, let alone parameterise. The resulting dynamics of these systems would likely be as inscrutable as the very systems that they were meant to be providing insight into. Therefore alternative strategies to attacking this neural modelling problem are required.

Because the goal of modelling complex systems, such as the brain, is the generation of insight into the functional significance of the emergent dynamics, some form of simplification is frequently of critical importance.

For systems comprised of a large number of elements such simplifications generally depend upon defining ensemble states (macrostates) that are in some sense analogous to thermodynamic variables that describe the overall properties of a gas (e.g. temperature, entropy and pressure). Thus from the perspective of modelling the EEG this means replacing individual cell properties and their interactions by a range of continuous functions that depend on some form of spatial averaging. Therefore neural, and in particular cortical, tissue is to be viewed as a spatial continuum, for it can be readily argued that the huge number of neurons in human cortex ($\approx 10^{10} - 10^{11}$), and the enormously high density of synaptic connections ($\approx 10^9 \text{mm}^{-3}$), make a continuum description of the dynamics of populations of neurons, at the spatial scale resolvable by the EEG (≈ 1 cm) compelling. By effecting such a coarse-graining it becomes possible to develop parsimonious theories of the dynamics of EEG which incorporate the bare minimum of physiological and anatomical detail, that despite their apparent biological simplicity are capable of producing a rich range of physiologically and be-

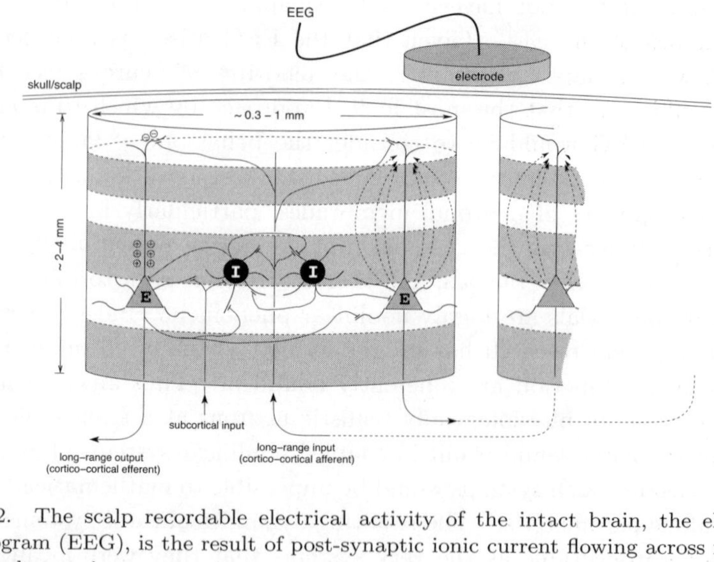

Fig. 7.2. The scalp recordable electrical activity of the intact brain, the electroencephalogram (EEG), is the result of post-synaptic ionic current flowing across neuronal cell membranes in response to temporally synchronous presynaptic neuronal activity. Electric field lines, drawn directed along the axis of the apical dendritic trees of excitatory (E) neurons (which comprise approximately 85% of all neurons in cortex) on the right, are the result of the inward flow of current due to excitatory synaptic activity in the distal portions of the apical dendrite. Such vertically oriented current dipoles make the greatest contributions to the surface-recordable electrocorticogram (ECoG) and EEG. Diagram not drawn to scale. Reproduced from Ref. 3.

haviourally plausible dynamical phenomena. The resulting theories, known variously as cortical field theories (CFT), neural field theories, mean field models (MFM) or mass action theories,[a] aim at being to brain function what the Hodgkin-Huxley equations are to single neuronal activity (see for example Ref. 2).

7.1.1. *Chapter outline*

The aim of this chapter is to provide an overview of the various macroscopic and mesoscopic approaches to understanding the dynamical complexity of brain function by modelling the EEG. Further, using a model familiar to the authors, we will illustrate how the methods of dynamical systems theory, nonlinear dynamics and the bifurcation of vector fields can be applied in order to theoretically investigate potential physiological and dynamical principles responsible for the emergence of dynamical complexity in brain function. It is important to emphasise that this chapter will illustrate the application of dynamical systems principles to understanding brain function based on modelling the EEG, focusing in particular on one theoretical approach. It is not intended as an overview of the *dynamical approach to neuroscience*[2] nor as a comprehensive review of the panoply of dynamical approaches to modelling brain function.[4]

Following a brief overview of the various macroscopic modelling approaches to modelling EEG we will discuss the various methods available to analyse the dynamical properties of these models. Because all current models of EEG are essentially nonlinear, we will focus particularly on the methods of nonlinear dynamics. This methodological introduction will then be followed by an in depth illustration of the application of these methods to the electrocortical model of Liley and co-workers. The focus on the model of Liley *et al* is not meant to imply that the theoretical mechanisms underlying cortical electrorhythmogenesis have been resolved. Indeed they still remain the subject of considerable speculation. Rather the reason for focusing on this theory is simply because it has been intensively studied using the methods and techniques of nonlinear dynamics.

[a]Finer distinctions may be drawn between these various types of approaches. Some may choose to characterise a mass action theory as that in which the constituting neural populations have been reduced to point masses with the corresponding neural activity remaining localised. In contrast a neural field theory neural activity is often characterised by spatially propagating fields of neural activity. In practise such distinctions are largely nominal, as most modern approaches to modelling EEG combine aspects of both, and therefore we will use the terms interchangeably.

7.2. Overview of Macroscopic Neural Modelling Approaches

In a cubic millimetre of mammalian cortex there is of the order of 20,000–100,000 neurons connected by about 3 kilometres of axonal fibre.[5] Because of these immense numbers it is easy to justify the construction of theories in which the constitutive neuronal networks of cortex are spatially continuous. We can then define a range of spatially continuous macroscopic variables that represent the bulk properties of a local region of cortex. Generally neural activity is averaged over roughly the extent of a cortical macrocolumn with the corresponding variables typically being the mean firing rates or the mean value of the cell (soma) membrane potential. The earliest models of neural mass action applied to the cortex that attempted to describe the spatial and temporal behaviour of these aggregate masses dealt mainly, if not exclusively, with excitatory interactions (e.g. Ref. 6). Later models incorporated inhibitory interactions, paying more attention to the anatomical topology of connections between these masses, and took into account the conversion of efferent axonal activity into afferent dendritic activity (and the converse process), dendritic integration, axonal dispersion and synaptic delays.

In particular two complementary macroscopic formulations have had a sustained influence on approaches to the large scale modelling of brain activity. The model of Wilson and Cowan[7,8] extended the early approaches[6] by considering two functionally distinct neuronal populations (excitatory and inhibitory) connected by all combinations of feed-forward and feed-back synaptic connectivity. The macroscopic state variables characterising the activity of the respective neuronal populations were the proportion of neurons of a particular type becoming active at a time t at x per unit time, i.e. the mean firing rate at time t and location x. The resulting formulation, known as the Wilson and Cowan equations, is in terms of a coupled pair of nonlinear integro-differential equations.

The other macroscopic formulation is attributable to Amari.[9,10] In contrast to that of Wilson and Cowan, which is an *activity based* model, the model of Amari[9,10] is a *voltage based* model, in which the macroscopic state variables are the mean soma membrane potential of suitably defined neuronal populations. While in his original formulation Amari considered only a single type of spatially distributed neuronal population interacting by local (proximal) excitation and distant (distal) inhibition, it can be trivially extended to the case of multiple interacting spatially continuous neuronal populations.

While both of these models have formed a staple diet for a range of biomathematical explorations, they have not been particularly successful in articulating the genesis of rhythmic activity in the EEG. This is attributable to a number of factors, the most important of which are those in which a range of mathematical simplifications are not adequately justified by an appeal to the physiology. For example both the models of Amari[9,10] and Wilson and Cowan[7,8] unrealistically assume that the effects of synaptic activity are felt instantaneously at the neuronal soma, whereas empirically these effects are well known to steadily increase to a maximum on a time scale typically characteristic of whether the synapse is excitatory or inhibitory. Nevertheless, physiologically more sophisticated approaches to modelling the bulk activity of neural tissue can be seen as extensions of the pioneering approaches of Wilson and Cowan[7,8] and Amari.[9,10] For this reason we will briefly outline these respective modelling approaches.

7.2.1. *The Wilson-Cowan neural field equations*

By assuming that cortical and thalamic neural tissue is comprised of two interacting, but functionally distinct, excitatory and inhibitory neuronal populations, Wilson and Cowan[7,8] were able to develop a continuum two state description of the dynamics of neural tissue. The state of their bulk neuronal population model neural tissue was defined in terms of the time-coarse-grained fraction of excitatory, $E(t)$, and inhibitory, $I(t)$, neurons firing per unit time. For point neural masses they were able to derive the following equations of motion for $E(t)$ and $I(t)$

$$\tau_E \frac{dE}{dt} = -E + (1 - r_E E) S_E[c_{EE}E(t) - c_{IE}I(t) + P(t)] \qquad (7.1)$$

$$\tau_I \frac{dI}{dt} = -I + (1 - r_I I) S_I[c_{EI}E(t) - c_{II}I(t) + Q(t)] \qquad (7.2)$$

where $\tau_{E,I}$ are nominally the membrane time constants of the respective neural populations and determine their characteristic response times to incoming activity. The respective neuronal absolute refractory periods are denoted by $r_{E,I}$. The connectivity coefficients $c_{EE,IE,EI,II} \geq 0$ quantify the respective neuronal population interactions, whereas the functions $S_{E,I}$ describe the relationship between neuronal population input (e.g. $c_{EE}E(t) - c_{IE}I(t) + P(t)$) and output in the absence of refractory effects. In general, because firing rates will be bounded below by a zero firing rate and above by a maximal firing rate, the $S_{E,I}$ are typically sigmoidal functions of their arguments i.e. $S_E \equiv 1/(1+\exp[-a(E-\theta_E)]$. $P(t)$ and $Q(t)$ define the external input to the excitatory and inhibitory sub-populations.

While no analytical solutions exist for these equations they can, like many other two-dimensional nonlinear systems, be analysed qualitatively in the phase plane. Because the lumped Wilson and Cowan formulation constitutes a planar (2-D) system, the only types of dynamics it is capable of supporting are steady states (fixed points) and limit cycles. By plotting the respective loci of points for which $dE/dt = 0$ and $dI/dt = 0$ and examining the nature of any intersections (steady states or fixed points) further insight can be gained into the qualitative properties of the emergent dynamics. The respective equations for the $dE/dt = 0$ and $dI/dt = 0$ isoclines are

$$c_{IE}I = c_{EE}E - S_E^{-1}\left(\frac{E}{1-r_E E}\right) + P \tag{7.3}$$

$$c_{EI}I = c_{II}E + S_I^{-1}\left(\frac{I}{1-r_I I}\right) - Q \tag{7.4}$$

where S_E^{-1} and S_I^{-1} are the inverse sigmoidal functions. Assuming that $S_{E,I}$ are indeed sigmoidal and $c_{EE,IE,EI,II} \geq 0$, the respective nullclines can be either monotone or cubic-like, and generically speaking there can be one, three or five steady states. Because there are an odd number of steady states, the number of stable and unstable steady states will be unequal. Typically there will be stable nodes/spirals, saddles and unstable nodes/spirals. In general, the stable manifold of the saddles will act as a boundary separating domains of stability associated with stable steady states. In addition to these fixed point attractors there will also exist limit cycle states, where topological considerations eliminate the possibility of chaotic solutions unless there is some periodic external driving.

A considerable body of work has been devoted to the analysis of equations 7.1 and 7.2, including phase plane analyses of the number, type and properties of the equilibria of the system, bifurcation analyses, and the behaviour of multiply coupled Wilson-Cowan type systems. While Wilson and Cowan made no special claims regarding the applicability of their model to understanding the EEG, they were able to numerically demonstrate the existence of sustained limit cycle activity and damped oscillatory behaviours in response to brief stimulating inputs for a range of suitably chosen parameters. The existence of these phenomena, while generic in planar systems, nevertheless suggested that neural populations models are relevant to understanding the genesis of evoked and spontaneous electroencephalographic phenomena such as the alpha rhythm and time locked, stimulus induced, average evoked potentials.

The spatially lumped Wilson and Cowan equations are easily extended to the case that the respective excitatory and inhibitory neuronal populations have a spatial extent. In this case $E(x,t)dx$ and $I(x,t)dx$ now define the fraction of neurons firing per unit time at time t within dx about x, and neuronal population input will now consist of spatially integrated activity weighted by some appropriately chosen, distance dependent, connectivity function. Numerical solutions to the resulting integro-differential equations reveal a range of additional phenomena that include the generation of propagating active transients, long lasting oscillatory responses and spatially inhomogeneous stable steady states. For an in depth review of this and related modelling approaches the reader is referred to Ref. 11.

7.2.2. *The lateral-inhibition type neural field theory of Amari*

In the Wilson and Cowan model, equations of motion for time averaged *neuronal firing rates* were derived. This and related models are therefore referred to as *activity based* models. However there also exists an alternative way of formulating mean field, or continuum models, referred to as *voltage based* models, which are arguably more pertinent to modelling, and thereby understanding, the genesis of EEG dynamics. In this modelling approach the resulting equations of motion describe the spatio-temporal evolution of the average membrane potential of neurons. It is empirically well established that the surface recorded EEG/ECoG is linearly related to perturbations in the average membrane potential of excitatory cortical (pyramidal) neurons.[12]

One of the biomathematically most influential *voltage based* continuum models of cortical dynamics is that of Amari.[9,10] In its most general form, this model considers, m distinct spatially distributed neuronal populations, in which the average membrane potential impulse response to incoming (axonal) input from other neuronal populations is $\propto \exp[-t/\tau]$. The resulting field equations can then be written as

$$\tau_i \frac{\partial u_i}{\partial t} = -u_i + \sum_{j=1}^{m} \int w_{ij}(x,x';t-t')f_j[u_j(x',t')]dx'dt' + s_i(x,t) \quad (7.5)$$

where $u_i(x,t)$ is the average membrane potential of neurons of type i at time t at x, $s_i(x,t)$ is external input and f_i is a nonlinear function that describes the average firing rate (action potential emission rate) of the i-th neuronal population as a function of u_i. The functions $w_{ij}(x,y;t)$ define the strength of connectivity between neuronal populations. In particular $w_{ij}(x,y;t)$ represents the magnitude of input to neurons of type i at x from neurons of type j at y at a time t before (thus incorporating the effects of conduction and synaptic delays).[b] In order to investigate the mathematical features and consequences of this theory, a number of simplifications are often made. Firstly the equations are reduced to a single scalar integro-differential equation by assuming a single, one dimensionally distributed, homogeneous neural population in which conduction and synaptic delays are negligible. Further, the nonlinear output function $f_i(u_i)$ is, for mathematical convenience, redefined to be the Heaviside step function, such that neurons fire at a maximum rate when the average membrane potential exceeds some threshold θ. Using these simplifications, equations 7.5 can be rewritten

$$\tau\frac{\partial u}{\partial t} = -u + \int w(x-y)H[u(y,t') - \theta]dx + s(x,t) \tag{7.6}$$

where $w(x-y) = w(x,y)$ by virtue of the homogeneity, and H is the Heaviside step function. The function $w(y)$ is often referred to as the *synaptic footprint*, and is often defined to be of lateral inhibition type such that it is excitatory ($w(y) > 0$ for some defined neighbourhood about the origin) for nearby neurons and inhibitory ($w(y) \leq 0$ for regions outside defined neighbourhood) for more distant neurons. This pattern of lateral inhibition connectivity is typically referred to as *Mexican hat connectivity*. Solutions to this scalar equation reveal, among other properties, that patterns of localised excitation, once evoked by stimulation, can be persistently retained in the deviations of u from rest after the stimulation has been removed. Such self-sustained states have been hypothesised to be the neuronal correlates of working memory. The lateral inhibition type connectivity is not a particularly effective model for modelling interacting populations of excitatory and inhibitory neurons, as the scalar fields generated are not capable of oscillating. Nevertheless the framework defined by equations 7.5

[b]In general there is an inconsistent use of subscripts in the literature. Some authors, typically from a physics background, choose to define the subscript order as *target, source*. Others, often from a more biological background, prefer to use *source, target* given we read from left to right. We have chosen not to enforce any consistency choosing to retain the authors original definitions.

is sufficiently general to enable the modelling of neural fields comprised of excitatory and inhibitory neurons. Doing so reveals, in addition to persistent (stationary) states of localised excitation, a range of phenomena that includes oscillatory waves, travelling waves and a range of active transient behaviours. For a detailed review of contemporary biomathematical explorations involving the Amari equations, the reader is encouraged to consult Ref. 13.

7.2.3. *EEG specific mean field models*

The modelling frameworks developed by Wilson and Cowan, and Amari, can be conceived as forming the basis for more elaborate theoretical approaches to modelling the EEG. While the majority of mean-field theories of EEG model the dynamics of at least two cortical neuronal populations (excitatory and inhibitory), details of the topology of connectivity, and the underlying physiological assumptions regarding the properties of the neural masses, can vary substantially. In this section we aim to summarise the essential features of a number of well-known theories in order to reveal the physiological degrees of freedom that need to be considered in order to generate theories of EEG having sufficient dynamical complexity and empirical merit. For further details the reader should consult Refs. 4, 14 and 15.

7.2.3.1. *Freeman's neural mass action and the K-set hierarchy*

Motivated by the desire to explain the electrocortical dynamics of the olfactory bulb and pre-pyriform cortex, Freeman[12] developed a hierarchy of neural interactedness—the well-known K set hierarchy, in which functionally differentiated populations of neurons are schematised to interact over progressively larger physical scales. The purpose of the K set hierarchy was to facilitate a systems oriented description of electrical dynamics of cortex. The simplest form of neural set that Freeman considered was the non-interactive set or KO set. Members of this set have a common source of input and a common sign of output (excitatory or inhibitory), but *do* not interact synaptically or by any other means with co-members. At this level the characteristic form of the neuronal population response to incoming activity is specified. Unlike the first-order response of the Wilson and Cowan (equations 7.1 and 7.2) or Amari models (equation 7.5) Freeman argued on the basis of detailed experiment that these population responses (or in his terminology *"pulse-to-wave"* conversion) could be described by

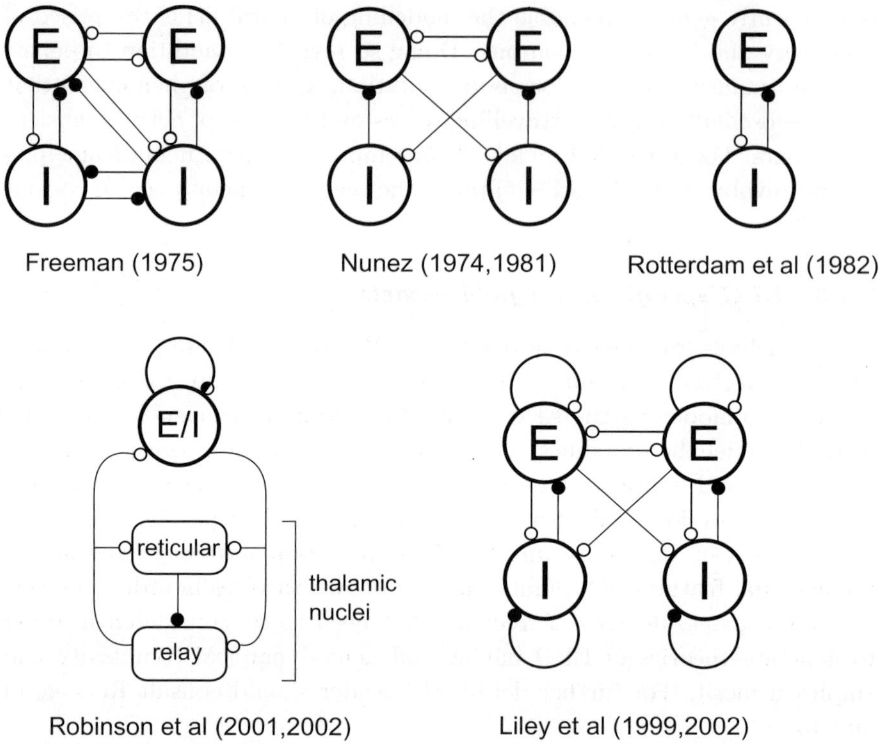

Fig. 7.3. Schematic outline of the connection topologies of a number of mean-field approaches to modelling the dynamics of EEG. All approaches consider the bulk interactions between two functionally differentiated neuronal populations. 'E' stands for excitatory, 'I' for inhibitory neuronal populations. Open circles represent excitatory connections, filled circles inhibitory ones.

third order linear, time invariant, differential equations.

The K-set hierarchy is next extended to include KO sets in which there is a non-zero level of functional interaction between members of the set. This defines the KI set, and is broadly divided into two types—the mutually excitatory KI_e set and the mutually inhibitory KI_i set. When there exists dense functional interaction between two KI sets, a KII set is formed. All possible interactions are allowed to occur between member KI sets and, locally, neocortex can be viewed as one or many interconnected KII sets. The K-set hierarchy can be extended to $KIII$ and KIV sets which nominally correspond to cortical areas and regions. The KII set of Freeman is equivalent physiologically and anatomically to the topology of cortex considered by Wilson and Cowan. Mathematically, the KII set is defined by 4 nonlinearly coupled sets of third order differential equations. The nonlinear couplings define how the induced population response to incoming synaptic activity is transduced into a neural population firing rate output. Freeman refers to the corresponding nonlinear function as the *"wave-to-pulse"* conversion function and has argued that such a function is an asymmetric sigmoid of the form $f(v) \propto \exp(-a\exp[-bv])$.[16]

This theoretical approach of Freeman has been widely applied to understanding bulk electrorhythmogenesis in paleocortex (olfactory bulb) and neocortex with experimentally constrained numerical simulations revealing a range of physiologically relevant and complex behaviour that includes the generation of gamma band (> 30 Hz), oscillatory activity and putative chaos[17].

7.2.3.2. *Nunez's global model of electrorhythmogenesis*

Broadly speaking, connectivity in cortex can be divided into short and long range. Short range connectivity typically occurs over scales of the order of millimetres, and represents the connectivity of excitatory and inhibitory neurons with each other due to the local branching of their unmyelinated axons within cortex. Because the corresponding axonal fibres always stay within cortex, such connectivity is typically referred to as *intracortical connectivity.* Long range connectivity, in contrast, occurs over scales of many centimetres in the human brain, and is due exclusively to the long range axons of excitatory (pyramidal) neurons. Because separated regions of cortex are synaptically connected with these long range fibres, such connectivity is referred to as *cortico-cortical connectivity.* However, due to the large distances over which these axons travel, they are expected to be associated with significant conduction delays. Nunez[18,19] has argued that these conduction delays are the principle determinants of the spectral features of human EEG. On this basis Nunez constructed a *global model of electrorhythmogenesis*, which in one form can be written as

$$h_E(x,t) = \int w_{EE}(v,x,x')E(x',t-\frac{|x-x'|}{v})\,dv\,dx' + p_E(x,t) \quad (7.7)$$

$$h_I(x,t) = \int w_{IE}(x,x')I(x',t)\,dx + p_I(x,t) \quad (7.8)$$

where $h_{E,I}$ are defined to be the number of *"active"* excitatory and inhibitory synapses per unit volume, respectively. $E(x',t)$ and $I(x',t)$ are respectively the number of excitatory and inhibitory action potential firings per unit time, whereas p_E and p_I are arbitrary external inputs. The velocity dependent propagation of excitatory activity by cortico-cortical fibres onto excitatory neurons is defined by $w_{EE}(v, x, x')$ and implicitly incorporates the possibility that the fibres connecting various regions have a distribution of conduction velocities. In contrast w_{IE} defines the intracortical distribution of local inhibitory activity onto excitatory neurons, which is independent of conduction velocity. These equations are closed by assuming particular forms for $E(x,t) \equiv E(h_E(x,t), h_I(x,t))$ and $I(x,t) \equiv I(h_E(x,t), h_I(x,t))$.

Equations 7.7 and 7.8 can be linearised and transformed into the Fourier domain to define a range of theoretical *electrocortical* dispersion relationships, which can then be compared with an empirical frequency-wavenumber spectra obtained using multi-electrode EEG recordings. Further, by incorporating a number of additional assumptions, equations 7.7 and 7.8 can be transformed into a set of nonlinear partial differential equations[20] which admit analytical and numerical solutions that reproduce a range of empirical electrocortical phenomena.

7.2.3.3. *Lopes da Silva's local model of the alpha rhythm*

The ability of the global resonance model of Nunez[18,19,21,22] to be a dynamically relevant theory for the genesis of rhythmic activity in the EEG rests on the assumption that the neuronal population response to incoming spike activity is, on the time scale of any conduction delays, effectively instantaneous. However, the varying time scales of neurotransmitter mediated neuronal interaction and the existence of the subsequent finite dendritic cable delays virtually eliminates the possibility that this will ever be the case. Indeed, experiment suggests that the response of the neuron membrane potential (and by inference the average membrane potential of a neuronal population) to incoming pre-synaptic spikes is at the very least second order: membrane potential rises to a peak and then decays away (in both cases with characteristic time courses).[23] These *"impulse"* responses are referred respectively to as excitatory postsynaptic potentials (EPSP) or inhibitory postsynaptic potentials (IPSP) depending on whether the spike arose from an excitatory or inhibitory neuron, and can have characteristic time courses of the same order or greater than any axonal conduction delays.

On this basis Lopes da Silva *et al*[24,25] constructed a bulk model of the

EEG in which lumped or spatially distributed populations of excitatory and inhibitory neurons synaptically interacted with each other via EPSPs and IPSPs having the form $PSP(t) = V_{PSP} t \exp[-t/\tau_{PSP}]$, and where the relationship between mean neuronal population firing rate was a nonlinear (sigmoidal) function of the neurons' average membrane potential. Thus, unlike the approaches of Wilson and Cowan[7,8] and Amari,[10] the response of a neuronal population to an incoming spike is not instantaneous. The linear analysis of the suitably parameterised model reveals a strong resonance in the alpha band range (8 − 13 Hz), as well as exhibiting a range of characteristics that are in qualitative agreement with experimental data on alpha rhythmic activity recorded from the dog.[26]

7.2.3.4. *Robinson's thalamocortical reverberation model*

Prior to Robinson *et al*[27] all bulk models of cortical electrorhythmogenesis had assumed that the EEG emerged through the reverberant interactions between at least two spatially distributed, functionally differentiated, cortical or thalamic, neuronal populations. However there exists significant reciprocal connectivity between cortex and the subcortical structure that determines and controls its input, the thalamus. Robinson *et al*[27] have argued that the inclusion of such cortico-thalamic feedback in a bulk or mean-field theory is crucial in order to plausibly model the essential dynamical features of normal (e.g. alpha rhythm) and abnormal (e.g. spike wave epilepsy) EEG activity.[28,29]

7.2.3.5. *Liley's coupled macrocolumnar continuum theory*

The theory of Liley *et al*[3,30–32] incorporates many of the features of the above models as well as including a few additional features aimed at achieving greater physiological veracity. It includes the effects of long range axonal conduction delays, the second/third order postsynaptic *"impulse"* response associated with neurotransmitter kinetics and the dendritic cable properties, as well as the effects of the synaptic reversal potentials which make the amplitude of the respective PSPs dependent on the ongoing postsynaptic/somatic membrane potential. Since we will illustrate the dynamical potential of mean field approaches to modelling brain function using this theory we describe it in more detail than the previously discussed approaches.

In this theory cortical activity is locally described by the mean soma membrane potentials of a spatially distributed excitatory neuronal population, h_e, and a spatially distributed inhibitory neuronal population, h_i. The connection with physiological measurement is through h_e, which uncontroversially is assumed to be linearly related to the scalp recordable EEG.[12] Excitatory and inhibitory neuronal populations are modelled as single passive RC compartments into which all PSPs terminate. Thus the response of the mean soma membrane potential h_k ($k = e, i$) to synaptic inputs I_{lk} is given by

$$\tau_k \frac{\partial h_k}{\partial t} = h_k^r - h_k + \sum_{l=e,i} \frac{h_{lk}^{eq} - h_k}{|h_{lk}^{eq} - h_k^r|} I_{lk} \tag{7.9}$$

where h_k^r is the resting mean soma membrane potential, τ_k the mean membrane time constant and I_{jk} the PSP inputs. Double subscripts indicate first source and then target, thus, for example, I_{ei} indicates PSP inputs from an excitatory to an inhibitory population. Note the PSP inputs, which correspond to transmitter activated postsynaptic channel conductance, are weighted by the respective ionic driving forces $h_{jk}^{eq} - h_k$, where h_{jk}^{eq} are the respective *synaptic reversal potentials*. This follows the *conductance* based approaches typically used to model networks of synaptically interacting networks of individual model neurons.[33,34] The PSPs I_{lk} can be traced to two essential sources: short-range and long-range feedback and feedforward excitation (I_{ee}, I_{ei}) and local feedforward and feedback inhibition (I_{ie}, I_{ii}). The time course of the PSP is described by a critically damped oscillator driven by the mean rate of incoming excitatory or inhibitory axonal pulses.[c] Thus for EPSPs and IPSPs we have respectively

$$\left(\frac{\partial}{\partial t} + \gamma_{ek}\right)^2 I_{ek} = \Gamma_{ek}\gamma_{ek} e \left[N_{ek}^{\beta} S_e(h_e) + p_{ek} + \phi_{ek}\right] \tag{7.10}$$

$$\left(\frac{\partial}{\partial t} + \gamma_{ik}\right)^2 I_{ik} = \Gamma_{ik}\gamma_{ik} e \left[N_{ik}^{\beta} S_i(h_i) + p_{ik}\right] \tag{7.11}$$

where the terms in square brackets on the RHS correspond to the different sources of incoming axonal pulses: $N_{lk}^{\beta} S_l(h_l)$ are axonal pulses arriving from local feedback or feedforward excitation or inhibition, p_{lk} represent extracortical (thalamic) sources and ϕ_{ek} are those axonal pulses arriving from the exclusively excitatory cortico-cortical (long range) fibre

[c]This is the differential form for the well-known *alpha function* ($t \exp[-at]$) used in dendritic cable theory to model the time course of individual PSPs.[35] Such a form has been used by Lopes da Silva *et al* (see 7.2.3.3) in their theory of EEG generation.

systems. The N_{lk}^{β} quantify the strength of anatomical population connectivity, whereas the Γ_{lk} are peak PSP amplitudes of type l on cells of type k occurring at time $1/\gamma_{lk}$ after the arrival of the presynaptic spike. The functions S_k convert the mean membrane potential of the neuron populations to an equivalent mean firing rate, and are given by the sigmoid $S_k(h_k) = S_k^{max}\{1 + \exp[\sqrt{2}(h_k - \mu_k)/\sigma_k]\}^{-1}$. The transmission of axonal pulses ϕ_{ek} is assumed to occur by long range fibre systems having a single conduction velocity, v_{ek}, and an exponential fall off in the strength of connectivity with increasing distance between source and target populations

$$\left(\frac{\partial}{\partial t} + v_{ek}\Lambda_{ek}\right)^2 \phi_{ek} - \frac{3}{2}\nabla^2\phi_{ek} = N_{ek}^{\alpha}v_{ek}^2\Lambda_{ek}^2 S_e(h_e) \qquad (7.12)$$

where $1/\Lambda_{ek}$ is the characteristic scale for the fall off in cortico-cortical connectivity. This damped wave equation (or telegraph equation) is not unique to this theory. In particular Robinson *et al*[27,36] and Jirsa and Haken[20] have both used an inhomogeneous wave equation in order to spatiotemporally propagate axonal activity.

Collectively equations 7.9 -7.12 are referred to as the Liley equations, and are capable of reproducing the main features of spontaneous human EEG. In particular, autonomous limit cycle and chaotic oscillatory activity in the alpha band (8 – 13 Hz) is easily produced, as we will see shortly.

7.3. Linearisation and Parameter Space Search

One of the major advantages in constructing computational models of cortical activity is clearly the possibility of applying existing analytical and numerical methods to extract valuable information on known (and unknown) dynamical regimes associated with the brain. This represents the essence of theoretical computational modelling in neurosciences. Since there exists comprehensive and detailed analysis for the model by Liley described in the previous section, it is worthwhile discussing some of the techniques that have been used in the investigation of diverse topics such as the generation of alpha rhythm, epilepsy, anaesthesia and the dynamical complexity of the brain.

A straightforward approach to study the behaviour of complex systems, especially in the exploratory phases of characterisation of a generic nonlinear model, is linearisation. Linearisation investigates small disturbances around fixed points of the system, i.e. around state variables $\vec{Z} = \vec{Z}^*$, which in our case are spatiotemporally constant solutions of Equations 7.9-7.12. For hyperbolic fixed points, i.e. points for which all eigenvalues have nonzero real part, the Hartman-Grobman theorem states that a linear expansion in $\vec{z}$ with $\vec{Z} = \vec{Z}^* + \vec{z}$ captures the essential local dynamics.[37] For example, referring to the state variables introduced above as part of the Liley equations (Eqs. 7.9 -7.12), we can define a state vector

$$\vec{Z} \equiv (h_e, h_i, I_{ee}, I_{ei}, I_{ie}, I_{ii}, \phi_{ee}, \phi_{ei})^T \ , \tag{7.13}$$

and determine the fixed points by

$$\begin{aligned}
h_k^* &= h_k^{\mathrm{r}} + \frac{h_{ek}^{\mathrm{eq}} - h_k^*}{|h_{ek}^{\mathrm{eq}} - h_k^{\mathrm{r}}|} I_{ek}^* + \frac{h_{ik}^{\mathrm{eq}} - h_k^*}{|h_{ik}^{\mathrm{eq}} - h_k^{\mathrm{r}}|} I_{ik}^* \ , \\
\phi_{ek}^* &= N_{ek}^\alpha S_k^* = \frac{N_{ek}^\alpha S_k^{\max}}{1 + \exp\left(-\sqrt{2}\frac{h_k^* - \mu_k}{\sigma_k}\right)} \ , \\
I_{ek}^* &= \Gamma_{ek} \mathrm{e} \left[N_{ek}^\beta S_e^* + p_{ek} + \phi_{ek}^*\right] / \gamma_{ek} \ , \\
I_{ik}^* &= \Gamma_{ik} \mathrm{e} \left[N_{ik}^\beta S_i^* + p_{ik}\right] / \gamma_{ik} \ ,
\end{aligned} \tag{7.14}$$

which, after substitution, reduce to just two equations in h_e^* and h_i^*. If multiple solutions exist, it is customary to consider 'default' fixed point $\vec{Z}^{*,r}$ by choosing the h_e^* closest to rest h_e^{r}. The small perturbation vector can be defined as follows:

$$\vec{z} \equiv \vec{a} \times \exp(\lambda t) \times \exp\left(\mathrm{i}\vec{k} \cdot \vec{x}_{\mathrm{cort}}\right) \ , \tag{7.15}$$

and expanded linearly in components $[\vec{a}]_m$. For example, the equation for ϕ_{ee} becomes

$$\left(\frac{1}{v_{ee}^2}\lambda^2 + \frac{2\Lambda_{ee}}{v_{ee}}\lambda + k^2 + \Lambda_{ee}^2\right)[\vec{a}]_7 = N_{ee}^\alpha \Lambda_{ee}^2 \frac{S_e^{\max}\sqrt{2}\upsilon}{\sigma_e(1+\upsilon)^2}[\vec{a}]_1 \ , \tag{7.16}$$

with $\upsilon \equiv \exp[-\sqrt{2}(h_e^* - \mu_e)/\sigma_e]$. Treating all PDEs (i.e Eqs. 7.9 -7.12) in a similar fashion, we end up with an equation set

$$\sum_j \mathsf{B}_{ij}(\lambda, k)[\vec{a}]_j = 0, \quad \text{with } i, j = 1, \ldots, 8 \ .$$

In matrix notation $\mathsf{B}(\lambda, k)\vec{a} = 0$, which allows us to write the condition for the existence of non-trivial solutions as

$$\det \mathsf{B}(\lambda, k) = 0 \, .$$

However, searching for roots $\lambda(k)$ of the above equation is efficient only in special cases. Instead it is more appropriate to introduce auxiliary real variables $Z_{9,\ldots,14} = \partial Z_{3,\ldots,8}/\partial t$, to eliminate second order time derivatives. Our example Eq. 7.16 then becomes

$$[\vec{a}]_{15} = \lambda [\vec{a}]_7 \, ,$$

$$\left(\frac{1}{v_{ee}^2}\lambda + \frac{2\Lambda_{ee}}{v_{ee}}\right)[\vec{a}]_{15} + \left(k^2 + \Lambda_{ee}^2\right)[\vec{a}]_7 N_{ee}^{\alpha}\Lambda_{ee}^2 \frac{S_e^{\max}\sqrt{2}\upsilon}{\sigma_e(1+\upsilon)^2}[\vec{a}]_1 \, .$$

Again, for all PDEs, we can write a new but equivalent form

$$\sum_j \mathsf{B}_{ij}(\lambda, k)[\vec{a}]_j = \sum_j \left[\mathsf{A}_{ij}(k) - \lambda\delta_{ij}\right][\vec{a}]_j = 0, \quad \text{with } i, j = 1, \ldots, 14 \, ,$$

with the Kronecker δ_{ij}. In matrix notation $\mathsf{A}(k)\vec{a} = \lambda\vec{a}$, hence $\lambda(k)$ solutions are eigenvalues. One of the advantages of this extension is that powerful algorithms are available[38] to solve the system above as

$$\sum_l \mathsf{A}_{il}\mathsf{R}_{lj} = \lambda_j \mathsf{R}_{ij} \, , \quad \sum_l \mathsf{L}_{il}\mathsf{A}_{lj} = \lambda_i \mathsf{L}_{ij} \, , \quad \lambda_j \sum_l \mathsf{L}_{il}\mathsf{R}_{lj} = \lambda_i \sum_l \mathsf{L}_{il}\mathsf{R}_{lj},$$

with $i, j, l = 1, \ldots, 14$, and all quantities are functions of k. The λ_j denote 14 eigenvalues with corresponding right $[\vec{r}_j]_i = \mathsf{R}_{ij}$ (columns of R) and left $[\vec{l}_j]_i = \mathsf{L}_{ji}$ (rows of L) eigenvectors. The third equation implies orthogonality for non-degenerate eigenvalues $\sum_l \mathsf{L}_{il}\mathsf{R}_{lj} = \delta_{ij}n_j$ and one can orthonormalise $\mathsf{LR} = \mathsf{RL} = \mathbb{1}$. For spatial distributions of perturbations different k-modes will generally mix quickly with time. Because this is a completely general method to determine the stability of spatial eigenmodes, it can be readily (and indeed has — see e.g. Ref. 36) applied to other mean field formulations.

Knowledge of the fixed points and linear responses also provides useful information for the numerical simulation of mean field theories formulated as PDEs. For example, using the Liley equations (Eqs. 7.9 -7.12), one can model[32,39] the cortical sheet as square, connected at the edges to form a torus, and discretise space ($N \times N$ samples of length d_s) and time ($t = nt_s$ with $n = 0, 1, \ldots$). The most direct way to rewrite the PDEs in suitable form to evolve the equations numerically is to introduce 'Euler forward' time derivative and five-point Laplacian formulae

$$\frac{\partial}{\partial t} f(t) \to \frac{f_{n+1} - f_n}{t_s} = \dot{f}_n \ ,$$
$$\frac{\partial^2}{\partial t^2} f(t) \to \frac{f_{n+2} - 2f_{n+1} + f_n}{t_s^2} = \frac{\dot{f}_{n+1} - \dot{f}_n}{t_s} \ ,$$
$$\nabla^2 f(x,y) \to \frac{1}{d_s^2}\left(f^{i+1,j} + f^{i,j+1} - 4f^{i,j} + f^{i-1,j} + f^{i,j-1}\right) \ ,$$

with the resulting algebraic equations solved for the 'next time step' f_{n+1}, $\dot{f}_{n+1}$. Notice that the five-point Laplacian is particularly convenient for parallelisation,[32] since only one-point-deep edges of the parcellated torus need to be communicated between nodes. The 'Euler forward' formulae will converge slowly $\mathcal{O}(t_s^2, d_s^2)$ but *robustly*, which is important since the system dynamics can change drastically for different parameter sets. The Courant-Friedrichs-Lewy condition for a wave equation sets the upper bound of the time step t_s, i.e. $t_s^2 < d_s^2/v^2$. For instance, if we consider a maximum $v = 10\,\mathrm{m/s}$ and a spatial spacing of $d_s = 1$ mm, then $t_s < 10^{-4}$ s, so that in practise we choose $t_s = 5 \times 10^{-5}$ s. Using the previous solutions for the fixed point search, the entire cortex can be initialised to its (default) fixed point value $\vec{Z}(\vec{x}_{\mathrm{cort}}) = \vec{Z}^*$ at $t = 0$, whereas for parameter sets that have no fixed point in physiological range we can set $h_e(\vec{x}_{\mathrm{cort}}) = h_e^r$, $h_i(\vec{x}_{\mathrm{cort}}) = h_i^r$ and other state variables to zero. Sometimes it is advantageous to have no external inputs, so that any observed dynamics must be self-sustained. In this case some added spatial variation in $h_e(\vec{x}_{\mathrm{cort}})$ helps to excite $k \neq 0$ modes quickly.

One of the common, somewhat troublesome features of mean field brain dynamics modelling resides in the dimension of the parameter spaces considered. For 'physiological' parameters a wide range of model dynamics can be encountered, with the constraint that proper parameterisations should produce electroencephalographically plausible dynamics. In general, two approaches can be employed to generate such parameter sets. The first is to fit the model to real electroencephalographic data, although it should be noticed that there is still considerable uncertainty regarding the reliability, applicability and significance of using experimentally obtained data for fitting or estimating sets of nonlinear ordinary differential equations.[40] Alternatively one can explore the physiologically admissible multi-dimensional parameter space in order to identify parameter sets that give rise to 'suitable' dynamics, for instance those showing a dominant alpha rhythm.

In regards to the Liley model, one could stochastically or heuristically explore the parameter space by solving the full set of spatiotemporal equations. However, the computational costs of this approach are forbidding at this point in time. Alternatively, the parameter space of a simplified model, e.g., spatially homogeneous without the Laplacian in (7.12), can be searched. Linearisation is again very useful in this context. In fact, if the defining system can be approximated by linearisation, then one can estimate the spatiotemporal dynamics merely from the resulting eigensystem. Such an analysis is exceedingly rapid compared with the direct solution of the equations. Then, parameter sets randomly sampled from the physiologically admissible parameter space can be simply tested for physiological consistency. Finally, following Ref. 32, which shows how one can model plausible EEG recorded from a single electrode, the *noise fluctuation* power spectrum $S(\omega)$ for the Liley model (Eqs. 7.9 -7.12) can be estimated by

$$S(\omega) = \frac{1}{2\pi} \int dk\, k \left| \Psi(k) \left\{ \mathsf{R} \cdot \mathrm{diag} \left[\frac{1}{i\omega - \lambda_n(k)} \right] \cdot \mathsf{L} \cdot \vec{p} \right\}_1 \right|^2 , \qquad (7.17)$$

and then evaluated for physiological veracity. The left and right eigen-matrices, L and R, have been defined above, here $\mathsf{LR} = \mathbb{1}$ and $\Psi(k)$ is the electrode point spread function,[31] and $\vec{p} \equiv (0, 0, p_{ee}, 0, p_{ei}, 0, p_{ie}, 0, p_{ii}, 0, 0, 0, 0, 0)^T$.

The obvious drawback of the procedure described above is that nonlinear solutions of potential physiological relevance are inevitably missed. However, 'linear' parameter sets can be continued in one and two dimensions to reveal a plethora of electroencephalographically plausible nonlinear dynamical behaviour, as we will see in the next section.

7.4. Characteristics of the Model Dynamics in the Liley Equations: Multistability, Chaos, Bifurcations and Emergence

The computational approach is clearly of little value unless we try to connect quantitative predictions with physiologically and neurologically relevant phenomena. For example, numerical solutions to equations (7.9-7.12) for a range of physiologically admissible parameter values reveal a large array of deterministic and noise driven dynamics at alpha band frequencies.[30,31,41,42] In fact, alpha band activity appears in three distinct dynamical scenarios: as linear noise driven, limit cycle or chaotic oscillations. Furthermore, another interesting quantitative indication given by the model is the prediction that reverberant activity between inhibitory neuronal populations is causally central to the alpha rhythm. In particular it is found that the strength and form of inhibitory→ inhibitory synaptic interactions appears to be the most sensitive determinants of the frequency and damping of emergent alpha band activity. Indeed this sensitivity to inhibitory modulation has been exploited to explain the effects that a range of sedative and anaesthetic agents have on the EEG[32] as many of the most commonly used anaesthetic and sedative agents are thought to selectively target inhibition. Taken together these results, and a range of physiologically pertinent results obtained through other mean field modelling approaches (see Ref. 4 for a detailed review) all underscore the relevance of the mean field paradigm to understanding brain dynamics. Thus, at least in principle, it is possible to conceive of characterising the complex changes in dynamics in regards to cognition[40] or in a range of central nervous system diseases such as epilepsy.[43]

On the other hand, the model also shows a richness of dynamical scenarios that stimulate conjectures and call for explanations at different neurobiological levels, both theoretically and experimentally. One such example is the so-called "multistability", or the coexistence of different dynamical regimes for a unique set of physiologically relevant parameters. In particular, it has been shown[44] that in the 10-dimensional local reduction of the Liley model ($\phi_{ek} = 0$), two limit cycle attractors and one chaotic attractor are simultaneously present in a two-dimensional plane of the ten-dimensional volume of initial conditions. One multiperiodic small-amplitude limit cycle possesses the largest spectral peak, lying within the alpha band ($8 - 13$ Hz) of the EEG, whereas the other large-amplitude cycle may correspond to epilepsy. The basin of attraction for different initial conditions in h_e and h_i is depicted in Figure 7.4, and delay-embedded attractors with their corresponding power spectra are in Figures 7.5-7.6.

The origin of multistable dynamics in the Liley model can be explained with the aid of a one-parameter bifurcation analysis in the excitatory-excitatory pulse density p_{ee}, using the continuation software AUTO[45] (see the Appendix). Generally speaking, bifurcation analysis is instrumental in revealing regions of parameter space where abrupt changes in dynamics occur.[46] Mathematically these abrupt changes correspond to bifurcations, whereas physically they resemble phase transition phenomena in ordinary matter.

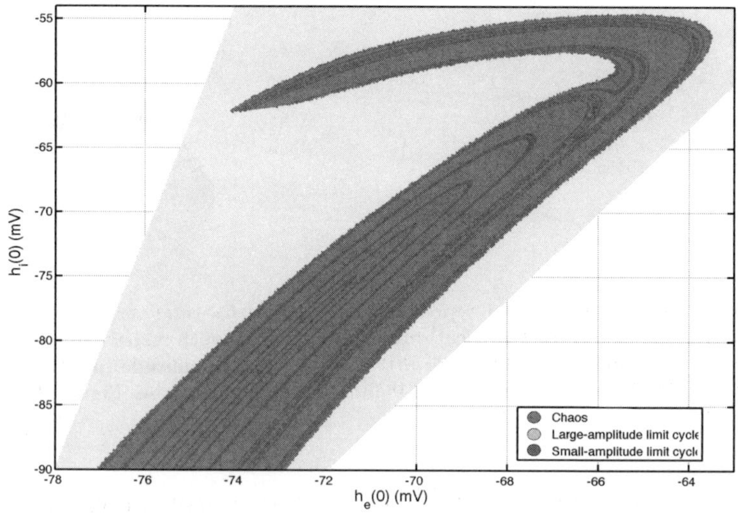

Fig. 7.4. Initial-condition map for the parameter set under investigation. Large-amplitude limit cycles are depicted by light grey points, chaotic sets are represented by an intermediate grey shade, and small-amplitude limit cycles are represented by dark grey points. Unshaded areas represent regions of initial condition space where no numerical integrations were performed but where only large-amplitude limit cycles are expected to exist. There are 594,656 chaotic points (about 39.6% of the points sampled), 37,815 small-amplitude limit cycle points (about 2.5% of the points sampled) and 867,529 points for the large-amplitude limit cycle cases (about 57.8% of the points sampled) out of a total of 1,500,000 points sampled. The parameters for the model which generate this initial-condition map can be found in Ref. 44. Figure reproduced from Ref. 44.

As illustrated in Figure 7.7, the system loses stability through a supercritical Hopf bifurcation (hb), which gives birth to oscillatory behaviour at $p_{ee} \approx 6.9$. As p_{ee} increases, a series of period doubling bifurcations (pd) for the periodic orbits occur, resulting in a cascade (pdc) which gives rise to chaos at $p_{ee} \approx 9.387$. There also exists a further subcritical Hopf bifurcation at very large values of the pulse density (not shown in the Figure), from which an unstable cycle emerges. The continuation of the latter shows that, after a saddle-node orbit point, this cycle becomes stable and heads back to the region of small p_{ee}. This periodic branch, which eventually ends in a homoclinic trajectory (hm), indeed contains the large-amplitude attractor depicted in Figures 7.5a) and 7.6a) at $p_{ee} = 9.43$. Homoclinics represent the tangency between the stable and the unstable components of the manifold at a fixed point (saddle-node), which is reached in the limit of infinite pos-

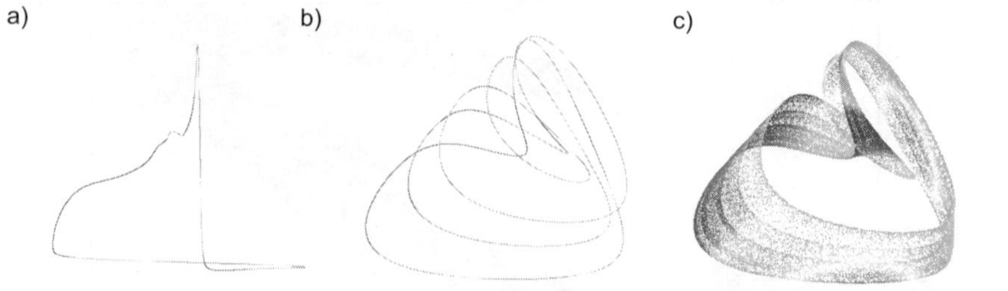

Fig. 7.5. Delay-time-embedded representations of the h_e time series of the model, shaded according to the relative local rate of change of h_e with respect to time. a) low frequency, high amplitude, limit cycle attractor b) small amplitude, alpha frequency, limit cycle attractor c) alpha frequency dominant chaotic attractor. Figure reproduced from Ref. 44.

itive and negative times. On the other hand, the small-amplitude periodic attractor described in 7.5b) and 7.6c) is nested in the "sea"of chaos and its continuation is shown in Figure 7.8. Even though it crosses unstable segments of the fast orbits involved in the doubling cascade, it does not belong to any of them. Due to the stiffness of the system, it is not possible to conclude whether the branch is isolated or may be originated from a branch point for a cascading generation greater than the sixth (see Figure 7.8). Nonetheless, the existence of so-called "isolae"[47] of low-amplitude orbits is ruled out since the branch is not closed: after stability is lost through saddle-node (snlc) and period doubling (pd) points, the unstable segment terminates in a homoclinic bifurcation at $p_{ee} \approx 10.22$.

A general feature of the Liley model is that variations in p_{ee} (excitatory input to excitatory neurons) and p_{ei} (excitatory to inhibitory) result in the system producing a range of dynamically differentiated alpha activities. It is clear that if p_{ei} is much larger than p_{ee}, a stable equilibrium is the unique state of the EEG model and driving the equations in this state with white noise typically produces sharp alpha resonances.[31] When one increases p_{ee}, this equilibrium loses stability in a Hopf bifurcation and periodic motion sets in with a frequency of about 11 Hz. As we have seen, for still larger p_{ee} the fluctuations can become irregular and the limiting behaviour of the model is governed by a chaotic attractor. In general, the different dynamical states can be distinguished by computing the largest Lyapunov exponent (LLE), which is negative for equilibria, zero for (quasi)-periodic fluctuations, and positive for chaos. For instance, in the previously discussed multistable example, the chaotic attractor was found to have a

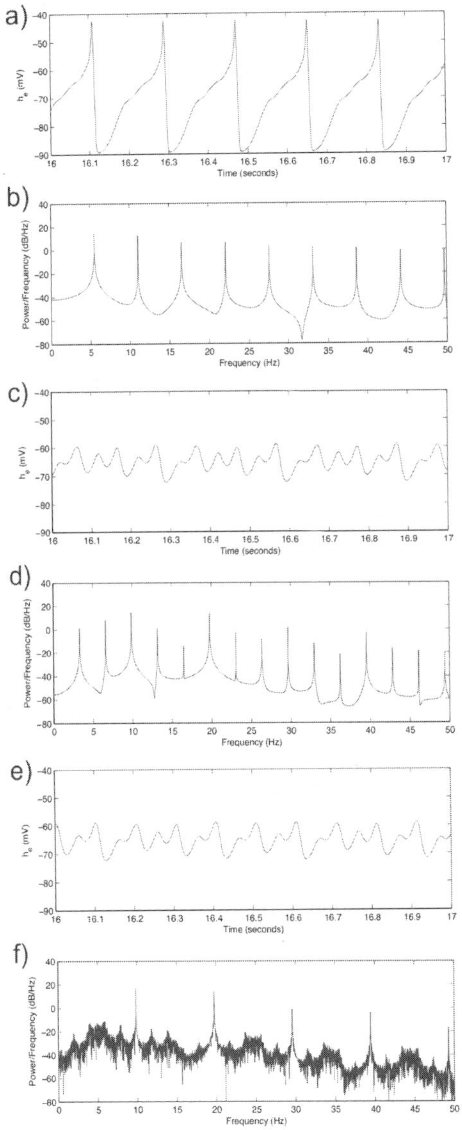

Fig. 7.6. Time series and power spectra for the attractors investigated, using the same h_e time series as per Figure 7.5. Note the presence in panels a) and b) of waves of large amplitude, low frequency and of a spiky, angular nature, which, as we have suggested, could broadly correspond to epileptic activity. Panels c) and d) show low amplitude activity with a dominant peak at approximately 10 Hz (in the alpha band) and a further dominant harmonic at approximately 20 Hz, which corresponds to the beta band of the EEG spectrum. Subharmonics and higher harmonics also exist. Panels e) and f) correspond to the chaotic dynamics of the system. Panel f) in particular illustrates the broadband spectral features which are so familiar in chaotic dynamics, with dominant peaks at approximately 10 Hz and higher harmonics. Panel e) illustrates the low amplitude, aperiodic nature of the time series. Figure reproduced from Ref. 44.

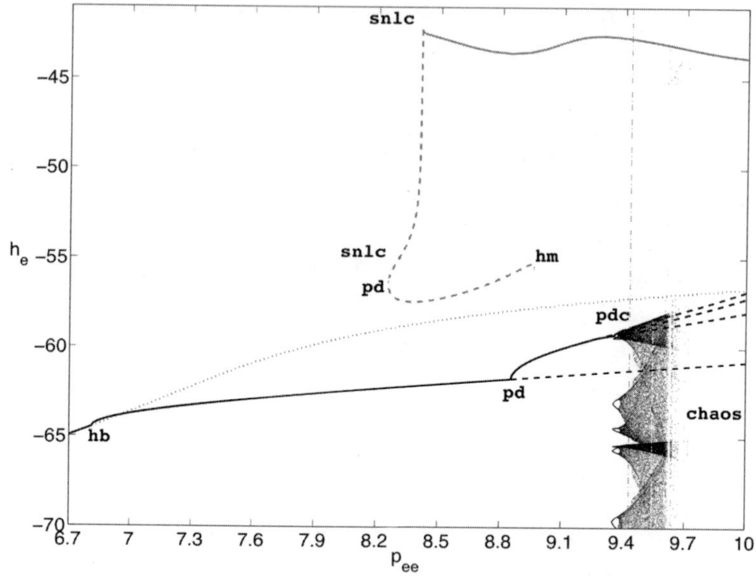

Fig. 7.7. One-parameter bifurcation diagram showing the absolute maxima of h_e in terms of p_{ee} for steady state (dotted black), low-amplitude periodic (solid black) and large-amplitude periodic (grey) regimes. Stable (unstable) branches are continuous (dashed) and the vertical(dash-dot) line indicates the value of p_{ee} at which multistability occurs. Chaos (in black) ensues after a period doubling cascade (pdc) of orbits originating from a supercritical Hopf bifurcation (hb) and is represented through a Poincaré section of the relative maxima and minima of h_e. Notice the stable large-amplitude orbits interspersed with chaos for $p_{ee} \approx 9.66$ (i.e. the black points that coincide at the top with the cycles in grey) and the very small window of stability between a saddle-node (snlc) and a period doubling (pd) point for the same large-amplitude branch before it ends in a homoclinic bifurcation (hm). Figure reproduced from Ref. 44.

moderate value of the largest Lyapunov exponent (3.4 s^{-1} base e) with an associated Kaplan-Yorke (Lyapunov) dimension of 2.086.

For a different parameter set which still retains physiological relevance, bifurcation analysis[48] indicates that the boundary of the chaotic parameter set is formed by infinitely many saddle-node and period-doubling bifurcations, as shown in Figure 7.9(a). All these bifurcations converge to a narrow wedge for negative, and hence un-physiological, values of p_{ee} and p_{ei}, literally pointing to the crucial part of the diagram where a *Shilnikov saddle-node homoclinic bifurcation* takes place. It is worthwhile discussing this finding in some detail.

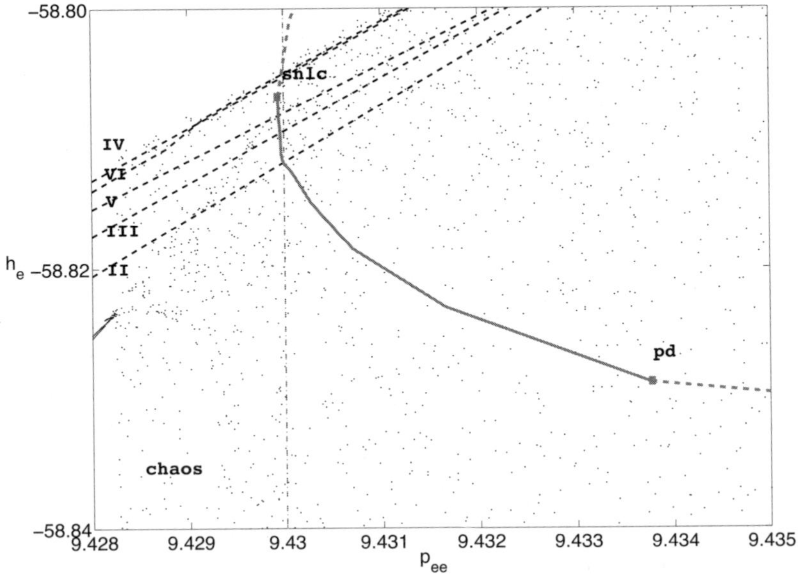

Fig. 7.8. The periodic branch that corresponds to the small-amplitude attractor (grey line) is hidden in the chaotic region (black dots) and crosses the unstable segments of cascading orbits (black dashed lines). The branch then loses stability through a period doubling (pd) and a saddle-node point (snlc), terminating in two separate homoclinic bifurcations (not shown). The very limited window of stability for the orbits just includes $p_{ee} = 9.43$ (vertical thin black dash-dot line). The generation of cascading orbits is indicated in Roman numerals, according to their appearance with increasing p_{ee}. Also note that, within a period, the small-amplitude orbits in the cascade have a number of maxima (or minima) that are always powers of two: this is not the case for the small-amplitude attractor which has six (see Figure 7.6) and thus cannot be directly generated through doubling. Figure reproduced from Ref. 44.

Figure 7.9(b) is a codimension two plot, which shows a sketch of the bifurcation diagram at the tip of the wedge. The interplay among bifurcations is particularly interesting. The line with the cusp point of saddle-nodes c separates regions with one and three equilibria, and the line of Hopf bifurcations terminates there at the Bogdanov–Takens point bt, where the system presents a double zero eigenvalue. The point gh is a generalised Hopf point, where the Hopf bifurcation changes from sub- (hard) to supercritical (soft). The line which emanates from bt represents a homoclinic bifurcation, which coincides with the line of saddle-node bifurcations on an open interval, where it denotes orbits homoclinic to a saddle node. In the normal form, this interval is bounded by the points n_1 and n_2, at which points the homoclinic orbit does not lie in the local centre manifold. While the normal form is two-dimensional and only allows for a single orbit homoclinic to the saddle-node equilibrium, the high dimension of the macrocolumnar EEG model ($\phi_{ek} = 0$) allows for several orbits homoclinic to the saddle node. If we consider the numerical continuation of the homoclinic along the saddle node curve, starting from n_1 as shown in Figure 7.9(b), it actually overshoots n_2 and folds back at t_1, where the centre-stable and centre-unstable manifolds of the saddle node have a tangency. In fact, the curve of homoclinic orbits folds several times before it terminates at n_2, and this creates an interval, bounded by t_1 and t_2, in which up to four homoclinic orbits coexist. This, according to Shilnikov's theorem, signals the existence of infinitely many periodic orbits. This is the hallmark of chaos, which emerges from the left wedge of the saddle-nodes in Figure 7.9. The reader wishing to grasp the essence of the geometry of the Shilnikov saddle-node homoclinic bifurcation should consult Ref. 49.

It is important to understand that in contrast to the homoclinic bifurcation of a saddle focus, commonly referred to as the Shilnikov bifurcation, this route to chaos has not been reported before in the analysis of any mathematical model of a physical system. Also, while the Shilnikov saddle node bifurcation occurs at negative, and thus unphysiological, values of p_{ee} and p_{ei}, it nevertheless organises the qualitative behaviour of the EEG model in the biologically meaningful parameter space. It is essential to remark that this type of organisation persists in a large part of the parameter space: for example, if a third parameter is varied, the codimensional two points c, bt and gh collapse onto a degenerate Bogdanov–Takens point of codimension three, which represents an organising centre controlling the qualitative dynamics of an even larger part of the parameter space.

A further surprising feature in the rich dynamical texture of the model is the fact that parameter sets that have been chosen to give rise to physiologically realistic behaviour in one domain, can produce a range of unexpected, but physiologically plausible, activity in another. For example, parameters can be chosen to accurately model eyes-closed alpha and the surge in total EEG power that has been observed to occur during anaesthetic induction.[32] Among other conditions, sets are required to have a sharp alpha resonance (full-width half-maximum > 5) and moderate mean excitatory and inhibitory neuronal firing rates $< 20/\mathrm{s}$. Surprisingly, a large fraction of these sets also turns out to produce limit cycle (nonlinear) gamma band activity under mild parameter perturbations.[42] This is particularly rele-

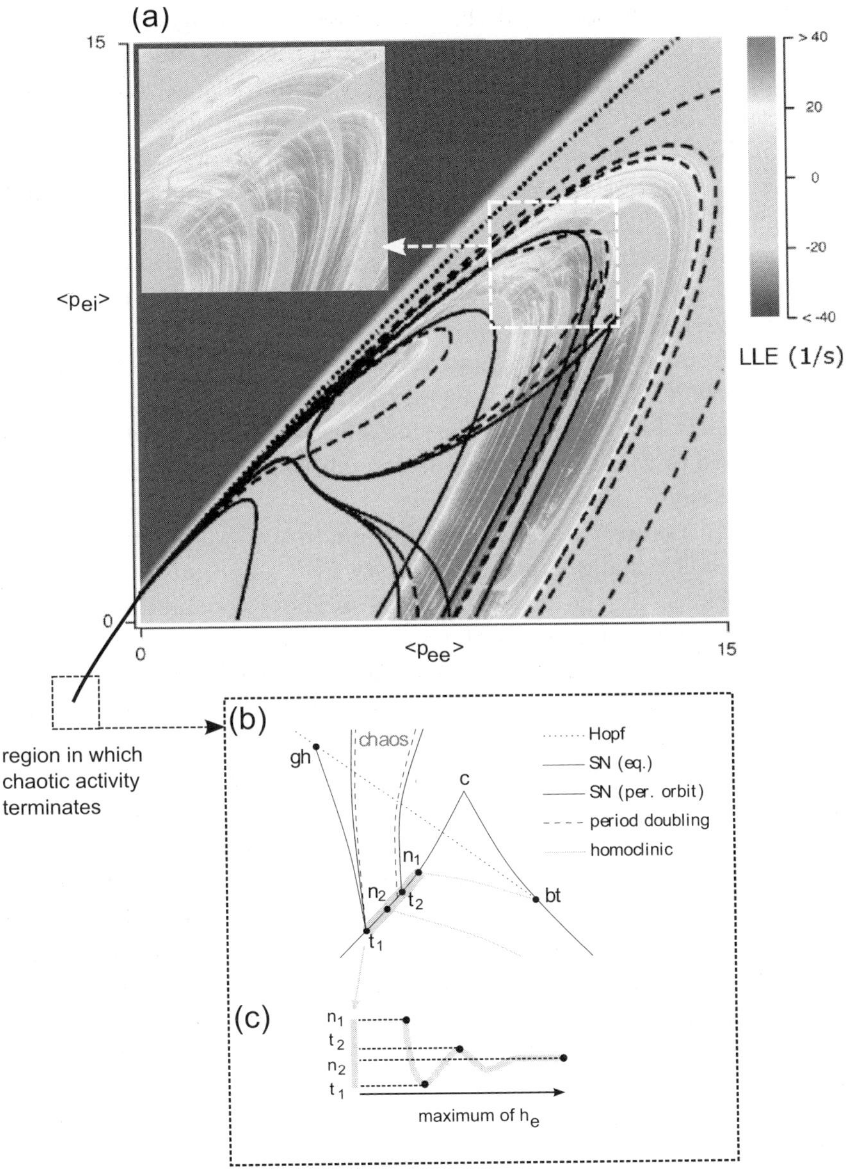

Fig. 7.9. (a) The largest Lyapunov exponent (LLE) of the dynamics of a simplified local model ($\phi_{ek} = 0$) for a physiologically plausible parameter set exhibiting robust (fat-fractal) chaos.[41] Superimposed is a two parameter continuation of saddle-node and period-doubling bifurcations. The leftmost wedge of chaos terminates for negative values of the exterior forcings, p_{ee} and p_{ei}. (b) Schematic bifurcation diagram at the tip of the chaotic wedge. bt = Bogdanov–Takens bifurcation, gh = generalised Hopf bifurcation, and SN = saddle node. Between t_1 and t_2 multiple homoclinic orbits coexist and Shilnikov's saddle node bifurcation takes place. (c) Schematic illustration of the continuation of the homoclinic orbit between points n_1 and t_1. Figure adapted from Refs. 48 and 41.

vant since gamma band (> 30 Hz) oscillations are thought to be one of the distinctive and necessary features present in cognitive functioning.[50] In turn, it appears that the existence of weakly damped, noise driven, linear alpha activity can be associated with limit cycle 40 Hz activity, and that transitions between these two dynamical states can occur.

Bifurcation analysis of this parameter set again reveals non trivial findings. Figure 7.10 illustrates a bifurcation diagram (column 11 of Table V in Ref. 32, see also Table 1 in Ref. 42) for the spatially homogeneous reduction $\nabla^2 \to 0$ of (7.12). The choice of bifurcation variables is motivated by two observations: i) differential increases in $\Gamma_{ii,ie}$ have been shown to reproduce a shift from alpha to beta band activity, similar to what is seen in the presence of low levels of $GABA_A$ agonists such as benzodiazepines;[51] and ii) the dynamics of linearised solutions for the case when $\nabla^2 \neq 0$ are particularly sensitive to variations of parameters affecting inhibitory→inhibitory neurotransmission,[31] such as N_{ii}^{β} and p_{ii}.

Specifically, Figure 7.10 illustrates the results of a two parameter bifurcation analysis for changes in the inhibitory PSP amplitudes via $\Gamma_{ie,ii} \to r\Gamma_{ie,ii}$ and changes in the total number of inhibitory→inhibitory connections via $N_{ii}^{\beta} \to kN_{ii}^{\beta}$. The parameter space has physiological meaning only for positive values of r and k. The saddle-node bifurcations of equilibria have the same structure as for the 10-dimensional homogeneous reduction discussed previously, in that there are two branches joined at a cusp point. Furthermore, we have two branches of Hopf bifurcations, the one at the top being associated with the birth of alpha limit cycles and the other with gamma limit cycles. This former line of Hopf points enters the wedge shaped curve of saddle-nodes of equilibria close to the cusp point and has two successive tangencies in fold-Hopf points (fh). The fold-Hopf points are connected by a line of tori or Neimark–Sacker bifurcation, corresponding to orbits with oscillations at two incommensurate frequencies. The same curve of Hopf points then ends in a Bogdanov–Takens (bt) point, from which a line of homoclinics emanate. Contrary to the previous example, this line of homoclinics does not give rise to a Shilnikov saddle-node bifurcation, but is instead involved in a different scenario leading to complex behaviour (including chaos), called a *homoclinic doubling cascade.* Essentially, a cascade of period doubling bifurcations which is responsible for the birth of chaos also collides with lines of homoclinics and, as a consequence, not only are infinitely many periodic orbits created, but so are infinitely many homoclinic connections.[53] All these periodic and homoclinic orbits coexist with a stable equilibrium, reinforcing the idea that multistability is a frequent

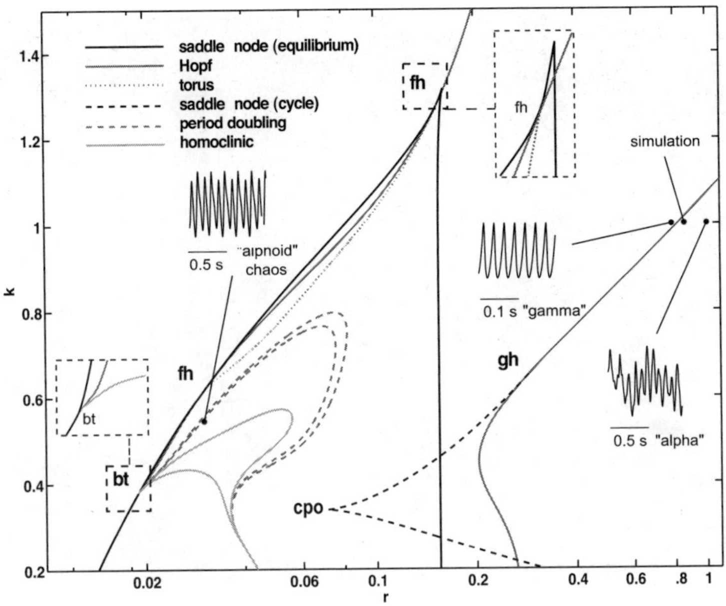

Fig. 7.10. Partial bifurcation diagram for the spatially homogeneous model, $\nabla^2 \to 0$ in (7.12), as a function of the two scaling parameters k and r, defined by $\Gamma_{ie,ii} \to r\Gamma_{ie,ii}$ and $N_{ii}^{\beta} \to kN_{ii}^{\beta}$, respectively. The codimension two points have been labelled fh for fold-Hopf, gh for generalised Hopf and bt for Bogdanov–Takens. The right-most branch of Hopf points corresponds to the emergence of gamma frequency (≈ 37 Hz) limit cycle activity via a subcritical Hopf bifurcation above the point labelled gh. A homoclinic doubling cascade takes place along the line of homoclinics emanating from bt. Insets on the left show schematic blowups of the fh and bt points. Additional insets show time series of deterministic (limit cycle and chaos) and noise-driven dynamics for a range of indicated parameter values. The point labelled simulation represents the parameter values used for the numerical simulations of Figure 7.11. Figure reproduced from Ref. 52.

and common feature in our model. The second line of Hopf bifurcations in the gamma frequency range (> 30 Hz) does not interact with the lines of saddle nodes in the relevant portion of the parameter space. Both branches of Hopf points change from sub- to supercritical at gh around $r^* = 0.27$, so that bifurcations are "hard" for $r > r^*$ in either case. These points are also the end points of folds for the periodic orbits, and the gamma frequency ones form a cusp (cpo) inside the wedge of saddle nodes of equilibria.

There are many perspectives and developments that can be drawn from

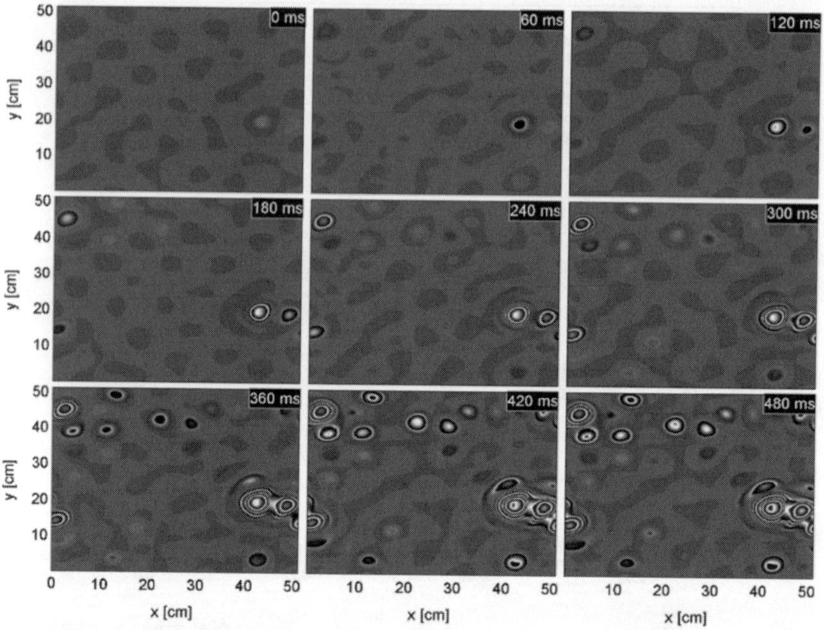

Fig. 7.11. Numerical solutions of the equations 7.9–7.12 for a human-sized cortical torus with $r = 0.875$. h_e is mapped every 60 ms (grayscale: -76.9 mV black to -21.2 mV white). For $r = 0.875$ the linearisation becomes unstable for a range of wavenumbers around 0.6325/cm. From random h_e synchronised gamma activity emerges and spatial patterns with a high degree of phase correlation (gamma 'hotspots') form with a frequency consistent with the predicted subcritical Hopf bifurcations of the spatially homogeneous equations, compare Figure 7.10. Figure reproduced from Ref. 42.

the discussions above. Firstly, an analysis of the bifurcation diagrams for general parameter sets known to be associated with emergent activity and a rise in power spectrum is currently under way. This should give a deeper understanding of the role of inhibition for the birth of coherent gamma oscillations and elucidate some of the typical dynamical mechanisms and bifurcation patterns that can support a cognitive-like response in our model. Secondly, the detection, investigation and continuation of particular oscillatory structures named mixed mode oscillations[54,55] represents another active field of our current research. It is believed that the alternation of spike-like bursts with subthreshold periodic behaviour might be central in waking mechanism for anaesthetised brains, such as the so-called "burst suppression".[56] Finally, it is of great importance to estimate the contribution of inhomogeneous solution of Equation 7.12 to all the phenomena we reviewed so far. This calls for a bifurcation analysis of PDEs rather than ODEs, which is a still rather unexplored and exciting field in its own right.

7.5. Conclusion

Currently the most serious rival to a symbolic model of cognition is that of connectionism. Initially inspired by neurophysiology, connectionism argues that cognition arises out of the emergent activity of multilayered networks of non-linear nodes or elements (corresponding to primitive neurons) with variously positively and negatively weighted interconnections. Symbolic representation is abandoned and instead replaced by invariant patterns of activity that may be understood as corresponding to, among other things, particular perceptual events. Such perceptual events may correspond to memory retrieval, sensory categorization or any other compositional feature of cognition. However training such a network to perform a certain task does not ensure that the resulting network is a physiologically, and thereby a psychologically/behaviourally, plausible model of the task. While networks comprised of a finite number of physiologically plausible neuron-like elements can be constructed, their relevance as a bottom-up frame work for understanding cognition is restricted in the following ways: i) the computational tractability of brain-sized networks ii) the vast size of the parameter and state spaces needed to meaningfully instantiate network dynamics and iii) the difficulties in comparing network states with quantitative measures of behaviour.

Fortunately, from a physiological perspective (e.g. Freeman 1975) the methodological problems posed by connectionistic brain models can be overcome by various schemes in which some form of spatial and/or temporal coarse-graining is used to transform an enumerable (discrete) network of many many millions of individual connections and neurons into continuous networks in which the activity, and thus the state, is continuously distributed in space. This approach accords well with the generally held belief that behaviour emerges out of the cooperative behaviour of populations of neurons, and is consonant with the practical limitations of all forms of non-invasive neuro-imaging (fMRI, EEG/ECoG or MEG) in which some form of spatial and/or temporal averaging of neuronal activity occurs. In section 7.2 we reviewed the major theoretical frameworks underlying this approach, and in particular discussed a number of major theories of EEG electrorhythmogenesis that utilised, in broad terms, one or a number of these frameworks.

Because of their non-linear integral, differential or integro-differential form the resulting mean field theories are expected to exhibit a range of complicated spatio-temporal dynamics, which are expected to be of relevance in understanding the global features of brain dynamics and brain function and thus human cognition. In order to illustrate the potential and relevance of this approach we focused on the mean field formulation of Liley *et al* and showed that considerable dynamical complexity emerges from relatively straightforward physiological predicates. While this theory has some experimental support, so do a number of other competing approaches (for example see e.g. Refs. 27, 36, 22 and 21), and thus the patterns of complex activity produced, and their routes, must be regarded as physiologically speculative.

Regardless of which theory emerges as the more fruitful in terms of accounting for the physiological genesis and significance of rhythmic EEG activity, all imply the existence, at some level, of non-linear patterns of activity. The detection of nonlinear patterns of activity in real EEG will therefore be of crucial importance in the evaluation of the various modelling approaches to cortical electrorhythmogenesis. However there are significant challenges in the detection of weak non-linearity (let alone chaotic activity) in noisy signals such as scalp recorded EEG. While the use of the surrogate data technique is now virtually standard in the identification of weak non-linearity[57] experience has shown that it has to be judiciously applied to signals in which there is a significant periodic component (such as the alpha rhythm),[58,59] and in which non-linearity and non-stationarity are predicted to coexist.[52]

Acknowledgments

The authors would like to thank Mathew Dafilis, Brett Foster and Ingo Bojak for useful discussions, and the careful reading of a number of earlier drafts. This work was supported by the Australian Research Council through grants DP0559949 and DP0879137, and by COSNet through an overseas travel grant awarded to FF.

Appendix A. Continuation Software and Selected Problems

Several well-designed, efficient and free software packages are available for the study of dynamical systems. The most relevant are collated and presented in a comprehensive website to which the reader is redirected (www.dynamicalsystems.org), which also contains references, pictures and interesting general information for the dynamical systems' aficionados. At least three packages are worth being mentioned for their popularity and usefulness: XPPAUT, AUTO and MATCONT.[60] The first one represents a computational laboratory with extensive built-in capabilities for the study of ordinary differential equations and maps. A host of numerical integrators are present, alongside tools for the calculation of Lyapunov exponents, the determination of nullclines and fixed points, the realisation of Poincaré and Ruelle plots, animations and movies just to cite a few. It is our strong advice to the reader who feels interested in these themes to consider Ref. 61, where the features of the software are illustrated with the help of a number of worked out popular dynamical models, examples and exercises. XPPAUT also contains some basic features of AUTO, a software designed for continuation and boundary value problems (BVP) for differential and algebraic equations. AUTO uses the methods of collocation points and pseudo-arclength continuation to efficiently compute codimension one, two (and in some cases three) points and track and extend homoclinic and heteroclinic orbits. It also solves integral and differential BVPs and can be used for applications like manifold computation or for the extension of objects like point(cycle)-to-cycle connections.[62] AUTO is essentially a command-line type of application, which interfaces very well with C and FORTRAN, works seamlessly with systems with many degrees of freedom and is accompanied with sleek plotting and animating interfaces. MATCONT and its companions Cl_MatCont and Cl_MatCont_for_maps are also very interesting tools for bifurcation analysis and integration of dynamical systems and maps, and are built as add-ons of the popular engineering software MATLAB. One of the advantages of MATCONT over AUTO is that the former has built-in routines for the automatic detection of all codimension one and two points, and its graphic user interface leads to a somehow immediate "feel" of operational knowledge. Both packages are also filled with demos and ready-to-go very instructive examples.

Let us conclude this appendix with a couple of selected open questions. There are, of course, much more still unsolved problems and possible developments in reference to the results presented in the previous sections, which can be tackled with the software above described. The adventurous reader can feel free to play with the following two, and possibly email us in case something interesting comes up.

- It still unknown what the fate of the Shilnikov saddle-node bifurcation will exactly be for different combinations of two parameters. The reader might want to use either AUTO or MATCONT to reproduce the results in Figure 7.9 for p_{ee} and p_{ei} with the set given in Ref. 48, and then change the continuation variables to see how the plot changes. Is there still a Shilnikov point or is it destroyed? Are you able to show that there is the appearance of a generalised Bogdanov–Takens point due to the coalescence of the codimension two bifurcations in three parameters?
- It has been shown in Ref. 63 that the 14D model which includes the cortico-cortical connections can be extended to simulate the effect of anaesthetic agents on the cerebral cortex, through modifications in the postsynaptic response. In particular, it would be interesting to understand how anaesthetic concentration affects the results shown in Figure 7.10, which originate only from variations in $\Gamma_{ii,ie}$ and N_{ii}^{β}. So, in terms of the rate constant and the amplitude of the inhibitory postsynaptic potential, what will happen, for example, to the homoclinic doubling cascade previously shown? How will the emergent 40 Hz activity be influenced?

References

1. B. L. Foster, I. Bojak, and D. T. J. Liley, Population based models of cortical drug response — insights form anaesthesia, *Cognitive Neurodyn.* (2008). in press.
2. E. M. Izhikevich, *Dynamical Systems in Neuroscience: The Geometry of Excitability and Bursting.* (MIT Press, Cambridge, MA, 2007).
3. D. T. J. Liley and I. Bojak, Understanding the transition to seizure by modeling the epileptiform activity of general anesthetic agents, *J. Clin. Neurophsiol.* **22**, 300–313, (2005).
4. G. Deco, V. K. Jirsa, P. A. Robinson, M. Breakspear, and K. Friston, The dynamic brain: from spiking neurpons to neural masses and cortical fields, *PLoS Comput. Biol.* **4**(8), e1000092, (2008).
5. V. Braitenberg and A. Schüz, *Cortex: Statistics and geometry of neuronal connectivity.* (Springer, New York, 1998), 2nd edition.
6. R. L. Beurle, Poperties of a mass of cells capable of regenerating pulses, *Trans. Roy. Soc. (Lond) B.* **240**, 55–94, (1956).
7. H. R. Wilson and J. D. Cowan, Excitatory and inhibitory interactions in localised populations of model neurons, *Biophys. J.* **12**, 1–24, (1972).
8. H. R. Wilson and J. D. Cowan, A mathematical theory of the functional dynamics of cortical and thalamic nervous tissue, *Kybernetik.* **13**, 55–80, (1973).

9. S. Amari, Homogeneous nets of neuron-likme elements, *Biol. Cybern.* **17**, 211–220, (1975).
10. S. Amari, Dynamics of pattern formation in lateral-inhibition type neural fields, *Biol. Cybern.* **27**, 77–87, (1977).
11. B. Ermentout, Neural networks as spatio-temporal pattern-forming systems, *Rep. Prog. Phys.* **61**, 353–430, (1998).
12. W. J. Freeman, *Mass Action in the Nervous System.* (Academic Press, New York NY, 1975).
13. S. Coombes, Waves, bumps, and patterns in neural field theories, *Biol. Cybern.* **93**, 91–108, (2005).
14. M. Breakspear and V. K. Jirsa. Neuronal dynamics and brain connectivity. In eds. V. K. Jirsa and A. R. McIntosh, *Handbook of Brain Connectivity*, pp. 3–64. Springer-Verlag, Berlin, (2007).
15. J. J. Wright and R. R. Kydd, The electroencephalogram and cortical neural networks, *Network.* **3**, 341–362, (1992).
16. W. J. Freeman, Nonlinear gain mediating cortical stimulus-response relations, *Biol. Cybern.* **33**, 237–247, (1979).
17. W. J. Freeman, Simulation of chaotic EEG patterns with a dynamic model of the olfactory system, *Biol. Cybern.* **56**, 139–150, (1987).
18. P. L. Nunez, The brain wave equatiomn: a model for the EEG, *Math. Biosci.* **21**, 279–297, (1974).
19. P. L. Nunez, *Electric fields of the brain: The neurophysics of EEG.* (Oxford University Press, New York, 1981), 1st edition.
20. V. K. Jirsa and H. Haken, Field theory of electromagnetic brain activity, *Phys. Rev. Lett.* **77**, 960–963, (1996).
21. P. L. Nunez, Generation of human EEG by a combination of long and short range neocortical interactions, *Brain Topography.* **1**, 199–215, (1989).
22. P. L. Nunez, *Neocortical Dynamics and Human EEG Rhythms.* (Oxford University Press, New York, 1995).
23. E. R. Kandel, J. H. Schwartz, and T. M. Jessell, Eds., *Principles of Neural Science.* (McGraw-Hill, New York, 2000), 4th edition.
24. F. H. Lopes da Silva, A. Hoeks, H. Smits, and L. H. Zetterberg, Model of brain rhythmic activity, *Biol. Cybern.* **15**(1), 27–37, (1974).
25. A. van Rotterdam, F. H. Lopes da Silva, J. van den Ende, M. A. Viergever, and A. J. Hermans, A model of the spatial-temporal characteristics of the alpha rhythm, *Bull. Math. Biol.* **44**, 283–305, (1982).
26. F. H. Lopes da Silva and W. Storm van Leeuwen. The cortical alpha rhythm in dog: the depth and surface profile of phase. In eds. M. A. B. Brazier and H. Petsche, *Architectonics of the Cerebral Cortex*, pp. 319–33. Raven Press, New York, (1978).
27. P. A. Robinson, C. J. Rennie, J. J. Wright, H. Bahramali, E. Gordon, and D. L. Rowe, Prediction of electroencephalographic spectra from neurophysiology, *Phys. Rev. E.* **63**, 021903, (2001).
28. S. Rodrigues, J. R. Terry, and M. Breakspear, On the genesis of spike-wave activity in a mean-field model of human corticothalamic dynamics, *Phys. Lett. A.* **355**, 352–357, (2006).

29. M. Breakspear, J. A. Roberts, J. R. Terry, S. Rodrigues, , M. N, and P. A. Robinson, A unifying explanation of generalised seizures via the bifurcation analysis of a dynamical brain model, *Cerebral Cortex.* **16**(9), 1296–313, (2006).
30. D. T. J. Liley, P. J. Cadusch, and J. J. Wright, A continuum theory of electro-cortical activity, *Neurocomp.* **26-27**, 795–800, (1999).
31. D. T. J. Liley, P. J. Cadusch, and M. P. Dafilis, A spatially continuous mean field theory of electrocortical activity, *Network: Comput. Neural Syst.* **13**, 67–113, (2002). See also.[64]
32. I. Bojak and D. T. J. Liley, Modeling the effects of anesthesia on the electroencephalogram, *Phys. Rev. E.* **71**, 041902, (2005).
33. M. L. Hines and N. T. Carnevale, NEURON: a tool for neuroscientists, *The Neuroscientist.* **7**, 123–135, (2001).
34. J. Bower and D. Beeman, *The Book of GENESIS: Exploring Realistic Neural Models with the GEneral NEural SImulation System.* (Springer-Verlag, New York, 1998), 2nd edition.
35. H. C. Tuckwell, *Introduction to Theoretical Neurobiology.* vol. Volume 1, Linear Cable Theory and Dendritic Structure, (CUP, Cambridge, 1988).
36. P. A. Robinson, C. J. Rennie, and J. J. Wright, Propagation and stability of waves of electrical activity in the cerebral cortex, *Phys. Rev. E.* **56**, 826–840, (1997).
37. S. H. Strogatz, *Nonlinear dynamics and chaos.* (Westview Press, Cambridge, MA, 2000).
38. W. H. Press, B. P. Flannery, S. A. Teukolsky, and W. T. Vetterling, *Numerical Recipes: The Art of Scientific Computing.* (Cambridge University Press, Cambridge (UK) and New York, 1992), 2nd edition.
39. I. Bojak, D. T. J. Liley, P. J. Cadusch, and K. Cheng, Electrorhythmogenesis and anaesthesia in a physiological mean field theory, *Neurocomp.* **58-60**, 1197–1202, (2004).
40. C. Uhl and R. Friedrich. Spatio-temporal modeling based on dynamical systems theory. In ed. C. Uhl, *Analysis of Neurophysiological Brain Functioning*, pp. 274–306. Springer-Verlag, Berlin, (1999).
41. M. P. Dafilis, D. T. J. Liley, and P. J. Cadusch, Robust chaos in a model of the electroencephalogram: Implications for brain dynamics, *Chaos.* **11**, 474–478, (2001).
42. I. Bojak and D. T. J. Liley, Self-organized 40 Hz synchronization in a physiological theory of eeg, *Neurocomp.* **70**, 2085–2090, (2007).
43. F. H. Lopes da Silva, W. Blanes, S. N. Kalitzin, J. Parra, P. Suffczyński, and D. N. Velis, Dynamical diseases of brain systems: Different routes to epileptic seizures, *IEEE Trans. Biomed. Eng.* **50**, 540–548, (2003).
44. M. P. Dafilis, F. Frascoli, P. J. Cadusch, and D. T. J. Liley, Chaos and generalised multistability in a mesoscopic model of the electroencephalogram, *Physica D* **238**, 1056–1060 (2009).
45. E. J. Doedel, B. E. Oldeman, A. R. Champneys, F. Dercole, T. Fairgrieve, Y. Kuznetsov, R. Paffenroth, B. Sandstede, X. Wang, and C. Zhang, *Computer code AUTO-07P: continuation and bifurcation software for ordinary*

differential equations. (Concordia University, Canada, 2007).
46. Y. A. Kuznetsov, *Elements of Applied Bifurcation Theory.* (Springer-Verlag, New York, 2004).
47. E. J. Doedel, H. B. Keller, and J. P. Kernevez, Numerical analysis and control of bifurcation problems: (i) bifurcation in finite dimensions, *Int. J. Bif. Chaos.* **1**, 493–520, (1991).
48. L. van Veen and D. T. J. Liley, Chaos via Shilnikov's saddle-node bifurcation in a theory of the electroencephalogram, *Phys. Rev. Lett.* **97**, 208101, (2006).
49. L. P. Shilnikov and A. Shilnikov, Shilnikov saddle-node bifurcation, *Scholarpedia.* **3**(4), 4789, (2008).
50. G. Buzsáki, *Rhythms of the Brain.* (Oxford University Press, New York, 2006).
51. D. T. J. Liley, P. J. Cadusch, M. Gray, and P. J. Nathan, Drug-induced modification of the system properties associated with spontaneous human electroencephalographic activity, *Phys. Rev. E.* **68**, 051906, (2003).
52. D. T. J. Liley, I. Bojak, M. P. Dafilis, L. van Veen, F. Frascoli, and B. Foster. Bifurcations and state changes in the human alpha rhythm: theory and experiment. In eds. A. Steyn-Ross and M. Steyn-Ross, *Modeling Phase Transitions in Brain Activity.* Springer-Verlag, Berlin, (2009).
53. B. E. Oldeman, B. Krauskopf, and A. R. Champneys, Death of period doublings: Locating the homoclinic doubling cascade, *Physica D.* **146**, 100–120, (2000).
54. H. Rotstein, M. Wechselberger, and N. Kopell, Canard induced mixed-mode oscillations in a medial entorhinal cortex layer ii stellate cell model, *SIAM J. Appl. Dyn. Syst.* **7**, 1582–1611, (2008).
55. I. Erchova and D. J. McConigle, Rhythms of the brain: an examination of mixed mode oscillation approaches to the analysis of neurophysiological data, *Chaos.* **18**, 015115–015115–14, (2008).
56. H. S. Lukatch, C. E. Kiddoo, and M. B. MacIver, Anesthetic-induced burst suppression EEG activity requires glutamate-mediated excitatory synaptic transmission, *Cereb. Cortex.* **15**, 1322–1331, (2005).
57. H. Kantz and T. Schreiber, *Nonlinear time series analysis.* (Cambridge University Press, New York, 2003), 2nd edition.
58. M. Small and C. K. Tse, Applying the method of surrogate data to cyclic time series, *Physica D.* **164**, 187–201, (2002).
59. J. Theiler, P. S. Linsay, and D. M. Rubin. Detecting nonlinearity in data with long coherence times. time series prediction: forecasting the future and understanding the past. In eds. A. S. Weigend and N. A. Gershenfeld, *Proceedings of SFI Studies in Complexity*, vol. XV, pp. 429–55. Addison-Wesley, Reading, CA, (1993).
60. A. Dhooge, W. Govaerts, and Y. A. Kuznetsov, Numerical analysis and control of bifurcation problems: (i) bifurcation in finite dimensions, *ACM Trans. Math. Software.* **1**, 141–164, (2003).
61. B. Ermentrout, *Simulating, analyzing, and animating dynamical systems: a guide to XPPAUT for researchers and students.* (SIAM, Philadelpha, PA, 2002).

62. E. J. Doedel, B. W. Kooi, G. A. K. V. Voorn, and Y. A. Kuznetsov, Continuation of connecting orbits in 3d-odes: (i) point-to-cycle connections, *Int. J. Bif. Chaos.* **18**, 1889–1903, (2008).
63. I. Bojak and D. T. J. Liley, Modeling the effects of anesthesia on the electroencephalogram, *Phys Rev E.* **71**(4 Pt 1), 041902, (2005).
64. D. T. J. Liley, P. J. Cadusch, and M. P. Dafilis, Corrigendum: A spatially continuous mean field theory of electrocortical activity, *Network: Comput. Neural Syst.* **14**, 369, (2003).

Chapter 8

Jaynes' Maximum Entropy Principle, Riemannian Metrics and Generalised Least Action Bound

Robert K. Niven[1] and Bjarne Andresen[2]

[1]*School of Aerospace, Civil and Mechanical Engineering, The University of New South Wales at ADFA, Canberra, ACT, 2600, Australia.* *r.niven@adfa.edu.au*

[2]*Niels Bohr Institute, University of Copenhagen, Copenhagen Ø, Denmark.* *andresen@fys.ku.dk*

The set of solutions inferred by the generic maximum entropy (MaxEnt) or maximum relative entropy (MaxREnt) principles of Jaynes – considered as a function of the moment constraints or their conjugate Lagrangian multipliers – is endowed with a Riemannian geometric description, based on the second differential tensor of the entropy or its Legendre transform (negative Massieu function). The analysis provides a generalised *least action bound* applicable to all Jaynesian systems, which provides a lower bound to the cost (in generic entropy units) of a transition between inferred positions along a specified path, at specified rates of change of the control parameters. The analysis therefore extends the concepts of "finite time thermodynamics" to the generic Jaynes domain, providing a link between purely static (stationary) inferred positions of a system, and dynamic transitions between these positions (as a function of time or some other coordinate). If the path is unspecified, the analysis gives an absolute lower bound for the cost of the transition, corresponding to the geodesic of the Riemannian hypersurface. The analysis is applied to (i) an equilibrium thermodynamic system subject to mean internal energy and volume constraints, and (ii) a flow system at steady state, subject to constraints on the mean heat, mass and momentum fluxes and chemical reaction rates. The first example recovers the *minimum entropy cost* of a transition between equilibrium positions, a widely used result of finite-time thermodynamics. The second example leads to a new *minimum entropy production principle*, for the cost of a transition between steady state positions of a flow system.

Contents

8.1. Introduction

Jaynes' *maximum entropy principle* (MaxEnt) and its extension, the *maximum relative entropy principle* (MaxREnt), based on the principles of inductive (probabilistic) rather than deductive reasoning, arguably constitutes one of the most important tools for the solution of indeterminate problems of all kinds.[1–7] In this method, one maximises the entropy function of a system – a measure of its statistical spread over its parameter space – subject to the set of constraints on the system, to determine its "least informative" or "most probable" probability distribution.[1,2,7] By a series of generic "Jaynes relations", this can then be used to calculate the macroscopic properties of the system, providing the best (inferred) description of the system, subject to all that is known about the system. Since its inception half a century ago, the MaxEnt and MaxREnt principles have been successfully applied to the analysis of a diverse range of systems, including in thermodynamics (its first and foremost application), solid and fluid mechanics, mathematical biology, transport systems, networks, economic, social and human systems.[1–9]

The aim of this study is to examine a valuable extension to Jaynes' generic approach, by endowing the set of solutions inferred by Jaynes' method – considered as a function of the set of moment constraints and/or their conjugate Lagrangian multipliers – with a Riemannian geometric interpretation, using a metric tensor furnished directly by Jaynes' method. The analysis leads to a generalised *least action bound* applicable to all

Jaynesian systems, which provides a lower bound for the cost (in generic entropy units) of a transition between different inferred positions of the system. The analysis therefore extends the concepts of "finite time thermodynamics", developed over the past three decades,[10–43] to the generic Jaynes domain. The analysis reveals a deep, underlying connection between the essentially static manifold of stationary positions predicted by Jaynes' method, and lower bounds for the cost of dynamic transitions between these positions.

The manuscript proceeds as follows. In §8.2, the theoretical principles of Jaynes' MaxEnt and MaxREnt methods are discussed, followed by an appraisal of a generalised free energy (generalised potential) concept associated with Jaynes' method. In §8.3, the concepts of a Riemannian metric, arc length and action sums and integrals are developed in a generic Jaynesian context, leading to a generic least action bound for transitions on the manifold of Jaynes solutions. Considerations of minimum path lengths, involving calculation of the geodesic in Riemannian space, are also discussed. In §8.4, the foregoing principles are applied to (i) an equilibrium thermodynamic system subject to mean internal energy and volume constraints, and (ii) a flow system at steady state, subject to constraints on the mean heat, mass and momentum fluxes and chemical reaction rates. The first example (§8.4.1) recovers the *minimum entropy cost* of a transition between equilibrium positions, a widely used result of finite-time thermodynamics. The second example (§8.4.2) leads to a new *minimum entropy production principle*, for the cost of a transition between steady state positions of a flow system. The analyses reveal the tremendous utility of Jaynes' MaxEnt and MinXEnt methods augmented by the least action bound, for the analysis of probabilistic systems of all kinds.

8.2. Jaynes' Generic Formulation (MaxREnt)

8.2.1. *Theoretical principles*

The usefulness of Jaynes' method for statistical inference arises from its *generic* formulation, first expounded by Jaynes and other workers in the context of information theory,[1–7] but which can be reinterpreted using a combinatorial framework (the "Boltzmann principle").[44–51] In consequence, the method can be applied to *any* probabilistic system involving the allocation of entities to categories; this includes – but is not restricted to – thermodynamic systems. For maximum generality, it is useful to include source or prior probabilities q_i associated with each category $i = 1, ..., s$, to give the maximum relative entropy (MaxREnt) or minimum cross-entropy (MinXEnt) principle. In the event of equal q_i, this reduces to the special case of Jaynes' maximum entropy (MaxEnt) principle.[1–7]

The MaxREnt method proceeds as follows. To infer the "least informative" or "most probable" distribution of a probabilistic system, we wish to identify its observable *realization* or *macrostate* of maximum probability $\mathbb{P}$. This is equivalent to maximising the following dimensionless function, chosen for several "nice" mathematical properties:[44,45]

$$\mathfrak{H} = \frac{1}{N} \ln \mathbb{P}, \tag{8.1}$$

For a system of N distinguishable entities allocated to s distinguishable categories, it can be shown that the distribution is governed by the multinomial distribution $\mathbb{P} = N! \prod_{i=1}^{s} q_i^{n_i}/n_i!$, where n_i is the occupancy of the ith category and $N = \sum_{i=1}^{s} n_i$. In the asymptotic limit $N \to \infty$, (8.1) reduces to the relative entropy function[2] (the negative of the Kullback-Leibler function[52,53]):

$$\mathfrak{H} = -\sum_{i=1}^{s} p_i \ln \frac{p_i}{q_i} \tag{8.2}$$

where $p_i = n_i/N$ is the frequency or probability of occupancy of the ith category. Maximisation of (8.2) is subject to the normalisation constraint and any moment constraints on the system:

$$\sum_{i=1}^{s} p_i = 1, \tag{8.3}$$

$$\sum_{i=1}^{s} p_i f_{ri} = \langle f_r \rangle, \quad r = 1, ..., R, \tag{8.4}$$

where f_{ri} is the value of the property f_r in the ith category and $\langle f_r \rangle$ is the mathematical expectation of f_{ri}. Applying Lagrange's method of undetermined multipliers to (8.2)-(8.4) gives the stationary or "most probable" distribution of the system (denoted *):

$$\begin{aligned} p_i^* &= q_i \exp\left(-\lambda_0 - \sum_{r=1}^{R} \lambda_r f_{ri}\right) = \frac{1}{Z} q_i \exp\left(-\sum_{r=1}^{R} \lambda_r f_{ri}\right), \\ Z &= e^{\lambda_0} = \sum_{i=1}^{s} q_i \exp\left(-\sum_{r=1}^{R} \lambda_r f_{ri}\right) \end{aligned} \tag{8.5}$$

where λ_r is the Lagrangian multiplier associated with the rth constraint, Z is the partition function and $\lambda_0 = \ln Z$ is the Massieu function.[4] In thermodynamics, the constraints $\langle f_r \rangle$ are usually taken to represent conserved quantities, and thus correspond to extensive variables (e.g. internal energy, volume and numbers of particles), whilst the multipliers λ_r emerge as functions of the intensive variables of the system (e.g. temperature, pressure and chemical potentials). It is useful to preserve this distinction between extensive and intensive variables, even beyond a thermodynamic context.

By subsequent analyses,[1–7,54] one can derive the maximum relative entropy $\mathfrak{H}^*$ and the derivatives of $\mathfrak{H}^*$ and λ_0 for the system:

$$\mathfrak{H}^* = \lambda_0 + \sum_{r=1}^{R} \lambda_r \langle f_r \rangle \tag{8.6}$$

$$\frac{\partial \mathfrak{H}^*}{\partial \langle f_r \rangle} = \lambda_r \tag{8.7}$$

$$\frac{\partial^2 \mathfrak{H}^*}{\partial \langle f_m \rangle \partial \langle f_r \rangle} = \frac{\partial \lambda_r}{\partial \langle f_m \rangle} = g_{mr} \in \mathbf{g} \tag{8.8}$$

$$\frac{\partial \lambda_0}{\partial \lambda_r} = -\langle f_r \rangle \tag{8.9}$$

$$\frac{\partial^2 \lambda_0}{\partial \lambda_m \partial \lambda_r} = \langle f_r f_m \rangle - \langle f_r \rangle \langle f_m \rangle = -\frac{\partial \langle f_r \rangle}{\partial \lambda_m} = -\gamma_{mr} \in -\boldsymbol{\gamma} \tag{8.10}$$

The second derivatives of λ_0 in (8.10) express the dependence of each constraint on each multiplier, and therefore give the "capacities" or "susceptibilities" of the system (e.g. in thermodynamics, they define the heat capacity, compressibility, coefficient of thermal expansion and other material properties[12,55,56]). Their matrix $\boldsymbol{\gamma}$, the variance-covariance matrix of the constraints (with change of sign), is equal to the inverse of the matrix $\mathbf{g}$ of second derivatives of $\mathfrak{H}^*$ in (8.8), yielding the generic Legendre transformation between the $\mathfrak{H}^*(\langle f_1 \rangle, \langle f_2 \rangle, ...)$ and $\lambda_0(\lambda_1, \lambda_2, ...)$ descriptions of the system:[2]

$$\mathbf{g}\boldsymbol{\gamma} = \mathbf{I}, \tag{8.11}$$

where $\mathbf{I}$ is the identity matrix.[2] From (8.8) or (8.10) and the equality of mixed derivatives, we also obtain the generic reciprocal relations $\partial \langle f_r \rangle / \partial \lambda_m = \partial \langle f_m \rangle / \partial \lambda_r$ for the system.

Jaynes also showed that the incremental change in the relative entropy can be expressed as:[1]

$$d\mathfrak{H}^* = \sum_{r=1}^{R} \lambda_r \Big(d\langle f_r \rangle - \langle df_r \rangle \Big) = \sum_{r=1}^{R} \lambda_r \delta Q_r \tag{8.12}$$

where $\delta W_r = \langle df_r \rangle = \sum_{i=1}^{s} p_i^* df_{ri}$ and $\delta Q_r = \sum_{i=1}^{s} dp_i^* f_{ri}$ can be identified, respectively, as the increments of "generalised work" and "generalised heat" associated with a change in the rth constraint, and $\delta(\cdot)$ indicates a path-dependent differential. Eq. (8.12) gives a "generalised Clausius equality",[57] applicable to all multinomial systems in the asymptotic limit.

It is again emphasised that the above relations (8.5)-(8.12) apply to *any* probabilistic system of multinomial form, in the asymptotic limit. Although originally derived in thermodynamics, the above-mentioned quantities need not be interpreted as thermodynamic constructs, but have far broader application. Furthermore, the relations (8.5)-(8.12) apply to the stationary position of any multinomial probabilistic system. The derivatives (8.7)-(8.10) therefore relate to transitions of the system between different stationary positions, or in other words, to paths on the manifold of stationary positions. Whilst the lack of inclusion of non-stationary positions may seem unnecessarily restrictive, such geometry provides a sufficient foundation for most of engineering and chemical equilibrium thermodynamics. As will be shown, it is also useful for the analysis of many other systems of similar probabilistic structure.

8.2.2. *The generalised free energy concept*

It is instructive to insert (8.12) into the differential of (8.6) and rearrange in the form:

$$\begin{aligned} d\phi = -d\lambda_0 = -d\ln Z &= \sum_{r=1}^{R} \lambda_r \delta W_r + \sum_{r=1}^{R} d\lambda_r \langle f_r \rangle \\ &= -d\mathfrak{H}^* + \sum_{r=1}^{R} \lambda_r d\langle f_r \rangle + \sum_{r=1}^{R} d\lambda_r \langle f_r \rangle \end{aligned} \tag{8.13}$$

The negative Massieu function $-\lambda_0$ is therefore equivalent to a potential function ϕ which captures all possible changes in the system, whether they be in the entropy, constraints or multipliers. For constant multipliers, it simplifies to the weighted sum of generalised work on the system. It thus provides a dimensionless analogue of the free energy concept used in thermodynamics. For constant multipliers, $\phi|_{\{\lambda_r\}}$ also provides a measure of the dimensionless "availability", or the available "weighted generalised work", which can be extracted from a system. By extension of the principles of equilibrium thermodynamics, we can thus adopt the potential ϕ as a measure of distance from the stationary state. The system will converge towards a position of minimum ϕ, representing the balance between maximisation of entropy within the system $\mathfrak{H}^*$, and maximisation of the entropy generated and exported to the rest of the universe by the transfer of generalised heats δQ_r (see[63] for further discussion). The advantage of Jaynes' generic formulation is that ϕ can be defined for *any* multinomial probabilistic system, and is not restricted to thermodynamic systems.[1–4,7]

Returning to the second derivatives in the last section, we see that λ_0 can be replaced by $-\phi$ in (8.9)-(8.10). The latter provides a clean (non-negative) Legendre transformation between matrices $\mathbf{g}$ and $\boldsymbol{\gamma}$, and thus between the $\mathfrak{H}^*(\langle f_1\rangle, \langle f_2\rangle, ...)$ and $\phi(\lambda_1, \lambda_2, ...)$ representations of a system.

8.3. Riemannian Geometric Concepts

8.3.1. *Generalised Riemannian metrics and arc lengths*

Since the time of Gibbs,[58,59] examination of the geometry of the manifold of stationary positions has been of tremendous interest to scientists and engineers. In thermodynamics, this has typically involved analysis of the concave hypersurface defined by the Euler relation $S(\tilde{X}_1, \tilde{X}_2, ...)$, where S is the thermodynamic entropy and $\tilde{X}_r$ are the extensive variables, or alternatively of its Legendre transform, the convex hypersurface $\psi(Y_1, Y_2, ...)$ or $F(Y_1, Y_2, ...)$, where $\psi = F/T$ is a Planck potential function, F is a free energy and Y_r are the intensive variables.[55,56] Such interpretations have led to major advances in the understanding and analysis of thermodynamic processes and cycles.[58,59] However, adoption of the Jaynes MaxEnt framework (§8.2) permits a rather different insight, based on a Riemannian geometric interpretation. As will be evident from the previous discussion, this interpretation extends well beyond "mere" thermodynamics, forming a natural adjunct of Jaynes' generic formulation (§8.2).

Consider the R-dimensional hypersurface parameterised by the constraints $\{\langle f_r\rangle\}$ or their conjugate Lagrangian multipliers $\{\lambda_r\}$, representing the hypersurface of stationary states within the $(R+1)$-dimensional space given by $(\mathfrak{H}^*, \{\langle f_r\rangle\})$ or $(\phi, \{\lambda_r\})$. If the R parameters are linearly independent, the matrices of the second derivatives $\mathbf{g}$ (8.8) or $\boldsymbol{\gamma}$ (8.10) are positive definite (i.e. $\boldsymbol{x}^\top \mathbf{g} \boldsymbol{x} > 0$ or $\boldsymbol{x}^\top \boldsymbol{\gamma} \boldsymbol{x} > 0$ for any non-zero vector $\boldsymbol{x}$[60–62]). The matrices $\mathbf{g}$ or $\boldsymbol{\gamma}$ can therefore be adopted as Riemannian metric tensors associated with the stationary state hypersurface defined by $\{\langle f_r \rangle\}$ or $\{\lambda_r\}$, and used to interpret its geometric properties. Indeed, even if the R parameters are not always independent, whence $\mathbf{g}$ or $\boldsymbol{\gamma}$ are positive semidefinite (i.e. $\boldsymbol{x}^\top \mathbf{g} \boldsymbol{x} \geq 0$ or $\boldsymbol{x}^\top \boldsymbol{\gamma} \boldsymbol{x} \geq 0$ for $\boldsymbol{x} \neq \mathbf{0}$), the latter can still be adopted as pseudo-Riemannian metric tensors on the stationary hypersurface. This representation was first proposed by Weinhold,[10–14] and its implications in terms of a least action bound were subsequently developed, largely within a thermodynamic context, by Salamon, Berry, Andresen, Nulton and co-workers[15–36] and also by Beretta,[37–39] Diósi and co-workers,[40] Crooks and Feng[41,42] and Brody and Hook.[43] Some theoretical aspects of the adopted Riemannian formulation are discussed in Appendix A. It must be noted that the Riemannian formulation replaces – it cannot be used in conjunction with – the traditional convex or concave hypersurface interpretation normally used in thermodynamics and information theory.[30]

Firstly, the Riemannian geometric interpretation provides an *intrinsic differential* or *line element* (its square, a *metric*) with which to measure distances along a specified path on the manifold[60,62a]:

$$ds_{\mathfrak{H}^*} = \sqrt{d^2\mathfrak{H}^*} = \sqrt{\sum_{m,r=1}^{R} d\langle f_m \rangle \, g_{mr} \, d\langle f_r \rangle} = \sqrt{d\boldsymbol{f}^\top \mathbf{g} \, d\boldsymbol{f}}, \tag{8.14}$$

$$ds_{\phi} = \sqrt{d^2\phi} = \sqrt{\sum_{m,r=1}^{R} d\lambda_m \, \gamma_{mr} \, d\lambda_r} = \sqrt{d\boldsymbol{\Lambda}^\top \boldsymbol{\gamma} \, d\boldsymbol{\Lambda}}. \tag{8.15}$$

where $\boldsymbol{f} = [\langle f_1 \rangle, \langle f_2 \rangle, ..., \langle f_s \rangle]^\top$ and $\boldsymbol{\Lambda} = [\lambda_1, \lambda_2, ..., \lambda_s]^\top$. Integration between points a and b along a path on the manifold, defined by the set of increments $d\boldsymbol{f}$ or $d\boldsymbol{\Lambda}$, gives the *arc length* along that path between those points:[17,41,60,62]

$$L_{\mathfrak{H}^*} = \int_a^b ds_{\mathfrak{H}^*} = \int_a^b \sqrt{\sum_{m,r=1}^{R} d\langle f_m \rangle \, g_{mr} \, d\langle f_r \rangle} = \int_a^b \sqrt{d\boldsymbol{f}^\top \mathbf{g} \, d\boldsymbol{f}}, \tag{8.16}$$

$$L_{\phi} = \int_a^b ds_{\phi} = \int_a^b \sqrt{\sum_{m,r=1}^{R} d\lambda_m \, \gamma_{mr} \, d\lambda_r} = \int_a^b \sqrt{d\boldsymbol{\Lambda}^\top \boldsymbol{\gamma} \, d\boldsymbol{\Lambda}}. \tag{8.17}$$

The shortest such path is known as the *geodesic*. An infinite number of other paths on the manifold are also possible, of longer arc length, as also given

[a]Strictly, this line element is not a first fundamental form in Riemannian geometry;[60,62] its use as a distance measure is discussed in Appendix A.

by (8.16) or (8.17). If the manifold is parameterised by some parameter ξ – which can, but need not, correspond to time t – the arc lengths can be expressed in continuous form as:

$$L_{\mathfrak{H}^*} = \int_0^{\xi_{max}} \sqrt{\sum_{m,r=1}^{R} \frac{d\langle f_m \rangle}{d\xi} g_{mr} \frac{d\langle f_r \rangle}{d\xi}}\, d\xi = \int_0^{\xi_{max}} \sqrt{\dot{\boldsymbol{f}}^\top \mathbf{g}\, \dot{\boldsymbol{f}}}\, d\xi, \tag{8.18}$$

$$L_{\phi} = \int_0^{\xi_{max}} \sqrt{\sum_{m,r=1}^{R} \frac{d\lambda_m}{d\xi} \gamma_{mr} \frac{d\lambda_r}{d\xi}}\, d\xi = \int_0^{\xi_{max}} \sqrt{\dot{\boldsymbol{\Lambda}}^\top \boldsymbol{\gamma}\, \dot{\boldsymbol{\Lambda}}}\, d\xi \tag{8.19}$$

where the overdot indicates differentiation with respect to ξ.

The symmetry of the Legendre transformation (8.11) also permits a further insight. From (8.8) and (8.10), the metrics g_{mr} or γ_{mr} within the intrinsic differentials (8.14)-(8.15) can be substituted respectively by $\partial\lambda_r/\partial\langle f_m\rangle$ or $\partial\langle f_r\rangle/\partial\lambda_m$, to give:

$$ds_{\mathfrak{H}^*} = \sqrt{\sum_{m,r=1}^{R} d\langle f_m\rangle \frac{\partial \lambda_r}{\partial\langle f_m\rangle} d\langle f_r\rangle} = \sqrt{\sum_{r=1}^{R} d\lambda_r\, d\langle f_r\rangle} = \sqrt{d\boldsymbol{\Lambda}\cdot d\boldsymbol{f}} \tag{8.20}$$

$$ds_{\phi} = \sqrt{\sum_{m,r=1}^{R} d\lambda_m \frac{\partial\langle f_r\rangle}{\partial \lambda_m} d\lambda_r} = \sqrt{\sum_{r=1}^{R} d\lambda_r\, d\langle f_r\rangle} = \sqrt{d\boldsymbol{\Lambda}\cdot d\boldsymbol{f}} \tag{8.21}$$

In consequence, the intrinsic differentials are equal, $ds = ds_{\mathfrak{H}^*} = ds_\phi$, and so too are the arc lengths:

$$L = L_{\mathfrak{H}^*} = L_\phi = \int_0^{\xi_{max}} \sqrt{\dot{\boldsymbol{\Lambda}}\cdot\dot{\boldsymbol{f}}}\, d\xi, \tag{8.22}$$

From a Riemannian geometric perspective, it therefore does not matter whether one examines a system using its $\mathfrak{H}^*(\langle f_1\rangle, \langle f_2\rangle, ...)$ or $\phi(\lambda_1, \lambda_2, ...)$ representation. The above identities – touched on by several workers[27,34,40,41] – are not surprising, since the Legendre transforms $\mathfrak{H}^*$ and ϕ both have the character of entropy-related quantities, respectively indicating the (generic) entropy of a system and the capacity of a system to generate (generic) entropy.[63] The quantity $d\boldsymbol{\Lambda}\cdot d\boldsymbol{f}$ therefore expresses the second differential of generic entropy produced due to incremental changes in $\boldsymbol{\Lambda}$ and $\boldsymbol{f}$ (a generalised force-displacement or fluctuation-response relation). For all changes, $d\boldsymbol{\Lambda}\cdot d\boldsymbol{f} \geq 0$ must be valid, to preserve a positive definite metric (whence $\dot{\boldsymbol{\Lambda}} \cdot \dot{\boldsymbol{f}} \geq 0$);[12] this is in sympathy with a generalised form of the second law of thermodynamics, namely "*each net mean increment of (generic) entropy produced along a path must be positive*".

One further consideration arises from the recognition that most probabilistic systems involve *quantised* phenomena, which can only be approximated by the above continuous representation. For a system capable only of discrete increments in line elements $\Delta s_{\mathfrak{H}^*}$ or Δs_{ϕ} associated with a minimum dissipation parameter $\Delta\xi$ (e.g. a minimum dissipation time if $\xi = t$), the arc lengths are more appropriately given as:[25]

$$L_{\mathfrak{H}^*} = \sum_{\upsilon=1}^{M} \Delta s_{\mathfrak{H}^*,\upsilon} = \sum_{\upsilon=1}^{M} \sqrt{\Delta \boldsymbol{f}_{\upsilon}^{\top}\, \mathbf{g}_{\upsilon}\, \Delta \boldsymbol{f}_{\upsilon}} = \sum_{\upsilon=1}^{M} \sqrt{\dot{\boldsymbol{f}}_{\upsilon}^{\top}\, \mathbf{g}_{\upsilon}\, \dot{\boldsymbol{f}}_{\upsilon}}\, \Delta\xi_{\upsilon}, \tag{8.23}$$

$$L_{\phi} = \sum_{\upsilon=1}^{M} \Delta s_{\phi,\upsilon} = \sum_{\upsilon=1}^{M} \sqrt{\Delta \boldsymbol{\Lambda}_{\upsilon}^{\top}\, \gamma_{\upsilon}\, \Delta \boldsymbol{\Lambda}_{\upsilon}} = \sum_{\upsilon=1}^{M} \sqrt{\dot{\boldsymbol{\Lambda}}_{\upsilon}^{\top}\, \gamma_{\upsilon}\, \dot{\boldsymbol{\Lambda}}_{\upsilon}}\, \Delta\xi_{\upsilon} \tag{8.24}$$

where υ is the index of each increment. The last terms in (8.23)-(8.24) invoke the finite difference forms $\dot{\boldsymbol{f}}_{\upsilon} = \Delta \boldsymbol{f}_{\upsilon}/\Delta\xi_{\upsilon}$ or $\dot{\boldsymbol{\Lambda}}_{\upsilon} = \Delta\boldsymbol{\Lambda}_{\upsilon}/\Delta\xi_{\upsilon}$, strictly valid only in the limits $\Delta\xi_{\upsilon} \to 0$. The two discrete length scales (8.23)-(8.24) are again equivalent, but there will most likely be some discrepancy between their values due to their finite difference formulation.

8.3.2. *Generalised action concepts and least action bound*

A Riemannian geometry can also be examined from a different perspective,[15,17,25,33,41] discussed with reference to Figure 8.1; the following analysis largely follows that given by Nulton and coworkers,[25] converted into generic form. Although applied to $\mathfrak{H}^*$, an analogous derivation can be given for the ϕ representation. Consider a system on the manifold of stationary positions, subject to displacements $\{\Delta\langle f_r\rangle\}$ in its stationary position. The modified (generic) entropy $\mathfrak{H}^*(\{\langle f_r\rangle + \Delta\langle f_r\rangle\})$ of the system can be ex-

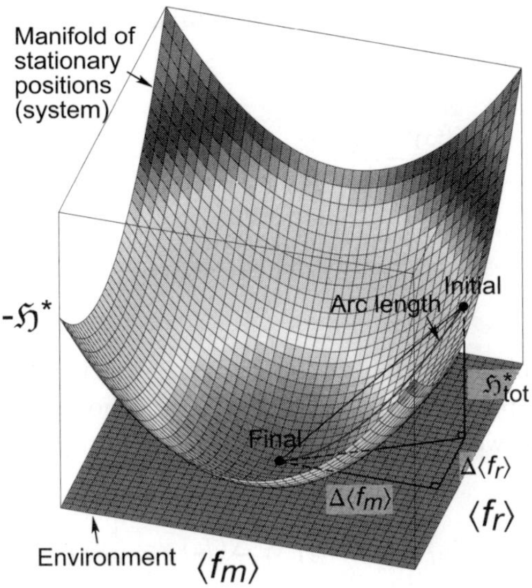

Fig. 8.1. Illustration of Riemannian geometry concepts, for a two-constraint system represented by $\mathfrak{H}^*(\langle f_m\rangle, \langle f_r\rangle)$ (for convenience the environment is shown as horizontal).

panded in a Taylor series about $\mathfrak{H}^*(\{\langle f_r\rangle\})$:

$$\begin{aligned}\mathfrak{H}^*(\{\langle f_r\rangle{+}\Delta\langle f_r\rangle\}) &= \mathfrak{H}^*(\{\langle f_r\rangle\}) + \sum_{r=1}^{R} \lambda_r|_{\{\langle f_r\rangle\}} \Delta\langle f_r\rangle \\ &+ \frac{1}{2!}\sum_{m,r=1}^{R} \left.\frac{\partial^2 \mathfrak{H}^*}{\partial\langle f_m\rangle \partial\langle f_r\rangle}\right|_{\{\langle f_r\rangle\}} \Delta\langle f_m\rangle \Delta\langle f_r\rangle \\ &+ \frac{1}{3!}\sum_{m,r,\ell=1}^{R} \left.\frac{\partial^3 \mathfrak{H}^*}{\partial\langle f_\ell\rangle \partial\langle f_m\rangle \partial\langle f_r\rangle}\right|_{\{\langle f_r\rangle\}} \Delta\langle f_\ell\rangle \Delta\langle f_m\rangle \Delta\langle f_r\rangle + \ldots\end{aligned} \tag{8.25}$$

where use is made of (8.7). The corresponding change in entropy of the "reservoir" or "environment" of constant $\{\lambda_r^{env}\}$, by which this change is effected, is given (exactly) by:[17,25]

$$\mathfrak{H}^{env}(\{\langle f_r\rangle{+}\Delta\langle f_r\rangle\}) = \mathfrak{H}^{env}(\{\langle f_r\rangle\}) + \sum_{r=1}^{R} \lambda_r^{env}|_{\{\langle f_r\rangle\}} \Delta\langle f_r\rangle^{env} \tag{8.26}$$

At the stationary state, $\lambda_r = \lambda_r^{env}$, whilst from the constraints (conservation laws), $\Delta\langle f_r\rangle = -\Delta\langle f_r\rangle^{env}$.[25] Addition of (8.25)-(8.26) thus yields the total change in the entropy of the system and environment for the step process:

$$\begin{aligned}\Delta\mathfrak{H}^* &= \frac{1}{2!}\sum_{m,r=1}^{R} \left.\frac{\partial^2 \mathfrak{H}^*}{\partial\langle f_m\rangle \partial\langle f_r\rangle}\right|_{\{\langle f_r\rangle\}} \Delta\langle f_m\rangle\Delta\langle f_r\rangle \\ &+ \frac{1}{3!}\sum_{m,r,\ell=1}^{R} \left.\frac{\partial^3 \mathfrak{H}^*}{\partial\langle f_\ell\rangle\partial\langle f_m\rangle \partial\langle f_r\rangle}\right|_{\{\langle f_r\rangle\}} \Delta\langle f_\ell\rangle\Delta\langle f_m\rangle\Delta\langle f_r\rangle + ...\end{aligned} \tag{8.27}$$

Provided the manifold is smooth, continuous, continuously differentiable (i.e., there are no phase changes in the neighbourhood) and the step sizes $\{\Delta\langle f_r\rangle\}$ are small, we can neglect the higher order terms in (8.27), giving:

$$\Delta\mathfrak{H}_\upsilon^* \approx \frac{1}{2}\sum_{m,r=1}^{R} \Delta\langle f_m\rangle_\upsilon \, g_{mr,\upsilon}|_{\{\langle f_r\rangle\}} \, \Delta\langle f_r\rangle_\upsilon = \frac{1}{2}\Delta\boldsymbol{f}_\upsilon{}^\top \, \mathbf{g}_\upsilon \, \Delta\boldsymbol{f}_\upsilon \tag{8.28}$$

where the subscript denotes the υth equilibration step. In the ϕ representation, the analogous form is obtained (in this case, giving the loss in ϕ):

$$-\Delta\phi_\upsilon \approx \frac{1}{2}\sum_{m,r=1}^{R} \Delta\lambda_{m,\upsilon} \, \gamma_{mr,\upsilon}|_{\{\lambda_r\}} \, \Delta\lambda_{r,\upsilon} = \frac{1}{2}\Delta\boldsymbol{\Lambda}_\upsilon{}^\top \, \boldsymbol{\gamma}_\upsilon \, \Delta\boldsymbol{\Lambda}_\upsilon \tag{8.29}$$

The total increase in entropy or decrease in potential of the system and environment subject to a M-step process is therefore:

$$\Delta\mathfrak{H}_{tot}^* = \sum_{\upsilon=1}^{M} \Delta\mathfrak{H}_\upsilon^* \approx \sum_{\upsilon=1}^{M} \frac{1}{2}\Delta\boldsymbol{f}_\upsilon{}^\top \, \mathbf{g}_\upsilon \, \Delta\boldsymbol{f}_\upsilon = \sum_{\upsilon=1}^{M} \frac{1}{2}\dot{\boldsymbol{f}}_\upsilon{}^\top \, \mathbf{g}_\upsilon \, \dot{\boldsymbol{f}}_\upsilon \, \Delta\xi_\upsilon\Delta\xi_\upsilon, \tag{8.30}$$

$$-\Delta\phi_{tot} = -\sum_{\upsilon=1}^{M} \Delta\phi_\upsilon \approx \sum_{\upsilon=1}^{M} \frac{1}{2}\Delta\boldsymbol{\Lambda}_\upsilon{}^\top \, \boldsymbol{\gamma}_\upsilon \, \Delta\boldsymbol{\Lambda}_\upsilon = \sum_{\upsilon=1}^{M} \frac{1}{2}\dot{\boldsymbol{\Lambda}}_\upsilon{}^\top \, \boldsymbol{\gamma}_\upsilon \, \dot{\boldsymbol{\Lambda}}_\upsilon \, \Delta\xi_\upsilon\Delta\xi_\upsilon \tag{8.31}$$

Recognising $\Delta\xi_\upsilon$ as the minimum dissipation parameter for the υth step (e.g. the minimum dissipation time if $\xi = t$), one such term may be factored out of each sum, to give the *mean minimum dissipation parameter* $\bar{\epsilon}_n$ for

$n \in \{\mathfrak{H}^*, \phi\}$.[17,25] This gives $\Delta \mathfrak{H}^*_{tot} = \bar{\epsilon}_{\mathfrak{H}^*} \mathcal{J}_{\mathfrak{H}^*}$ or $-\Delta \phi_{tot} = \bar{\epsilon}_\phi \mathcal{J}_\phi$, with:

$$\mathcal{J}_{\mathfrak{H}^*} = \sum_{\upsilon=1}^{M} \frac{1}{2} \dot{\boldsymbol{f}}_\upsilon^{\top} \, \mathbf{g}_\upsilon \, \dot{\boldsymbol{f}}_\upsilon \, \Delta\xi_\upsilon, \tag{8.32}$$

$$\mathcal{J}_\phi = \sum_{\upsilon=1}^{M} \frac{1}{2} \dot{\boldsymbol{\Lambda}}_\upsilon^{\top} \, \boldsymbol{\gamma}_\upsilon \, \dot{\boldsymbol{\Lambda}}_\upsilon \, \Delta\xi_\upsilon \tag{8.33}$$

The summands $1/2 \dot{\boldsymbol{f}}_\upsilon^{\top} \mathbf{g}_\upsilon \dot{\boldsymbol{f}}_\upsilon$ or $1/2 \dot{\boldsymbol{\Lambda}}_\upsilon^{\top} \boldsymbol{\gamma}_\upsilon \dot{\boldsymbol{\Lambda}}_\upsilon$ in (8.32)-(8.33) can be viewed as *generalised energy* terms, akin to the kinetic energy in mechanics, with the metric $\mathbf{g}_\upsilon$ or $\boldsymbol{\gamma}_\upsilon$ representing the "mass" and $\dot{\boldsymbol{f}}_\upsilon$ or $\dot{\boldsymbol{\Lambda}}_\upsilon$ the "velocity".[64] The terms $\mathcal{J}_{\mathfrak{H}^*}$ or $\mathcal{J}_\phi$ can then be interpreted as the discrete *generalised action* of the specified process,[41] again by analogy with mechanics[b]. From the previous considerations (§8.3), the two action sums are equivalent, although once again, discrepancies may emerge from their finite difference formulation.

From the discrete form of the Cauchy-Schwarz inequality:

$$\left(\sum_{\upsilon=1}^{M} {a_\upsilon}^2\right)\left(\sum_{\upsilon=1}^{M} {b_\upsilon}^2\right) \geq \left(\sum_{\upsilon=1}^{M} a_\upsilon b_\upsilon\right)^2 \tag{8.34}$$

with $a_\upsilon = \sqrt{\dot{\boldsymbol{f}}^{\top} \mathbf{g}_\upsilon \dot{\boldsymbol{f}}} \, \Delta\xi_\upsilon$ or $\sqrt{\dot{\boldsymbol{\Lambda}}^{\top} \boldsymbol{\gamma}_\upsilon \dot{\boldsymbol{\Lambda}}} \, \Delta\xi_\upsilon$ and $b_\upsilon = 1$, it can be shown that:[25]

$$\bar{\epsilon}_n \mathcal{J}_n \geq \frac{L_n^2}{2M} \tag{8.35}$$

Physically, the number of steps is equal to $M = \xi_{max}/\bar{\epsilon}_n$, whence (8.35) reduces to:[17,25]

$$\bar{\epsilon}_n \mathcal{J}_n \geq \frac{\bar{\epsilon}_n L_n^2}{2\,\xi_{max}} \qquad \text{or} \qquad \mathcal{J}_n \geq \frac{L_n^2}{2\,\xi_{max}} \tag{8.36}$$

Eqs. (8.35)-(8.36) can be considered a generalised *least action bound*,[41] applicable to all probabilistic systems amenable to analysis by Jaynes' method. Its physical interpretation is that it specifies the minimum cost or penalty, in units of dimensionless entropy per unit ξ, to move the system from one stationary position (at $\xi = 0$) to another (at $\xi = \xi_{max}$) along the given path at the specified rates $\dot{\boldsymbol{\Lambda}}$ and/or $\dot{\boldsymbol{f}}$. If the latter rates proceed infinitely slowly, the lower bound of the action is zero, indicating that

[b]Crooks[41] applies the terms "energy" and "action" interchangeably; we consider that the present definitions are more in keeping with those used in mechanics. Many authors include the $\bar{\epsilon}_n$ term within $\mathcal{J}_n$, but we here wish to preserve the mathematical structure of a generalised action principle.

the process can be conducted at zero cost; otherwise, it is necessary to "do generalised work" to move the system along the manifold of stationary positions within a finite parameter duration ξ_{max}.

The generalised least action bound thus provides a lower bound for the "transition cost" of a process (in entropy-related units). If the process is reversible, the cost would be zero, but no process can be reversible in practice. Identification of this minimum cost is of paramount importance: there is no point in undertaking expensive changes to the process, or initiating costly social or political changes, in the attempt to do better than the minimum predicted by (8.35)-(8.36). Taking a thermodynamic example, the method can be applied to determine the minimum cost of industrial processes such as work extraction from combustion, a question of fundamental importance to human society. Most thermodynamics and engineering textbooks give the Carnot limit as the theoretical limit of efficiency, but the limits imposed by finite time thermodynamics are more restrictive (see §8.4.1).

The generalised least action bound therefore emerges from the Riemannian geometry of the state space, and hence from somewhat different considerations than the principle of least action employed in mechanics.[64] We consider that the two principles are connected, but are unable to examine this topic further here. For further exploratory expositions, the reader is referred to the work of Crooks,[41] Caticha[65] and Wang.[67,68]

The above discrete sums (8.30)-(8.31) can also be presented in integral form. Consider a system represented by $\mathfrak{H}^*$, subjected to a finite change in the multipliers $\Delta\lambda_r$ due to movement of the reference environment. The incremental change in entropy is, again to first order (compare (8.27)):[17,25,33,41]

$$d\mathfrak{H}^* \approx \frac{1}{2!} \sum_{r=1}^{R} \Delta\lambda_r d\langle f_r \rangle \tag{8.37}$$

Substituting $\Delta\lambda_r = \sum\limits_{m=1}^{R} g_{mr} \Delta\langle f_m \rangle$ from (8.8), and assuming a first order decay process:

$$\langle \dot{f}_m \rangle = \frac{\langle f_m \rangle - \langle f_m \rangle_{env}}{\epsilon_{\mathfrak{H}^*}} = \frac{\Delta\langle f_m \rangle}{\epsilon_{\mathfrak{H}^*}} \tag{8.38}$$

where $\epsilon_{\mathfrak{H}^*}$ is a minimum dissipation parameter (reciprocal rate constant),

(8.37) yields:

$$d\mathfrak{H}^* = \frac{1}{2} \sum_{m,r=1}^{R} \langle \dot{f}_m \rangle \, g_{mr} \, d\langle f_r \rangle \, \epsilon_{\mathfrak{H}^*} \tag{8.39}$$

The total change in entropy $\Delta \mathfrak{H}^*_{tot} = \int_0^{\xi_{max}} d\mathfrak{H}^*$ is then obtained as:

$$\Delta \mathfrak{H}^*_{tot} = \int_0^{\xi_{max}} \frac{1}{2} \, \dot{\boldsymbol{f}}^\top \, \mathbf{g} \, \dot{\boldsymbol{f}} \, \epsilon_{\mathfrak{H}^*} \, d\xi = \bar{\epsilon}_{\mathfrak{H}^*} \int_0^{\xi_{max}} \frac{1}{2} \, \dot{\boldsymbol{f}}^\top \, \mathbf{g} \, \dot{\boldsymbol{f}} \, d\xi = \bar{\epsilon}_{\mathfrak{H}^*} \mathcal{J}_{\mathfrak{H}^*} \tag{8.40}$$

Similarly, in the ϕ representation, we obtain:

$$-\Delta \phi_{tot} = \int_0^{\xi_{max}} \frac{1}{2} \, \dot{\boldsymbol{\Lambda}}^\top \, \boldsymbol{\gamma} \, \dot{\boldsymbol{\Lambda}} \, \epsilon_\phi \, d\xi = \bar{\epsilon}_\phi \int_0^{\xi_{max}} \frac{1}{2} \, \dot{\boldsymbol{\Lambda}}^\top \, \boldsymbol{\gamma} \, \dot{\boldsymbol{\Lambda}} \, d\xi = \bar{\epsilon}_\phi \mathcal{J}_\phi \tag{8.41}$$

In the continuous representation, the process does not proceed by a series of finite steps; instead, the reference variables continuously move ahead of those of the system.[17,25,33] However, we still see the influence of a finite decay parameter ϵ_n, which on integration yields the mean parameter $\bar{\epsilon}_n$. Each $\mathcal{J}_n$ term above can be regarded as the *action integral* corresponding respectively to (8.32)-(8.33). Based on the integral form of the Cauchy-Schwarz inequality,[62] it can be shown that the integral actions also satisfy the least action bound (8.35)-(8.36), with L_n in integral form.[17,25,33,41]

Finally, for the least action bound (8.36) to achieve equality, the summands or integrands of the arc length L_n and action J_n must be constant. This gives the simple result that for slow processes with constant dissipation parameter $\epsilon_n = \bar{\epsilon}_n$, the minimum action (whence minimum in $\bar{\epsilon}_n J_n$) is attained by a process which proceeds at a constant speed $\dot{s}_n = \sqrt{\dot{\boldsymbol{f}}^\top \, \mathbf{g} \, \dot{\boldsymbol{f}}} = \sqrt{\dot{\boldsymbol{\Lambda}}^\top \, \boldsymbol{\gamma} \, \dot{\boldsymbol{\Lambda}}}$.[28,29,33] For a constant metric, this is equivalent to constant rates of change of the parameter vector $\dot{\boldsymbol{f}}$ and/or $\dot{\boldsymbol{\Lambda}}$. For systems with a variable dissipation parameter $\epsilon(\xi)$, it was first considered that the minimum is attained at the constant speed $ds/d\eta$, expressed in the "natural" parameter units $\eta = \xi/\epsilon$.[28,29,33,40] This however oversimplifies the minimisation problem, which is better handled within a discrete (stepwise) framework.[25,35] As discussed in §8.4.1, such principles have been widely applied to thermodynamic systems.

8.3.3. *Minimum path length principle*

The above discrete or continuous forms of the least action bound (8.35)-(8.36) are based on consideration of a specified path on the manifold of stationary positions, of arc length L_n. In many situations, we may wish to determine the path of minimum arc length $L_{n,min}$ – the *geodesic* – on the manifold of stationary positions. From the calculus of variations, this is given by the Euler–Lagrange equations:[66]

$$\frac{\partial \dot{s}_{\mathfrak{H}^*}}{\partial \boldsymbol{f}} - \frac{d}{d\xi}\frac{\partial \dot{s}_{\mathfrak{H}^*}}{\partial \dot{\boldsymbol{f}}} = \mathbf{0} \tag{8.42}$$

$$\frac{\partial \dot{s}_{\phi}}{\partial \boldsymbol{\Lambda}} - \frac{d}{d\xi}\frac{\partial \dot{s}_{\phi}}{\partial \dot{\boldsymbol{\Lambda}}} = \mathbf{0} \tag{8.43}$$

where $\dot{s}_{\mathfrak{H}^*} = \sqrt{\dot{\boldsymbol{f}}^\top \mathbf{g}\, \dot{\boldsymbol{f}}}$ and $\dot{s}_\phi = \sqrt{\dot{\boldsymbol{\Lambda}}^\top \boldsymbol{\gamma}\, \dot{\boldsymbol{\Lambda}}}$ are the integrands respectively of $L_{\mathfrak{H}^*}$ or L_ϕ (8.16)-(8.19). For two-dimensional parameters $\boldsymbol{f}, \boldsymbol{\Lambda} \in \mathbb{R}^2$, (8.42)-(8.43) can be reduced further in terms of the three unit normals to the surface, giving the curve(s) on the manifold for which the geodesic curvature vanishes.[43,60,62] Depending on the specified problem, a geodesic may not exist, or there may be multiple or il-defined solutions. Provided it does exist, a geodesic leads to the double minimisation principle:

$$\mathfrak{J}_n \geq \frac{L_n^2}{2\,\xi_{max}} \geq \frac{{L_{n,min}}^2}{2\,\xi_{max}} \tag{8.44}$$

where the right hand side indicates the absolute lower bound for the action, irrespective of path. This principle has been applied to thermodynamic systems, as will be discussed in §8.4.1.

8.4. Applications

As noted, the foregoing Riemannian geometric interpretation (§8.3) has mainly been presented within an equilibrium thermodynamics context,[15–43] although it has been applied to non-equilibrium thermodynamic and flow systems,[69–73] information coding[74] and in economics.[75] In the following sections, the utility of Riemannian geometric properties and the least action bound are demonstrated for two types of system: a thermodynamic system at equilibrium, and a flow system at steady state.

8.4.1. *Equilibrium thermodynamic systems*

The application of Riemannian geometric principles to equilibrium thermodynamic systems has constituted a major new development over the past three decades, forming an important plank of *finite-parameter* or (with $\xi = t$) *finite-time thermodynamics*.[21,34,35] Such analyses have progressed in four overlapping stages:

- The initial studies by Weinhold[10–14] and early work by Salamon, Andresen, Berry, Nulton and coworkers[15,17,18,20,25,28,29] all examined a manifold based on an internal energy representation $U(X_1, X_2, ...)$, as a function of extensive variables X_r, which include the thermodynamic entropy S. The resulting quantity $\bar{\epsilon}_U \mathcal{J}_U$ (in the present notation) was interpreted as an *availability* or *exergy* function, with (8.36) indicating the most efficient path (defined by the minimum amount of work or minimum loss of availability) required to move the equilibrium position of the system.[17,18] Such analyses complement the thermodynamic geometry used by Gibbs,[58,59] and fit well with the traditional heat-work framework of 19th century thermodynamics.
- Subsequently, following earlier pioneering works,[69,76] the entropy manifold $S(\tilde{X}_1, \tilde{X}_2, ...)$ was examined from a Riemannian perspective,[20,21,23,25,31–33,33,34,41,42] where $\tilde{X}_r$ are the new extensive variables, of course related to the $U(X_1, X_2, ...)$ representation by Jacobian transformation.[20] The quantity $\bar{\epsilon}_S \mathcal{J}_S$ was interpreted as a measure of *energy dissipation* or *entropy production*, again providing a measure of process efficiency. It was realised that the lower bound in (8.36) provides a formal, mathematical definition of the degree of irreversibility of a transition between equilibrium positions, with reversibility only for $\mathcal{J}_S = 0$ (a definition vastly preferable to the cumbersome word-play still used in thermodynamics references; see the scathing criticism by Truesdell[77]). However, the primacy of the entropy representation over that based on internal energy was not fully appreciated in these early studies. The applicability of Riemannian geometry in other contexts – based directly on the MaxEnt framework of Jaynes[1,2] – is hinted at by Levine,[27] but unfortunately was not developed further at the time, nor, to the authors' knowledge, in any subsequent studies.
- Several studies have considered an entropy representation based on a metric defined on a probability space $\{p_i\}$, either from the Boltzmann principle[76] or using a Shannon or relative entropy measure.[22–24,26,27,37–39,41–43]

Several authors[37–39,41–43]) have extended this analysis, to establish a connection with the Fisher information matrix[78] and an "entropy differential metric" of Rao.[79] The analysis is also intimately connected with paths in a space of square root probabilities, and thence to formulations of quantum mechanics.[37–39] These insights – not examined further here – demand further detailed attention; they may well furnish an explanation for the utility of extremisation methods based on the Fisher information function in many physical problems.[80]

- Finally, several workers realised that Riemannian geometric principles can be applied to Legendre-transformed representations, e.g. based on various forms of the free energy F (conjugate to U) or the negative Planck potential F/T (conjugate to S), as functions of the intensive variables (or functions thereof).[23,27,41,42] This approach offers particular advantages for the analysis of real thermodynamic systems, in which the control parameters tend to be intensive rather than extensive variables (the canonical ensemble), and for which the intensive variables do not exhibit sharp transitions or singularities associated with phase changes, as is the case for extensive variables.[41] Furthermore, the resulting metric is equivalent to the variance-covariance matrix of the constraints (8.10), and is therefore connected to fluctuation-dissipation processes within the system.

For completeness, we demonstrate – for a microcanonical thermodynamic system – how Riemannian geometric properties emerge as an inherent feature of Jaynes' MaxEnt formulation. Consider an isolated thermodynamic system, containing molecules of possible energy levels ϵ_i and volume elements V_j, subject to constraints on the mean energy $\langle U \rangle$ and mean volume $\langle V \rangle$. We consider the joint probability p_{ij} of a particle simultaneously occupying an energy level and volume element, giving the entropy function:

$$\mathfrak{H}_{eq} = -\sum_i \sum_j p_{ij} \ln p_{ij}, \tag{8.45}$$

where, without knowledge of any additional influences, we assume that each joint level ij is equally probable (hence the priors q_{ij} cancel out). Eq. (8.45) is maximised subject to the constraints:

$$\sum_i \sum_j p_{ij} = 1, \tag{8.46}$$

$$\sum_i \sum_j p_{ij} \epsilon_i = \langle U \rangle, \tag{8.47}$$

$$\sum_i \sum_j p_{ij} V_j = \langle V \rangle, \tag{8.48}$$

to give the equilibrium position:

$$p_{ij}^* = \frac{\exp(-\lambda_U U_i - \lambda_V V_j)}{\sum_i \sum_j \exp(-\lambda_U U_i - \lambda_V V_j)} = \frac{1}{Z} \exp(-\lambda_U U_i - \lambda_V V_j), \tag{8.49}$$

where Z is the partition function. From the existing body of thermodynamics, we can identify the Lagrangian multipliers as $\lambda_U = 1/kT$ and $\lambda_V = P/kT$, where k is the Boltzmann constant, T is absolute temperature and P is absolute pressure. Eq. (8.49) and Jaynes' relations (8.6)-(8.11) and (8.13) then reduce to:

$$p_{ij}^* = \frac{1}{Z} e^{-U_i/kT - PV_j/kT}, \tag{8.50}$$

$$S^* = k\mathfrak{H}_{eq}^* = k \ln Z + \frac{\langle U \rangle}{T} + \frac{P\langle V \rangle}{T} \tag{8.51}$$

$$k\,\boldsymbol{\Lambda}_{eq} = \left[\frac{\partial S^*}{\partial \langle U \rangle}, \frac{\partial S^*}{\partial \langle V \rangle}\right]^\top = \left[\frac{1}{T}, \frac{P}{T}\right]^\top \tag{8.52}$$

$$k\,\mathbf{g}_{eq} = \begin{bmatrix} \dfrac{\partial^2 S^*}{\partial \langle U \rangle^2}, & \dfrac{\partial^2 S^*}{\partial \langle U \rangle \partial \langle V \rangle} \\ \dfrac{\partial^2 S^*}{\partial \langle V \rangle \partial \langle U \rangle}, & \dfrac{\partial^2 S^*}{\partial \langle V \rangle^2} \end{bmatrix} = \begin{bmatrix} \dfrac{\partial}{\partial \langle U \rangle}\left(\dfrac{1}{T}\right), & \dfrac{\partial}{\partial \langle U \rangle}\left(\dfrac{P}{T}\right) \\ \dfrac{\partial}{\partial \langle V \rangle}\left(\dfrac{1}{T}\right), & \dfrac{\partial}{\partial \langle V \rangle}\left(\dfrac{P}{T}\right) \end{bmatrix} \tag{8.53}$$

$$\psi = k\phi_{eq} = -k \ln Z = -S^* + \frac{\langle U \rangle}{T} + \frac{P\langle V \rangle}{T} = \frac{G}{T} \tag{8.54}$$

$$\boldsymbol{f}_{eq} = \left[\frac{\partial \psi}{\partial (\frac{1}{T})}, \frac{\partial \psi}{\partial (\frac{P}{T})}\right]^\top = [\langle U \rangle, \langle V \rangle]^\top \tag{8.55}$$

$$\frac{\boldsymbol{\gamma}_{eq}}{k} = \begin{bmatrix} \dfrac{\partial^2 \psi}{\partial (\frac{1}{T})^2}, & \dfrac{\partial^2 \psi}{\partial (\frac{1}{T}) \partial (\frac{P}{T})} \\ \dfrac{\partial^2 \psi}{\partial (\frac{P}{T}) \partial (\frac{1}{T})}, & \dfrac{\partial^2 \psi}{\partial (\frac{P}{T})^2} \end{bmatrix} = \begin{bmatrix} \dfrac{\partial \langle U \rangle}{\partial (\frac{1}{T})}, & \dfrac{\partial \langle V \rangle}{\partial (\frac{1}{T})} \\ \dfrac{\partial \langle U \rangle}{\partial (\frac{P}{T})}, & \dfrac{\partial \langle V \rangle}{\partial (\frac{P}{T})} \end{bmatrix} \tag{8.56}$$

$$k\,\mathbf{g}_{eq} \frac{\boldsymbol{\gamma}_{eq}}{k} = \mathbf{I} \tag{8.57}$$

where S^* is the thermodynamic entropy at an equilibrium position, ψ is the negative Planck potential[81,82] (negative Massieu function[83]) and G is the Gibbs free energy. By Jacobian transformation of variables, using the following material properties (susceptibilities):[12,55,56]

$$\text{Heat capacity at constant pressure:} \qquad C_P = \left(\frac{\partial \langle H \rangle}{\partial T}\right)_P \tag{8.58}$$

$$\text{Isothermal compressibility:} \qquad \kappa_T = -\frac{1}{\langle V \rangle}\left(\frac{\partial \langle V \rangle}{\partial P}\right)_T \tag{8.59}$$

$$\text{Coefficient of thermal expansion:} \qquad \alpha = \frac{1}{\langle V \rangle}\left(\frac{\partial \langle V \rangle}{\partial T}\right)_P \tag{8.60}$$

where $\langle H \rangle = \langle U \rangle + P\langle V \rangle$ is the enthalpy, as well as the equality of cross-derivatives (Maxwell relation):

$$\frac{\partial \langle V \rangle}{\partial (\frac{1}{T})} = \frac{\partial \langle U \rangle}{\partial (\frac{P}{T})} \tag{8.61}$$

the ψ metric (8.56) reduces to[c]:

$$\frac{\boldsymbol{\gamma}_{eq}}{k} = T\langle V \rangle \begin{bmatrix} -\kappa_T P^2 + 2\,\alpha P T - \frac{C_P T}{\langle V \rangle}, & \kappa_T P - \alpha T \\ \kappa_T P - \alpha T, & -\kappa_T \end{bmatrix} \tag{8.62}$$

whence from (8.57):

$$k\mathbf{g}_{eq} = \frac{1}{T^2(\alpha^2 T \langle V \rangle - \kappa_T C_P)} \begin{bmatrix} \kappa_T, & \kappa_T P - \alpha T \\ \kappa_T P - \alpha T, & \kappa_T P^2 - 2\alpha P T + \frac{C_P T}{\langle V \rangle} \end{bmatrix} \tag{8.63}$$

Using (8.52), (8.55) and (8.62)-(8.63), the (dimensional) arc lengths (8.18)-(8.19) and action integrals (8.40)-(8.41) are obtained as:

$$\breve{L}_{S^*} = \int_0^{\xi_{max}} \sqrt{\dot{\boldsymbol{f}}_{eq}{}^{\top}\, k\mathbf{g}_{eq}\, \dot{\boldsymbol{f}}_{eq}}\, d\xi = \int_0^{\xi_{max}} \sqrt{\frac{N_{S^*}}{k^2 T^2}} d\xi \tag{8.64}$$

$$\breve{\mathcal{J}}_{S^*} = \int_0^{\xi_{max}} \frac{1}{2}\, \dot{\boldsymbol{f}}_{eq}{}^{\top}\, k\mathbf{g}_{eq}\, \dot{\boldsymbol{f}}_{eq}\, d\xi = \int_0^{\xi_{max}} \frac{N_{S^*}}{2k^2T^2} d\xi\,, \tag{8.65}$$

where $N_{S^*} \equiv -C_P \dot{T}^2 + 2\alpha \langle V \rangle T \dot{T} \dot{P} - \kappa_T \langle V \rangle T \dot{P}^2$, and

[c]The first variance is given erroneously by Callen.[55,56]

$$
\begin{aligned}
\breve{L}_\psi &= \int\limits_0^{\xi_{max}} \sqrt{\dot{\boldsymbol{\Lambda}}_{eq}^\top \frac{\boldsymbol{\gamma}_{eq}}{k} \dot{\boldsymbol{\Lambda}}_{eq}}\, d\xi \\
&= \int\limits_0^{\xi_{max}} \sqrt{\frac{N_\psi}{T^2\langle V\rangle(-C_P\kappa_T + \alpha^2 T\langle V\rangle)}}\, d\xi
\end{aligned}
\tag{8.66}
$$

$$
\begin{aligned}
\breve{\mathcal{J}}_\psi &= \int\limits_0^{\xi_{max}} \frac{1}{2} \dot{\boldsymbol{\Lambda}}_{eq}^\top \frac{\boldsymbol{\gamma}_{eq}}{k} \dot{\boldsymbol{\Lambda}}_{eq}\, d\xi \\
&= \int\limits_0^{\xi_{max}} \frac{N_\psi}{2T^2\langle V\rangle(-C_P\kappa_T + \alpha^2 T\langle V\rangle)} d\xi
\end{aligned}
\tag{8.67}
$$

where $N_\psi \equiv \langle V\rangle\langle\dot{U}\rangle\big[\kappa_T\langle\dot{U}\rangle + 2\langle\dot{V}\rangle(\kappa_T P - \alpha T)\big] + \langle\dot{V}\rangle^2\big[P\langle V\rangle(\kappa_T P - 2\alpha T) + C_P T\big]$.

Using (8.58)-(8.60), these two sets of measures can be shown to be equivalent. The above equations must be integrated along the particular thermodynamic path followed by the process, as defined by the velocities $\{\dot{T}, \dot{P}\}$ or $\{\langle\dot{U}\rangle, \langle\dot{V}\rangle\}$. For a process which follows a pre-determined path, e.g. an adiabatic, isothermal, isovolumetric or isopiezometric curve, this can be simplified by expressing the velocities (e.g. $\dot{P}$) as functions of one independent velocity (e.g. $\dot{T}$).

To comment on units: if the above quantities were calculated using the "pure" metrics $\mathbf{g}_{eq}$ or $\boldsymbol{\gamma}_{eq}$, in either case the line element ds_n, arc length L_n and the term $\bar{\epsilon}_n \mathcal{J}_n$ would be dimensionless (whence the action is in reciprocal ξ units). Use of the "natural" metric $k\mathbf{g}_{eq}$, as conducted here, gives the line element and arc length in $\sqrt{JK^{-1}}$ and the action in $JK^{-1}\xi^{-1}$, consistent with $\Delta S^*_{tot} = \bar{\epsilon}_{S^*} \mathcal{J}_{S^*}$ being in entropy units. In contrast, use of the "natural" metric $\boldsymbol{\gamma}_{eq}/k$ gives the line element and arc length in $\sqrt{KJ^{-1}}$ and the action in $KJ^{-1}\xi^{-1}$. The latter case can be rescued by use of a modified line element $d\breve{s}'_\psi = \sqrt{k\dot{\boldsymbol{\Lambda}}_{eq}^\top \boldsymbol{\gamma}_{eq}/k\, k\dot{\boldsymbol{\Lambda}}_{eq}}\, d\xi$ – as suggested by (8.52) and (8.56) – giving the line element and arc length in $\sqrt{JK^{-1}}$ and the action in $JK^{-1}\xi^{-1}$. Thus in both the S^* and ψ representations, the least action bound (8.36) can be used to determine the *minimum entropy cost* of a transition from one equilibrium position to another, along a specified path on the manifold of equilibrium positions. As noted earlier, for slow processes and constant ϵ_n, this is attained by a process which proceeds at a constant thermodynamic speed $\dot{s}$[28,29,33,40] (a more general result is available for rapid processes[84]). For variable ϵ_n and/or for stepwise phenomena, the process should be divided into individual steps placed at equal distances along the arc length traversed by the process, giving the so-called "equal thermodynamics distance" principle.[34–36] Such considerations have been applied to the optimisation of a wide variety of engineering and industrial batch and flow processes, including engine cycles, heat engines and pumps, chemical reactors, distillation towers and many other systems.

A final important point is that the minimum path length (double minimisation) principle (8.44) – involving calculation of the geodesic – has been applied to the analysis of equilibrium systems. In early work, this bound was established by applying the calculus of variations directly to particular thermodynamic problems, without use of a metric.[16,19] More recently, such lower bounds have been examined for particular thermodynamic systems.[34,43,100] In either case, for the entropy representation, this method yields an *absolute minimum entropy cost* $\Delta S^* \geq \Delta S^*_{min}$ for a transition between two equilibrium positions at particular rates of change, irrespective of the path. For cyclic or flow processes, this therefore gives a *minimum entropy production principle* $\dot{S} \geq \dot{S}_{min}$, providing one of the key concepts of finite-time (or finite-parameter) thermodynamics.

8.4.2. *Flow systems*

We now consider a flow system consisting of a control volume, subject to continuous flows of heat, particles and momentum, and within which chemical reactions may take place. A few workers have examined such non-equilibrium systems previously within a Riemannian context, including for the Onsager linear regime[69,70] and for extended irreversible thermodynamics.[71–73] A different perspective is provided here, based on a recent analysis of a flow system from a Jaynesian perspective.[63] This involves a probabilistic analysis of each infinitesimal element of the control volume, which experiences instantaneous values of the heat flux $\boldsymbol{j}_{Q,\mathcal{I}}$, mass fluxes $\boldsymbol{j}_{\mathcal{N}_c}$ of each species c, stress tensor $\boldsymbol{\tau}_{\mathcal{J}}$ and molar rate per unit volume $\hat{\dot{\xi}}_{\mathcal{L}_d}$ of each chemical reaction d, where the indices $\mathcal{I}, \mathcal{J}, \mathcal{L}_d, \mathcal{N}_c \in \{0, \pm 1, \pm 2, ...\}$. We therefore consider the joint probability $\pi_{\boldsymbol{I}} = \pi_{\mathcal{I},\mathcal{J},\{\mathcal{L}_d\},\{\mathcal{N}_c\}}$ of instantaneous fluxes through the element and instantaneous reactions within the element, giving the (dimensionless) "flux entropy" function:

$$\mathfrak{H}_{st} = -\sum_{\boldsymbol{I}} \pi_{\boldsymbol{I}} \ln \pi_{\boldsymbol{I}}. \tag{8.68}$$

Again assuming that each joint level $\boldsymbol{I}$ is equally probable, (8.68) is maximised subject to constraints on the mean values of the heat flux $\langle \boldsymbol{j}_Q \rangle$, mass fluxes $\langle \boldsymbol{j}_c \rangle$, stress tensor $\langle \boldsymbol{\tau} \rangle$ and molar reaction rates $\langle \hat{\dot{\xi}}_d \rangle$ through or within the element, as well as by the natural constraint (8.3). This gives the *steady state position* of the system:

$$\pi_{\boldsymbol{I}}^* = \frac{1}{\mathcal{Z}} \exp \left(-\boldsymbol{\zeta}_Q \cdot \boldsymbol{j}_{Q,\mathcal{I}} - \sum_c \boldsymbol{\zeta}_c \cdot \boldsymbol{j}_{\mathcal{N}_c} - \boldsymbol{\zeta}_{\boldsymbol{\tau}} : \boldsymbol{\tau}_{\mathcal{J}} - \sum_d \zeta_d \hat{\dot{\xi}}_{\mathcal{L}_d} \right) \tag{8.69}$$

where $\boldsymbol{\zeta}_Q, \boldsymbol{\zeta}_c, \boldsymbol{\zeta}_{\boldsymbol{\tau}}$ and ζ_d are the Lagrangian multipliers associated with the heat, particle, momentum and chemical reaction constraints, and $\mathcal{Z} = e^{\zeta_0}$ is the partition function. By a traditional control volume analysis,[85–88] the multipliers can be identified as:[63]

$$\boldsymbol{\zeta}_Q = -\frac{\theta \mathcal{V}}{k} \boldsymbol{\nabla} \left(\frac{1}{T} \right) \tag{8.70}$$

$$\boldsymbol{\zeta}_c = \frac{\theta \mathcal{V}}{k} \left[\boldsymbol{\nabla} \left(\frac{\mu_c}{M_c T} \right) - \frac{\boldsymbol{F}_c}{T} \right] \tag{8.71}$$

$$\boldsymbol{\zeta}_{\boldsymbol{\tau}} = \frac{\theta \mathcal{V}}{k} \boldsymbol{\nabla} \left(\frac{\boldsymbol{v}}{T} \right)^\top \tag{8.72}$$

$$\zeta_d = \frac{\theta \mathcal{V}}{k} \frac{A_d}{T} \tag{8.73}$$

where μ_c is the chemical potential of the cth constituent, M_c is the molar mass of the cth constituent, $\boldsymbol{F}_c$ is the specific body force on species c, $\boldsymbol{v}$ is the mass-average velocity, A_d is the chemical affinity of the dth reaction (< 0 for a spontaneous reaction), $\boldsymbol{\nabla}$ is the Cartesian gradient operator, and θ and $\mathcal{V}$ respectively are characteristic time and volume scales of the system. Generalising each component of the above multipliers as ζ_r and constraints as $\langle j_r \rangle$ with $r \in \{1, ..., R\}$, Jaynes' relations (8.6)-(8.11) and (8.13) reduce to:

$$\mathfrak{H}_{st}^* = \ln \mathcal{Z} + \sum_{r=1}^{R} \zeta_r \langle j_r \rangle = -\phi_{st} - \frac{\theta \mathcal{V}}{k} \hat{\dot{\sigma}} \tag{8.74}$$

$$\boldsymbol{\Lambda}_{st} = \left[\frac{\partial \mathfrak{H}_{st}^*}{\partial \langle j_1 \rangle}, ..., \frac{\partial \mathfrak{H}_{st}^*}{\partial \langle j_R \rangle} \right]^\top = [\zeta_1, ..., \zeta_R]^\top \tag{8.75}$$

$$\mathbf{g}_{st} = \begin{bmatrix} \dfrac{\partial^2 \mathfrak{H}_{st}^*}{\partial \langle j_1 \rangle^2} & \cdots & \dfrac{\partial^2 \mathfrak{H}_{st}^*}{\partial \langle j_1 \rangle \partial \langle j_R \rangle} \\ \vdots & \ddots & \vdots \\ \dfrac{\partial^2 \mathfrak{H}_{st}^*}{\partial \langle j_R \rangle \partial \langle j_1 \rangle} & \cdots & \dfrac{\partial^2 \mathfrak{H}_{st}^*}{\partial \langle j_R \rangle^2} \end{bmatrix} = \begin{bmatrix} \dfrac{\partial \zeta_1}{\partial \langle j_1 \rangle} & \cdots & \dfrac{\partial \zeta_R}{\partial \langle j_1 \rangle} \\ \vdots & \ddots & \vdots \\ \dfrac{\partial \zeta_1}{\partial \langle j_R \rangle} & \cdots & \dfrac{\partial \zeta_R}{\partial \langle j_R \rangle} \end{bmatrix} \tag{8.76}$$

$$\phi_{st} = -\ln \mathcal{Z} = -\mathfrak{H}^*_{st} + \sum_{r=1}^{R} \zeta_r \langle j_r \rangle = -\mathfrak{H}^*_{st} - \frac{\theta \mathcal{V}}{k} \hat{\dot{\sigma}} \tag{8.77}$$

$$\boldsymbol{f}_{st} = \left[\frac{\partial \phi_{st}}{\partial \zeta_1}, ..., \frac{\partial \phi_{st}}{\partial \zeta_R} \right]^{\top} = \left[\langle j_1 \rangle, ..., \langle j_R \rangle \right]^{\top} \tag{8.78}$$

$$\boldsymbol{\gamma}_{st} = \begin{bmatrix} \dfrac{\partial^2 \phi_{st}}{\partial \zeta_1^2} & \cdots & \dfrac{\partial^2 \phi_{st}}{\partial \zeta_1 \partial \zeta_R} \\ \vdots & \ddots & \vdots \\ \dfrac{\partial^2 \phi_{st}}{\partial \zeta_R \partial \zeta_1} & \cdots & \dfrac{\partial^2 \phi_{st}}{\partial \zeta_R^2} \end{bmatrix} = \begin{bmatrix} \dfrac{\partial \langle j_1 \rangle}{\partial \zeta_1} & \cdots & \dfrac{\partial \langle j_R \rangle}{\partial \zeta_1} \\ \vdots & \ddots & \vdots \\ \dfrac{\partial \langle j_1 \rangle}{\partial \zeta_R} & \cdots & \dfrac{\partial \langle j_R \rangle}{\partial \zeta_R} \end{bmatrix} \tag{8.79}$$

$$\mathbf{g}_{st}\, \boldsymbol{\gamma}_{st} = \mathbf{I} \tag{8.80}$$

where $\hat{\dot{\sigma}}$ can be identified as the local entropy production per unit volume (units of $JK^{-1}m^{-3}s^{-1}$). A flow system subject to constant flux and reaction rate constraints will therefore converge to a steady state position defined by a maximum in the flux entropy $\mathfrak{H}^*_{st}$ and a minimum in the flux potential ϕ_{st}. If these effects occur simultaneously, the system will converge to a position of maximum $\hat{\dot{\sigma}}$, therefore providing a conditional, local derivation of the *maximum entropy production* (MEP) principle,[63] which has been applied as a discriminator to determine the steady state of many non-linear flow systems.[89–97]

In Onsager's analysis of transport phenomena in the vicinity of equilibrium,[98,99] the fluxes and reaction rates are considered to be linear functions of the "forces" (the driving gradients and chemical affinities). In the present terminology, this would be written as:

$$\langle j_r \rangle = K \sum_m L^0_{rm} \zeta_m \tag{8.81}$$

where L^0_{rm} are the (constant) phenomenological coefficients at the zero-gradient position (i.e., at equilibrium) and $K = k/\theta\mathcal{V}$. In the present analysis, we do not claim linearity between $\langle j_r \rangle$ and ζ_m, nor consider that the system is "close to equilibrium", but simply adopt the partial derivatives $\partial\langle j_r \rangle/\partial\zeta_m$ within the metric $\boldsymbol{\gamma}_{st}$ (8.79) as a set of parameters (functions of ζ_m) with which to analyse the system. The present analysis therefore encompasses, but is not restricted to, Onsager's linear regime. The diagonal and many off-diagonal terms can readily be identified as functions of the conductivities (transport coefficients) and chemical reaction rate coefficents:[88]

$$\text{Heat conductivity:} \qquad \widetilde{\kappa}_{\imath\jmath} = -\frac{\partial \langle j_{Q\imath} \rangle}{\partial \left(\dfrac{\partial T}{\partial \jmath} \right)} \tag{8.82}$$

$$\text{Diffusion coefficient, species } c\text{:} \qquad \widetilde{D}^c_{\imath\jmath} = -\frac{\partial \langle j_{c\imath} \rangle}{\partial \left(\dfrac{\partial \hat{C}_c}{\partial \jmath} \right)} \tag{8.83}$$

$$\text{Viscosity coefficient:} \qquad \widetilde{\mu}_{\imath\jmath\kappa\ell} = -\frac{\partial \langle \tau_{\imath\jmath} \rangle}{\partial \left(\dfrac{\partial v_\kappa}{\partial \ell} \right)} \tag{8.84}$$

$$\text{Rate coefficient, reaction } d\text{:} \qquad \widetilde{k}_d = \frac{\partial \langle \hat{\dot{C}}_{cd} \rangle}{\partial \hat{C}_c} = \nu_{cd} M_c \frac{\partial \langle \hat{\dot{\xi}}_d \rangle}{\partial \hat{C}_c} \tag{8.85}$$

where $\hat{C}_c$ is the concentration of species c (units of kg m^{-3}; often used as a proxy for the chemical potential μ_c), $\langle \hat{\dot{C}}_{cd} \rangle$ is the mean rate of change of concentration of species c in the dth reaction (units of kg m^{-3} s^{-1}), ν_{cd} is the stoichiometric coefficient of species c in the dth reaction (positive if a product), and the indices $\imath, \jmath, \kappa, \ell \in \{x, y, z\}$. The remaining off-diagonal terms consist of the cross-process conductivity coupling coefficients and conductivity-reaction rate coefficients. The Riemannian metric $\boldsymbol{\gamma}_{st}$ can therefore be regarded as a function of the material properties or susceptibilities of a flow and chemical reactive system, in the same way that the Riemannian metric for an equilibrium system $\boldsymbol{\gamma}_{eq}$ is a function of its various susceptibilities, such as C_P, κ_T and α (§8.4.1). As with equilibrium systems, an abrupt change in a given component $\gamma_{st,rm}$ with ζ_m can be interpreted as the boundary of a phase change in the system. Notice also that symmetry of $\boldsymbol{\gamma}_{st}$ yields a set of Maxwell-like relations:[63]

$$\frac{\partial \langle j_r \rangle}{\partial \zeta_m} = \frac{\partial \langle j_m \rangle}{\partial \zeta_r} \tag{8.86}$$

These apply to all infinitesimal volume elements of a flow system, not merely those in the vicinity of equilibrium. Eqs. (8.86) considerably simplify the set of parameters needed for analysis, from R^2 to $\binom{R+1}{2}$ coefficients; further simplifications may be attainable in certain systems due to geometric and tensor symmetries.[88]

The above relations (8.74)-(8.79) can now be applied to develop a Riemannian description of a flow system on the manifold of steady state positions. In terms of the generalised derivatives, the (dimensionless) arc lengths (8.18)-(8.19) and action integrals (8.40)-(8.41) are obtained as:

$$\begin{aligned} L_{st} &= \int_0^{\xi_{max}} \sqrt{\dot{\boldsymbol{f}}_{st}^{\top} \mathbf{g}_{st} \dot{\boldsymbol{f}}_{st}}\, d\xi = \int_0^{\xi_{max}} \sqrt{\dot{\boldsymbol{\Lambda}}_{st}^{\top} \boldsymbol{\gamma}_{st} \dot{\boldsymbol{\Lambda}}_{st}}\, d\xi \\ &= \int_0^{\xi_{max}} \sqrt{\dot{\boldsymbol{\Lambda}}_{st} \cdot \dot{\boldsymbol{f}}_{st}}\, d\xi = \int_0^{\xi_{max}} \sqrt{\sum_{r=1}^{R} \frac{\partial \zeta_r}{\partial \xi} \frac{\partial \langle j_r \rangle}{\partial \xi}}\, d\xi \end{aligned} \tag{8.87}$$

$$\begin{aligned} \mathcal{J}_{st} &= \int_0^{\xi_{max}} \frac{1}{2} \dot{\boldsymbol{f}}_{st}^{\top} \mathbf{g}_{st} \dot{\boldsymbol{f}}_{st}\, d\xi = \int_0^{\xi_{max}} \frac{1}{2} \dot{\boldsymbol{\Lambda}}_{st}^{\top} \boldsymbol{\gamma}_{st} \dot{\boldsymbol{\Lambda}}_{st}\, d\xi \\ &= \int_0^{\xi_{max}} \frac{1}{2} \dot{\boldsymbol{\Lambda}}_{st} \cdot \dot{\boldsymbol{f}}_{st}\, d\xi = \frac{1}{2} \int_0^{\xi_{max}} \sum_{r=1}^{R} \frac{\partial \zeta_r}{\partial \xi} \frac{\partial \langle j_r \rangle}{\partial \xi}\, d\xi \end{aligned} \tag{8.88}$$

where, as shown, the two alternative $\mathfrak{H}_{st}^*$ and ϕ_{st} measures are equivalent. Once again, these equations must be integrated along the particular path taken between the initial and final steady state positions.

To comment on units: since the above quantities are calculated using the "pure" metrics $\mathbf{g}_{st}$ or $\boldsymbol{\gamma}_{st}$, the resulting line element ds_{st}, arc length L_{st} and the term $\bar{\epsilon}_{st}\mathcal{J}_{st}$ are dimensionless. Use of the "natural" metric $K\mathbf{g}_{st}$, for $K = k/\theta\mathcal{V}$, therefore gives the line element and arc length in $\sqrt{JK^{-1}m^{-3}s^{-1}}$ and the action in $JK^{-1}m^{-3}s^{-1}\xi^{-1}$, thereby giving $\bar{\epsilon}_{st}\mathcal{J}_{st}$ in units of entropy production per unit volume. Similarly, use of the "natural" metric $\boldsymbol{\gamma}_{st}/K$ in conjunction with the dimensional constraint vector $K\dot{\boldsymbol{\Lambda}}_{st}$ gives the line element and arc length in $\sqrt{JK^{-1}m^{-3}s^{-1}}$ and action in $JK^{-1}m^{-3}s^{-1}\xi^{-1}$, again giving $\bar{\epsilon}_{st}\mathcal{J}_{st}$ in units of entropy production per unit volume. The least action bound (8.36) therefore yields a *minimum entropy production principle*, which sets a lower bound for the entropy production associated with movement of a flow system from one steady state position to another along a specified path. From the previous analysis, this involves two separate minimisation principles:

- If the path is specified, the process of minimum entropy production will be one which proceeds at constant speed $\dot{s}$, assuming a slow process and a constant dissipation parameter ϵ. Alternately, if the dissipation parameter ϵ is not constant, the minimum entropy production process will be given by a constant arc length speed, in accordance with a steady state analogue of the "equal thermodynamic distance" principle.[25,34–36]
- If the path is not specified or can be varied, an absolute lower bound for

the entropy production is given by the geodesic in steady state parameter space, in accordance with the methods of §8.3.3.

Although they share a similar name, the minimum entropy production principle derived herein is quite different to that of Prigogine,[86] which concerns the selection of a steady state position relative to possible non-steady state positions, and which only applies to the Onsager linear regime. Similarly, it differs from the minimum entropy production principle obtained by the application of Riemannian geodesic calculations to the manifold of equilibrium positions, discussed at the end of §8.4.1.[16,19,34,43,100] The minimum principle derived herein is more general than both these principles, being applicable beyond the set of equilibrium positions, and also well outside the linear regime of non-equilibrium thermodynamics. In turn, it is based on the even broader generic formulation of the least action bound given herein, applicable to any system which can be analysed by Jaynes' method.

8.5. Conclusion

In this study, the manifold of stationary positions inferred by Jaynes' MaxEnt and MaxREnt principles – considered as a function of the moment constraints or their conjugate Lagrangian multipliers – is endowed with a Riemannian geometric description, based on the second differential tensor of the entropy or its Legendre transform (negative Massieu function) obtained from Jaynes' method. The analysis provides a generalised *least action bound* applicable to all Jaynesian systems, which provides a lower bound to the cost (in generic entropy units) of a transition between inferred positions along a specified path, at specified rates of change of the control parameters. The analysis therefore extends the concepts of "finite time thermodynamics", developed over the past three decades,[10–43] to the generic Jaynes domain, providing a link between purely static (stationary) inferred positions of a system, and dynamic transitions between these positions (as a function of time or some other coordinate). If the path is unspecified, the analysis gives an absolute lower bound for the cost of the transition, corresponding to the geodesic of the Riemannian hypersurface.

The analysis is then applied to (i) an equilibrium thermodynamic system subject to mean internal energy and volume constraints, and (ii) a flow system at steady state, subject to constraints on the mean heat, mass and momentum fluxes and chemical reaction rates. The first example recovers the *minimum entropy cost* of a transition between equilibrium positions, a widely used result of finite-time thermodynamics. The second example leads to a new *minimum entropy production principle*, for the cost of a transition between steady state positions of a flow system. The analyses reveal the tremendous utility of Jaynes' MaxEnt and MinXEnt methods augmented by the generalised least action bound, for the analysis of probabilistic systems of all kinds.

Acknowledgments

The first author thanks the European Commission for support as a Marie Curie Incoming International Fellow (FP6); The University of New South Wales, the University of Copenhagen and Technical University of Berlin for financial support; and Bob Dewar and Roderick Dewar for the opportunity to present this analysis at the 22nd Canberra International Physics Summer School, ANU, Canberra, December 2008.

Appendix 1: Riemannian Geometric Considerations

It is necessary to examine several salient features of the Riemannian geometric interpretation adopted herein.[60,62] Consider a hypersurface represented by the position vector $\boldsymbol{x} = [x_1, ..., x_n]^\top$, embedded within the n-dimensional space defined by the coordinates $(x_1, ..., x_n)$. For analysis, this hypersurface can be converted to the parametric representation $\boldsymbol{x}(\boldsymbol{u}) = [x_1(\boldsymbol{u}), ..., x_n(\boldsymbol{u})]^\top$, where $\boldsymbol{u} = [u_1, ..., u_{n-1}]^\top$ is the $(n-1)$-dimensional vector of parameters u_j, consisting of coordinates on the hypersurface. The *first fundamental form* of this geometry is defined by the metric:[60,62]

$$d\varsigma^2 = d\boldsymbol{x} \cdot d\boldsymbol{x} = \sum_{i=1}^{n-1} \sum_{j=1}^{n-1} a_{ij} du_i du_j = d\boldsymbol{u}^\top \mathbf{a}\, d\boldsymbol{u} \tag{A.1}$$

in which, by elementary calculus, the components of the tensor **a** can be shown to be:

$$a_{ij} = \frac{\partial \boldsymbol{x}}{\partial u_i} \cdot \frac{\partial \boldsymbol{x}}{\partial u_j} \tag{A.2}$$

Accordingly, **a** is symmetric. By Euclidean geometry, (A.1) can be used to calculate distances between two points a and b on the hypersurface $\boldsymbol{x}$, on the path defined by $\boldsymbol{u}$:

$$L_{\boldsymbol{x}} = \int_a^b d\varsigma = \int_a^b \sqrt{d\boldsymbol{u}^\top \mathbf{a}\, d\boldsymbol{u}} = \int_{\xi_a}^{\xi_b} \sqrt{d\dot{\boldsymbol{u}}^\top \mathbf{a}\, d\dot{\boldsymbol{u}}}\, d\xi \tag{A.3}$$

where the overdot indicates the derivative with respect to the path parameter ξ. The *second fundamental form* of the hypersurface is then defined by:[60,62]

$$-d\boldsymbol{x} \cdot d\boldsymbol{n} = \sum_{i=1}^{n-1}\sum_{j=1}^{n-1} b_{ij} du_i du_j = d\boldsymbol{u}^\top \mathbf{b}\, d\boldsymbol{u} \tag{A.4}$$

where $\boldsymbol{n}$ is the unit normal vector to the hypersurface. By differential calculus, it can be shown that:

$$b_{ij} = \frac{\partial \boldsymbol{x}}{\partial u_i \partial u_j} \cdot \boldsymbol{n} \tag{A.5}$$

The second fundamental form is not considered as a metric with which to calculate distances, but is used to examine the tangency and curvature properties of the manifold $\boldsymbol{x}$.[60,62]

In the present study, we wish to adopt the Jaynesian matrix $\mathbf{g}$ or $\boldsymbol{\gamma}$ as a Riemannian metric tensor for the calculation of arc lengths on the R-dimensional stationary state hypersurface, embedded in the $(R+1)$-dimensional space defined by $(\mathfrak{H}^*, \{\langle f_r \rangle\})$ or $(\phi, \{\lambda_r\})$. We therefore adopt the (somewhat peculiar) approach in which the coordinates $[x_2, ..., x_{R+1}]^\top$ are selected as the surface parameters $[u_1, ..., u_R]^\top$; i.e. with the hypersurface $\boldsymbol{x}_{\mathfrak{H}^*} = [\mathfrak{H}^*, \langle f_1 \rangle, ..., \langle f_R \rangle]^\top$ parameterised by $\boldsymbol{u}_{\mathfrak{H}^*} = \boldsymbol{f}$ and with $\boldsymbol{x}_\phi = [\phi, \lambda_1, ..., \lambda_R]^\top$ parameterised by $\boldsymbol{u}_\phi = \boldsymbol{\Lambda}$. Two necessary conditions for the use of $\mathbf{g}$ or $\boldsymbol{\gamma}$ as metric tensors is that they be symmetric and positive definite (or semi-definite); since they constitute Hessian matrices of the concave generic entropy $\mathfrak{H}^*$ or convex potential function ϕ, these conditions are satisfied, not only in thermodynamic applications but within the generic Jaynes formulation (with semi-definite behaviour only at singularities).[6,15] However, $\mathbf{g}$ and $\boldsymbol{\gamma}$ are related to a second, rather than a first, fundamental form.[15,31] For $\mathbf{g}$ or $\boldsymbol{\gamma}$ to be considered as metric tensors, they must be able to generate the first fundamental form of some position vector which describes the hypersurface. In mathematical terms, from (A.1):

$$ds^2_{\mathfrak{H}^*} = d\boldsymbol{f}^\top \mathbf{g}\, d\boldsymbol{f} = d\boldsymbol{u}_{\mathfrak{H}^*}{}^\top \mathbf{a}_{\mathfrak{H}^*}\, d\boldsymbol{u}_{\mathfrak{H}^*}, \tag{A.6}$$

$$ds^2_\phi = d\boldsymbol{\Lambda}^\top \boldsymbol{\gamma}\, d\boldsymbol{\Lambda} = d\boldsymbol{u}_\phi{}^\top \mathbf{a}_\phi\, d\boldsymbol{u}_\phi \tag{A.7}$$

From (8.8), (8.10), (8.13) and (A.2), taking advantage of tensor symmetries, the metric components must therefore satisfy:

$$g_{mr} = a_{\mathfrak{H}^*,mr} = \frac{\partial \boldsymbol{\omega}}{\partial \langle f_m \rangle} \cdot \frac{\partial \boldsymbol{\omega}}{\partial \langle f_r \rangle} = \frac{\partial^2 \mathfrak{H}^*}{\partial \langle f_m \rangle \partial \langle f_r \rangle} = \frac{\partial \lambda_r}{\partial \langle f_m \rangle} \tag{A.8}$$

$$\gamma_{mr} = a_{\phi,mr} = \frac{\partial \boldsymbol{\Omega}}{\partial \lambda_m} \cdot \frac{\partial \boldsymbol{\Omega}}{\partial \lambda_r} = \frac{\partial^2 \phi}{\partial \lambda_m \partial \lambda_r} = \frac{\partial \langle f_r \rangle}{\partial \lambda_m} \tag{A.9}$$

where $\boldsymbol{\omega}(\boldsymbol{f})$ and $\boldsymbol{\Omega}(\boldsymbol{\Lambda})$ are new R-dimensional position vectors, which from (A.10), are related by:

$$\mathbf{a}_{\mathfrak{H}^*} \, \mathbf{a}_{\phi} = \mathbf{I}. \tag{A.10}$$

In consequence, the metrics (8.14)-(8.15) and (8.20)-(8.21) and arc lengths (8.16)-(8.19) used herein are not measures of distance on the stationary state hypersurface defined by $\mathfrak{H}^*(\{\langle f_r \rangle\})$ or $\phi(\{\lambda_r\})$, but rather, on the transformed hypersurface given by $\boldsymbol{\omega}$ or $\boldsymbol{\Omega}$. In addition to the symmetry and positive definiteness conditions, it is therefore also necessary and sufficient that the hypersurface defined by $\boldsymbol{\omega}$ or $\boldsymbol{\Omega}$ exists within $\mathbb{R}^R$, is continuous and continuously differentiable – at least up to first order – except in the neighbourhood of singularities.

References

1. E.T. Jaynes, Information theory and statistical mechanics, Phys. Rev., **106**, 620-630 (1957).
2. E.T. Jaynes, Information theory and statistical mechanics, *in* Ford, K.W. (ed), Brandeis University Summer Institute, Lectures in Theoretical Physics, Vol. 3: Statistical Physics, Benjamin-Cummings Publ. Co., 1963, 181-218.
3. M. Tribus, Information theory as the basis for thermostatics and thermodynamics, J. Appl. Mech., Trans. ASME, **28**, 1-8 (1961).
4. M. Tribus, Thermostatics and Thermodynamics, D. Van Nostrand Co. Inc., Princeton, NJ, 1961.
5. J.E. Shore, R.W. Johnson, Axiomatic derivation of the principle of maximum entropy and the principle of minimum cross-entropy, IEEE Trans. Information Theory **IT-26**(1), 26-37 (1980).
6. J.N. Kapur, H.K. Kesevan, Entropy Optimization Principles with Applications, Academic Press, Inc., Boston, MA, 1992.
7. E.T. Jaynes (G.L. Bretthorst, ed.) Probability Theory: The Logic of Science, Cambridge U.P., Cambridge, 2003.
8. J.N. Kapur, H.K. Kesevan, *The Generalized Maximum Entropy Principle (with Applications)*, Sandford Educational Press, Waterloo, Canada, 1987.
9. J.N. Kapur, *Maximum-Entropy Models in Science and Engineering*, John Wiley, NY, 1989.
10. F. Weinhold, Metric geometry of equilibrium thermodynamics, J. Chem. Phys. **63**(6) 2479-2483 (1975).

11. F. Weinhold, Metric geometry of equilibrium thermodynamics. II. Scaling, homogeneity and generalized Gibbs-Duhem relations, J. Chem. Phys. **63**(6) 2484-2487 (1975).
12. F. Weinhold, Metric geometry of equilibrium thermodynamics. III. Elementary formal structure of a vector-algebraic representation of equilibrium thermodynamics, J. Chem. Phys. **63**(6) 2488-2495 (1975).
13. F. Weinhold, Metric geometry of equilibrium thermodynamics. IV. Vector-algebraic evaluation of thermodynamic derivatives, J. Chem. Phys. **63**(6) 2496-2501 (1975).
14. F. Weinhold, Metric geometry of equilibrium thermodynamics. V. Aspects of heterogeneous equilibrium, J. Chem. Phys. **65**(2) 559-564 (1976).
15. P. Salamon, B. Andresen, P.D. Gait, R.S. Berry, The significance of Weinhold's length, J. Chem. Phys. **73**(2) 1001-1002 (1980), erratum **73**(10) 5407 (1980).
16. P. Salamon, A. Nitzan, B. Andresen, R.S. Berry, Minimum entropy production and the optimization of heat engines, Phys. Rev. A **21** 2115-2129 (1980).
17. P. Salamon, R.S. Berry, Thermodynamic length and dissipated availability, Phys. Rev. Lett. **51**(13) 1127-1130 (1983).
18. P. Salamon, E. Ihrig, R.S. Berry, A group of coordinate transformations which preserve the metric of Weinhold, J. Math. Phys. **24**(10) 2515-2520 (1983).
19. B. Andresen, Finite-Time Thermodynamics, Physics Laboratory II, University of Copenhagen, Denmark, 1983.
20. P. Salamon, J. Nulton, E. Ihrig, On the relation between entropy and energy versions of thermodynamic length, J. Chem. Phys. **80**(1) 436-437 (1984).
21. B. Andresen, R.S. Berry, M.J. Ondrechen, P. Salamon, Thermodynamics for processes in finite time, Acc. Chem. Res. **17** 266-271 (1984).
22. P. Salamon, J.D. Nulton, R.S. Berry, Length in statistical thermodynamics, J. Chem. Phys. **82**(5) 2433-2436 (1985).
23. F. Schlögl, Thermodynamic metric and stochastic measures, Z. Phys. B **59** 449-454 (1985).
24. T. Feldman, B. Andresen, A. Qi, P. Salamon, Thermodynamic lengths and intrinsic time scales in molecular relaxation, J. Chem. Phys. **83**(11) 5849-5853 (1985).
25. J. Nulton, P. Salamon, B. Andresen, Q. Anmin, Quasistatic processes as step equilibrations, J. Chem. Phys. **83**(1) 334-338 (1985).
26. T. Feldman, R.D. Levine, P. Salamon, A geometrical measure for entropy changes, J. Stat. Phys. **42**(5/6) 1127-1134 (1986).
27. R.D. Levine, Geometry in classical statistical thermodynamics, J. Chem. Phys. **84**(2) 910-916 (1986).
28. J.D. Nulton, P. Salamon, Statistical mechanics of combinatorial optimization, Phys. Rev. A **37**(4) 1351-1356.
29. P. Salamon, J.D. Nulton, J.R. Harland, J. Pedersen, G. Ruppeiner, L. Liao, Simulated annealing with constant thermodynamic speed, Comput. Phys. Comm. **49** 423-428 (1988).

30. B. Andresen, R.S. Berry, R. Gilmore, E. Ihrig, P. Salamon, Thermodynamic geometry and the metrics of Weinhold and Gilmore, Phys. Rev. A **37**(3) 845-848 (1988).
31. B. Andresen, R.S. Berry, E. Ihrig, P. Salamon, Inducing Weinhold's metric from Euclidian and Riemannin metrics, Phys. Rev. A **37**(3) 849-851 (1988).
32. K.H. Hoffmann, B. Andresen, P. Salamon, Measures of dissipation, Phys. Rev. A **39**(7) 3618-3621 (1989).
33. B. Andresen, J.M. Gordon, Constant thermodynamic speed for minimizing entropy production in thermodynamic processes and simulated annealing, Phys. Rev. E **50**(6) 4346-4351 (1994).
34. P. Salamon, J.D. Nulton, The geometry of separation processes: A horse-carrot theorem for steady flow systems, Europhysics Letters **42**(5) 571-576 (1998).
35. P. Salamon, J.D. Nulton, G. Siragusa, T.R. Andersen, A. Limon, Principles of control thermodynamics, Energy **26** 307-319 (2001).
36. M. Schaller, K.H. Hoffmanm, G. Siragusa, P. Salamon, B. Andresen, Numerically optimized performance of diabatic distillation columns, Comput. Chem. Eng. **25** 1537-1548 (2001).
37. G.P. Beretta, A new approach to constrained-maximization nonequilibrium problems, in R.A. Gaggioli (ed.) Computer-Aided Engineering of Energy Systems: Second Law Analysis and Modeling, ASME Book H0341C-AES, **3**, 129-134 (1986).
38. G.P. Beretta, Dynamics of smooth constrained approach to maximum entropy, in M.J. Moran, E. Sciubba (eds), Second Law Analysis of Thermal Systems, ASME Book I00236, 17-24 (1987).
39. G.P. Beretta, Modeling non-equilibrium dynamics of a discrete probability distribution: General rate equation for maximal entropy generation in a maximum-entropy landscape with time-dependent constraints, Entropy **10** 160-182 (2008).
40. L. Diósi, K. Kulacsy, B. Lukács, A. Rácz, Thermodynamic length, time, speed, and optimum path to minimize entropy production, J. Chem. Phys. **105**(24) 11220-11225 (1996).
41. G.E. Crooks, Measuring thermodynamic length, Phys. Rev. Lett. **99** 100602 (2007).
42. E.H. Feng, G.E. Crooks, Far-from-equilibrium measurements of thermodynamic length, Phys. Rev. E **79** 012104 (2009).
43. D.C. Brody, D.W. Hook, Information geometry in vapour-liquid equilibrium, J. Phys. A: Math. Theor. **42** 023001 (33pp) (2009).
44. L. Boltzmann, Über die Beziehung zwischen dem zweiten Hauptsatze dewr mechanischen Wärmetheorie und der Wahrscheinlichkeitsrechnung, respective den Sätzen über das Wärmegleichgewicht, Wien. Ber. **76**, 373-435 (1877); English transl.: J. Le Roux (2002) 1-63 *http://www.essi.fr/~leroux/*.
45. M. Planck, Über das gesetz der Energieverteilung im Normalspektrum, Annalen der Physik **4**, 553-563 (1901).
46. I. Vincze, On the maximum probability principle in statistical physics, Progress in Statistics **2**, 869-895 (1974).

47. M. Grendar, M. Grendar, What is the question that MaxEnt answers? A probabilistic interpretation, *in* A. Mohammad-Djafari (ed.) MaxEnt 2000, Gif-sur-Yvette, France, 8-13 July 2000, AIP Conf. Proc. **568**, 83-94 (2001).
48. R.K. Niven, Exact Maxwell-Boltzmann, Bose-Einstein and Fermi-Dirac statistics, Phys. Lett. A **342**(4) 286-293 (2005).
49. R.K. Niven, Cost of s-fold decisions in exact Maxwell-Boltzmann, Bose-Einstein and Fermi-Dirac statistics, Physica A **365**(1) 142-149 (2006).
50. R.K. Niven, M. Grendar, Generalized classical, quantum and intermediate statistics and the Polya urn model, Physics Letters A **373** 621-626 (2009).
51. R.K. Niven, Combinatorial entropies and statistics, European Physics Journal B, in press.
52. S. Kullback, R.A. Leibler, On information and sufficiency, Annals Math. Stat. **22**, 79-86 (1951).
53. S. Kullback, Information Theory and Statistics, John Wiley, NY, 1959.
54. R.K. Niven, Combinatorial information theory: I. Philosophical basis of cross-entropy and entropy, *http://arxiv.org/abs/cond-mat/0512017 v5*, 2007.
55. H.B. Callen, Thermodynamics, John Wiley & Sons, NY, 1960.
56. H.B. Callen, Thermodynamics and an Introduction to Thermostatistics, 2nd ed., John Wiley & Sons, NY, 1985.
57. R. Clausius, Über verschiedene für die Anwendung bequeme Formen der Hauptgleichungen der mechanischen Wrmetheorie, Poggendorfs Annalen **125**, 335-400 (1865); English transl.: R.B. Lindsay *in* J. Kestin (ed.) The Second Law of Thermodynamics, Dowden, Hutchinson & Ross, PA (1976) 162-193.
58. J.W. Gibbs, A method of graphical representation of the thermodynamic properties of substances by means of surfaces, Trans. Connecticut Acad. **2**, 382-404 (1873).
59. J.W. Gibbs, On the equilibrium of heterogeneous substances, Trans. Connecticut Acad. **3**, 108-248 (1875-1876); 343-524 (1877-1878).
60. E. Kreysig, Differential Geometry, Dover Publ., NY, 1991.
61. R. Trasarti-Battistoni, Euclidean and Riemannian geometrical approaches to non-extensive thermo-statistical mechanics, arXiv:cond-mat/0203536v2 (2002).
62. D. Zwillinger, CRC Standard Mathematical Tables and Formulae, Chapman & Hall / CRC Press, Boca Raton, FL, 2003.
63. R.K. Niven, Derivation of the maximum entropy production principle for flow-controlled systems at steady state, Physical Review E, in press *http://arxiv.org/abs/0902.1568* (2009).
64. C. Lanczos, The Variational Principles of Mechanics, 3rd ed., University of Toronto Press, Toronto, 1966.
65. A. Caticha, C. Cafaro, CP954, in Knuth, K.H., Caticha, A., Center, J.L., Giffon, A., Rodrguez, C.C. (eds), AIP Conference Proceedings 954, 165-174 (2007).
66. R. Weinstock, Calculus of Variations, with Applications to Physics and Engineering, Dover Publ., NY, 1974.

67. Q.A. Wang, Maximum path information and the principle of least action for chaotic system, Chaos, Solitons and Fractals **23** 1253-1258 (2005).
68. Q.A. Wang, Maximum entropy change and least action principle for nonequilibrium systems, Astrophys. Space Sci. **305** 273-281 (2006).
69. G. Nathanson, O. Sinanoğlu, The geometry of near equilibrium irreversible thermodynamics, J. Chem. Phys. **72**(5) 3127-3129 (1980).
70. R. Gilmore, Le Châtelier reciprocal relations, J. Chem. Phys. **76**(11) 5551-5553 (1982).
71. S. Sieniutycz, R.S. Berry, Field thermodynamic potentials and geometric thermodynamics with heat transfer and fluid flow, Phys. Rev. A **43**(6) 2807-2818 (1991).
72. M. Chen, Symmetry transformations in extended irreversible thermodynamics, J. Math. Phys. **42**(6) 2531-2539 (2001).
73. M. Chen, On the intrinsic geometric structure of extended irreversible thermodynamics, J. Phys. A: Math. Gen. **36**4717-4727 (2003).
74. J.D. Flick, P. Salamon, B. Andresen, Metric bounds on losses in adaptive coding, Information Sci. **42** 239-253 (1987).
75. P. Salamon, J. Komlos, B. Andresen, J.D. Nulton, A geometric view of welfare gains with non-instantaneous adjustment, Math. Social Sci. **13** 153-163 (1987).
76. G. Ruppeiner, Thermodynamics: a Reimannian geometric model, Phys. Rev. A **20**(4) 1608-1613 (1979).
77. C. Truesdell, Rational Thermodynamics, McGraw-Hill, NY, 1969, chap. 7.
78. R.A. Fisher, On the mathematical foundations of theoretical statistics, Phil. Trans. Royal Soc. London A **222** 309-368 (1922).
79. C.R. Rao, Information and the accuracy attainable in the estimation of statistical parameters, Bull. Calcutta Math. Soc. **37** 81-91 (1945).
80. B.R. Frieden, Science from Fisher Information, An Introduction, 2nd ed., Cambridge U.P., 2004.
81. M. Planck, Treatise on Thermodynamics, Engl. transl., 3rd ed., Dover Publications, NY, 1945.
82. M. Planck, Introduction to Theoretical Physics, Vol. V: Theory of Heat, Engl. transl. H.L. Brose, Macmillan & Co., Ltd, 1932.
83. M. Massieu, Thermodynamique - Sur les fonctions caractéristiques des divers fluides, Comptes Rendus **69** 858-862; 1057-1061 (1869).
84. W. Spirkl, H. Ries, Optimal finite-time endoreversible processes, Phys. Rev. E. **52**(4) 3485-3489 (1995).
85. S.R. de Groot, P. Mazur, Non-Equilibrium Thermodynamics, Dover Publications, NY, 1984.
86. I. Prigogine, Introduction to Thermodynamics of Irreversible Processes, 3rd ed., Interscience Publ., NY.
87. H.J. Kreuzer, Nonequilibrium Thermodynamics and its Statistical Foundations, Clarendon Press, Oxford, 1981.
88. R.B. Bird, W.E. Stewart, E.N. Lightfoot, Transport Phenomena, 2nd ed., John Wiley & Sons, NY, 2002.
89. G.W. Paltridge, Global dynamics and climate - a system of minimum en-

tropy exchange, Quart. J. Royal Meteorol. Soc. **101**, 475-484 (1975).
90. G.W. Paltridge, The steady-state format of global climate, Quart. J. Royal Meteorol. Soc. **104**, 927-945 (1978).
91. G.W. Paltridge, Thermodynamic dissipation and the global climate system, Quart. J. Royal Meteorol. Soc. **107**, 531-547 (1981).
92. H. Ozawa, A. Ohmura, R.D. Lorenz, T. Pujol, The second law of thermodynamics and the global climate system: A review of the maximum entropy production principle, Rev. Geophys. **41**, article 4 (2003).
93. R.C. Dewar, Information theory explanation of the fluctuation theorem, maximum entropy production and self-organized criticality in non-equilibrium stationary states, J. Phys. A: Math. Gen. **36**, 631-641 (2003).
94. R.C. Dewar, Maximum entropy production and the fluctuation theorem, J. Phys. A: Math. Gen. **38**, L371-L381 (2005).
95. A. Kleidon, R.D. Lorenz (eds.) Non-equilibrium Thermodynamics and the Production of Entropy: Life, Earth and Beyond, Springer Verlag, Heidelberg, 2005.
96. L.M. Martyushev, V.D. Seleznev, Maximum entropy production principle in physics, chemistry and biology, Physics Reports **426**, 1-45 (2006).
97. S. Bruers, Classification and discussion of macroscopic entropy production principles, *arXiv:cond-mat/0604482v3*, 2007.
98. L. Onsager, Reciprocal relations in irreversible processes I, Phys. Rev. **37**, 405-426 (1931).
99. L. Onsager, Reciprocal relations in irreversible processes II, Phys. Rev. **38**, 2265-2279 (1931).
100. J.C. Schön, A thermodynamic distance criterion of optimality for the calculations of free energy changes from computer simulations, J. Chem. Phys. **105**(22) 10072 (1996).

Chapter 9

Complexity, Post-genomic Biology and Gene Expression Programs

Rohan B. H. Williams and Oscar Junhong Luo

Genome Biology Program,
John Curtin School of Medical Research,
Australian National University
rohan.williams@anu.edu.au

Gene expression represents the fundamental phenomenon by which information encoded in a genome is utilised for the overall biological objectives of the organism. Understanding this level of information transfer is therefore essential for dissecting the mechanistic basis of form and function of organisms. We survey recent developments in the methodology of the life sciences that is relevant for understanding the organisation and function of the genome and review our current understanding of the regulation of gene expression, and finally, outline some new approaches that may be useful in understanding the organisation of gene regulatory systems.

Contents

9.1. Introduction

Gene expression represents the fundamental phenomenon by which information encoded in a genome is utilised for the overall biological objectives of the organism. Understanding this level of information transfer is therefore essential for dissecting the mechanistic basis of form and function of organisms. Although the *mechanisms of eukaryotic gene expression have been under intensive study for several decades, our view of this phenomenon continues to grow in complexity.* This chapter will be organised into four major sections: first, we will provide a admittedly superficial view of the way in living organisms can be thought of as complex systems, using the conceptual framework common to all chapters in this volume. We then survey recent developments in the methodology of the life sciences that are relevant to understanding the organisation and function of the genome. Next, we review our current understanding of the regulation of gene expression, and finally, outline some new approaches that may be useful in understanding the organisation of gene regulatory systems. However, before discussing this specific topic in more detail, we will address some broader issues about the nature of organisms, and in particular how they relate to the unifying themes that underpin the study of complex systems.

9.1.1. *Organisms as complex systems*

That living systems, be it cells, organisms or populations, can be usefully considered as a complex system is largely self evident. In the present era, the consensus scientific perspective argues that living systems are entirely constructed of known matter and that the form and function of living systems are entirely consistent with known physical laws: albeit under a obviously daunting level of complexity and organisation from phenomenological, mechanistic and explanatory perspectives. While the term *organism* can refer to any living being (a cell, a plant or animal, or collections of organisms, perhaps of different species, exhibiting behaviour as a "superorganism"), we will tend to focus on multicellular animals, also known as *metazoans.* Organisms are clearly *open systems* and they are constantly exchanging matter and energy with their environment, both externally (*e.g.*

gas exchange through the respiratory system) or internally (*e.g.* transport of mRNA or proteins between the nucleus and cytoplasm of a cell). *Emergence* of behaviours or features that are not evident at lower organisational scales are ubiquitous in organisms: some examples include higher cognitive functions in humans and primates, and complex collective behaviour in lower eukaryotic organisms, such as slime moulds. Organisms are inherently *adaptive* systems in two senses: it can refer to the ability of the cell, tissue or organism to modify its function or behaviour in response, to say, a stimulus or different environment, but can also be used to describe an organism, or feature of organism, that has been subject to selection across evolutionary time scales. *History* is an essential and intrinsic aspect of living systems. Organisms, both at a level of individuals and a species, are entities that are highly organised in time[1] and they can be understood and studied at multiple time scales: firstly, there is the life history of an individual, from fertilisation, through development and ageing, until death.[3] Secondly *within* an individual's life history, functionally important temporal fluctuations in biochemical, physiological or behavioural state occur. Finally, organisms can be understood, in regards their origins and inter-relationship with other species, across evolutionary time. *Hierarchical structure* is common in organisms, often containing considerably redundancy and robustness—some examples include the evolution of central nervous systems, or the organisation of signalling pathways within a cell, or the "molecular machines"[2] associated with the control of gene expression. Organisms demonstrate many behaviours which are stunningly *predictable*: for example, the replication of DNA sequence in a cell via mitotic replication, or, perhaps most dramatically , the execution of developmental programs, in which a single fertilised germ cell develops into a complete organism. Aside from the complexity of living systems, a fundamental issue with *prediction* of biological systems, whether in "real-time" or in understanding their evolution, relates to the imprecision with which their environment can be measured or reconstructed. Therefore, *predictability*, at least, as understood by a physicist or control engineer, remains an elusive goal, but one that continues to exert a powerful intellectual attraction.[4,5] Organisms are inherently patterned across *multiple spatio-temporal scales*, and are highly *non*-disordered systems. Perhaps the most obvious example is the multicellular nature of animals and plants: collections of specific cell types, organised into highly structured and functional specialised compartment called tissues and organs, under the control of multiple regulatory systems (*e.g.* nervous, endocrine, immune systems).[3] Later we will examine the multi-scale organisation of the eukarotic nucleus, and its role in the regulation of gene expression. Organisms routinely exhibit *multiple* and *altered* states, which can be manifest on short– or long– term temporal scales, for example sleep versus waking. Some of these states are detrimental to the overall functioning of the organism: conditions that are called "disease" states by medical science, and that may be caused by myriad genetic and/or environmental perturbations, the latter being of both abiotic[6] or biotic origin (*e.g.* H1N1 swine influenza or *M. tuberculosis*).

9.2. An Overview of Genes and Gene Expression

9.2.1. *Parts: genes and their products*

For the purposes of our discussion, we will consider genes as "parts" in the complex system under study. This view is certainly arbitrary, and, depending on whether cellular, organismal or population–level phenomena are of primary interest, other biological entities (*e.g.* cells, organs, organisms or species) may be considered parts with equal validity. Modern biology, being largely dominated by molecular and cell level approaches, has been criticised from several perspectives as being too *gene-centric*.[7,8] Fundamentally though, it is difficult to argue that genes, and their protein products, do *not* play a central role in development and aging, including the genesis of disease processes. So what are genes and what do they do? Genes are fundamental material of heredity: the "instructions for life" inherited by an individual from each of its parents and encoded in an information-carrying biopolymer, *deoxyribonucleic acid* (DNA). On a functional level, genes specify, or encode (via a universal "alphabet": the so-called "genetic code"), the instructions for the manufacture of a biological-active biopolymer called a *protein*. Each gene encodes unique protein products.[a] Proteins are the fundamental building blocks for the structure and function of cells, and they are the most fundamental example of a *phenotype*. For example, one of the 20,000 or so genes in the human genome is called *leptin* which encodes for a specific type of signalling protein called a hormone. The *leptin* hormone is excreted by fat cells and is detected in the central nervous system, acting as a part of a body-wide signalling system involved in modulating energy intake.[11] Other types of genes may make proteins that build cell membranes, or act as intra-cellular signalling molecules, or

[a] Multiple proteins can be generated the same gene via the mechanism of alternative splicing.[9]

they may even regulate the expression of other genes, as is the case for a large class of genes that encode for DNA-binding regulatory proteins called *transcription factors*.[10] Proteins are not generated directly from DNA, but rather via an intermediary nucleic acid called *messenger ribonucleic acid* (mRNA), via a process called *transcription*[b] (**Figure 9.1**). Once the RNA form of a gene has been generated, it can be converted into protein form via a process called *translation* (**Figure 9.1**).

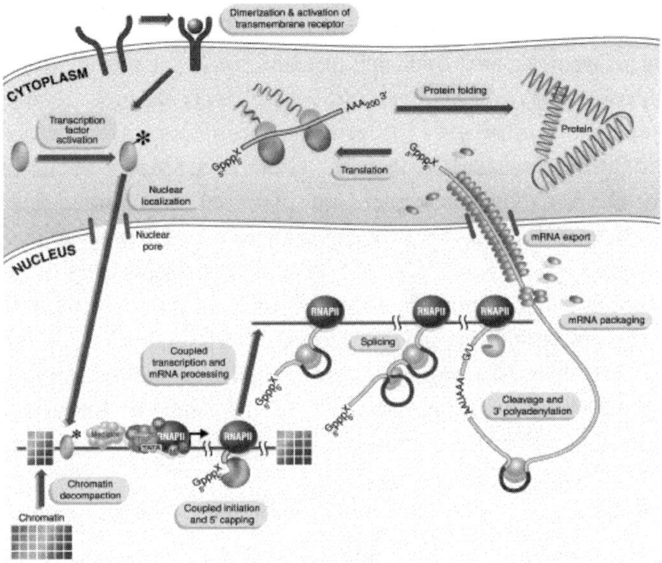

Fig. 9.1. A schematic representation of the mechanisms of gene expression, emphasising the interconnectedness of transcriptional and post-transcriptional mechanisms. Note that this figure pre-dates the widespread recognition that small regulatory RNAs play an important role in the post-transcriptional regulation of gene expression (see **Figure 9.3**). Reprinted from *Cell*, **108**, Orphanides and Reinberg, 439–451., (2002), with permission from Elsevier.

9.2.2. *The central dogma of molecular biology*

In the previous paragraph, the use of the term "information" refers to the sequence of amino acid residues (of which proteins are comprised) and the DNA and RNA sequence alphabets that code each amino acid (for example the amino arginine can be coded in DNA by the triplet sequence bases

[b]mRNA is not the only kind of RNA that is functional active in cells, but is the only RNA that encodes for proteins

AGA). The primacy of genetic information in determining the form and function of living systems was postulated in one of the major ideas of early molecular biology–the so-called *Central Dogma of Molecular Biology* named by the Nobel Prize winner F.H.C Crick (who determined the structure of the DNA molecule along with J.H. Watson). The Central Dogma postulates the nature of information transfer between the three biopolymers, DNA, mRNA and protein.[13] Its interpretation is often over-simplified, often being erroneously described as a "one-way" flow of information from genome to phenotype: actually the statement refers to which information transfers are *possible* or, as summarised by Crick himself "*once [genetic] information has got into a protein, it can't get back out again*".[13] In its most recent form, the Central Dogma is comprised of three types of information transfer: general, special and unknown (**Table 9.2.2**). General transfers occur in all cells, specifically DNA to DNA transfers are observable during *cell duplication*, DNA to RNA and RNA to protein transfers are observable during the process of *gene expression.* Certain special transfers are now seen as far more widespread than originally thought, for example *reverse transcription*, involving the conversion of information from RNA back into DNA is now known to occur in most organisms and is currently under intense study[14]). *Unknown transfers*, involving transfer of information from protein form back to nucleic acid form, have never been observed in any living cell.

Table 9.1. Genetic information transfer in the Central Dogma.

General	Special	Unknown
DNA→DNA *Replication*	RNA→RNA *Viral replication*	Protein→Protein
DNA→RNA *Transcription*	RNA→DNA *Reverse transcription*	Protein→DNA
RNA→Protein *Translation*	DNA→Protein	Protein→RNA

The Central Dogma refers to the transfer of genetic *information* itself, and does it not mean that on a mechanistic level, the the function and activity of genes cannot be influenced by extra-genomic factors: for example many genes are continuously expressed in feedback loops from upstream signalling pathways of intra- or extra-cellular origin, and other genes are expressed in response to specific external conditions (*e.g.* genes expressed following the activation of the immune response[12]). The transfer of genetic information should not be confused with the *mechanisms* by which these transfers are undertaken and the elegance of both the genetic code, and of the Central Dogma, belies the immense complexity of the actual mechanisms by which these information transfers are carried out in living cells (**Figure 9.1**), as well as the myriad control mechanisms for regulating these processes.[15,17]

9.2.3. *The text, and its reader*

It has become something of a modern cliché that the genome, specifically its DNA molecule, and more specifically genes (or other functionally important transcripts, such as non-coding RNA), encoded in that molecule specify the "instructions" that specify the form and function of an organism[7] . While that view is ultimately true, as all proteins within a cell must derive from either the cell's own genome (or those available from its inherited cellular context), this "text" of instructions needs a mechanism to "read" and interpret it correctly. This "reader" mechanism is comprised of collections of specialised regulatory proteins, estimated to number in the several hundreds.[15] In humans, only about 1% of DNA is comprised of sequences that actually encode for proteins, that is, what we would classically recognise as "genes"; whilst the remaining 99% is referred to as *non-coding DNA*.[c] The sequence regions immediately flanking the genes themselves constitute "controls" that the protein machinery of the "reader" can interact with to produce a viable messenger RNA transcript. More distant regions from genes, called enhancers, can also contain regulatory control sequences that can interact with the "reader" mechanism and modifiy or alter the expression of the associated gene.[19] The complexity of these controls varies widely across the the Tree of Life: bacteria and archea have relatively simple genomic organism, with minimal, well-organised regulatory DNA outside protein coding sequences; by contrast, higher eukaryotes like humans demonstrate stunningly complicated organisation of these regulatory sequences, reflecting their complex developmental programs, multicellularity and many modes of environmental interaction.[19]

These complex control sequences are required to permit the execution of developmental programs, that is, the process by which a fertilised egg moves through various development stages to become an adult organism. Understanding the mechanisms of development constitutes one of the major

[c]Sometimes referred to by the easily misinterpreted phrase: "junk DNA"

general problems of biology.[3] An outline of these processes is beyond the scope of this chapter but several features of such developmental processes deserve clarification. First, and perhaps most importantly for the rest of this chapter, while the execution of the entire process itself is stunningly deterministic, there is no simple relationship between genes and the outcome of their activity—or, in other words, *there is no gene that encodes for fingers, or a kidney or the autonomic nervous system*, rather these complex morphological and functional structures are specified by the precisely timed and co-ordinated action of the expression of many genes.[3] Once established, the ongoing function of differentiated cells in a given tissue is maintained by different sets of expressed genes.[20] Secondly, these developmental gene expression "programs" can be actively modulated by environmental (non-genetic) influences. The capacity for modulation of developmental programs can be readily observed in simple organisms—for example in sea urchins, where different environments can induce radically different morphological forms in the same species.[18] In mammalian species such environmental influences can even act *in utero* or *post natal*: for example, in the rat hippocampus (a brain region involved with long-term memory and navigation), the regulation of a gene encoding for *glucocorticoid receptor*, which can be modulated by different levels of by maternal care delievered by the mother, potentially resulting in upregulated expression glucocorticoid-mediated stress response in later life. This effects are mediated by DNA methylation, an example of so-called *epigenetic regulatory mechanisms* (see **Section 9.4**).

9.2.4. *The real world: genetic variation*

The complexity and interplay between multiple genetic elements in development processes can readily be observed when something goes wrong—there are documented mutations in >1000 human genes that can each cause stunningly abnormal human individuals[22] (these mutations are documented in the database *Mendelian Inheritence in Man*, or OMIM[23]). Although the vast majority of individuals of a given species contain the *same* complement of genes in their genomes there is functionally significant variation in the *structure* (coding content) of genes evident between any two individuals. For example, at a given location in a coding region of a gene, there may be a sequence variant in one individual which changes the sequence composition from an 'A' nucleotide to a 'C' nucleotide, which will lead to a different amino acid being produced in the protein for that gene encodes.

This modification may confer a different function to the activity of the protein (the case just provided is an example of a non-synonmous single nucleotide polymorphism, or SNP[24]). Such natural variation provides the "raw material" that natural selection will act on for evolution to occur, but will also contribute to inter-individual variation that will be evident within an individuals lifetime (irrespective of their reproductive success)—for example, why one smoker is more likely to develop lung cancer and another not. It is still very unclear which combinations of such common gene variants, observable in, say, >1% of members of a population, contribute the development of so-called "complex diseases" such as cancer, heart disease or obesity.[25]

9.3. Whole Genome Sequencing and the "Post-genomic" Era

9.3.1. *Genome sequencing and its influence*

Determining the base composition of nucleic acids was a major technical development of early molecular biology.[27] Subsequently, systematic identification and characterisation of the entire complement of genes in the genome of a given species was initiated as a research program in the early 1980s.[28] Despite the technical, methodological and organisational complexities of this process—and strong opposition from many who saw the entire project as a case of collecting data in the absence of a clear set of hypotheses—the first full genome sequence was generated for the bacteria *Haemophilus influenzae Rd* in 1995,[26] for the fruit fly *Drosophilia melanogaster* (a "model" organism extensively used in genetics and development) in 2000[29] and the draft whole genome of humans was generated in 2001.[30,31] At the time of writing, there are 581 whole genome sequencing projects being conducted in eukaryotic organisms, with 24 being completed, 264 being assembled and another 293 underway.[d] The rapid technical advances in sequencing has led to two major developments that are worth highlighting: the first is known as *meta-genomics*, or sequencing of DNA from all bacterial species in a given ecosystem (for example, the sequencing of sea water samples,[32] or bacteria in the human gut,[33] which has led to the identification of previously unknown species, and new insights into obesity, respectively), The second development has become known as *personal genomics*—the sequencing of the genomes of individual humans—the most

[d] `http://www.ncbi.nlm.nih.gov/genomes/leuks.cgi`

dramatic example of this is the *1000Genomes Project*,[e] in which the complete genomes of hundreds of related and unrelated human subjects will be completely sequenced. Building on recent catalogues of common variants[24] in multiple human populations,[f] this project will provide an unprecidented view of the nature of genetic variation between individuals of our species.

9.3.2. *Measurement in the post-genomic era*

Gene expression has been able to be *measured* since the development of a biochemical assay called the Northern blot in the 1970s, based on the phenomenon of nucleic acid hybridisation.[34] Later, more quantitative assays for measuring mRNA levels were developed based on the use of polymerase chain reaction (PCR) and via sequencing based approaches (SAGE: Serial Analysis of Gene Expression). A major breakthrough came in the mid–1990s, when a fortuitous combination of cheap robotics and optics, affordable laboratory computing and years of experience in recombinant DNA technology, resulted in the development of *expression microarrays*:[35] a methodology that permitted measurement of relative expression in all genes of a genome simultaneously.[g] At the time of writing, the same technology has been adapted as a genome-wide "read-out" method for many molecular biological assays, not just those that measure gene expression, including genome-wide surveys of histone modifications, DNA methylation and DNA accessibility[36] (see **Section 9.4**). These technologies have provided an unprecedented view of the function of the genome and several large-scale projects are employing these approaches to systematically discover regulatory elements in eukaryotic genomes.[37,38] More recently, DNA sequencing technology has been adapted to provide a read-out for these assays that is not dependent on the experimental complexities of hybridisation reactions, so called "next generation" sequencing.[39]

9.3.3. *Post-genomic biology as an information science?*

The aforementioned methodological developments that arose out of the sequencing projects have highlighted several facets of molecular and cell biology that were probably long overdue for examination and renewal: firstly, the new genome-wide technologies are *inherently data-rich*. The results of single gene expression assay, such as the Northern Blot mentioned above,

[e] `www.1000genomes.org`
[f] `www.hapmap.org`
[g] "Gene chips" is an equivalent but largely obsolete phrase for these devices

can be interpreted by examination of a single photographic image (or series of images in the context of an experiment with replicate observations), in contrast, a microoarray experiment would generate thousands of gene expression measurements for a single experiment sample, and a matrix of gene-by-sample measurements for a replicated experiment. By nature, such data requires extensive preprocessing to remove noise and systematic artefacts and extensive statistical analysis to obtain even preliminary results of any value. This interaction with the requisite statistical methodology has been traumatic, as most practising molecular and cell biologists have little or no training or experience in these areas, but also fruitful, as these new data have begun to stimulate new, far-ranging approaches to analysing so-called "high-dimensional" data.[40] Mostly, however, this new emphasis on *data* has resulted in importance being placed on the role of the information sciences. Like astronomy and astrophysics, the World Wide Web (WWW) has become an indispensable tool in modern biological research—with databases like National Center for Biotechnology Information (NCBI)[h] and Ensembl[i] (and a myriad of smaller, more specific collections[41]) depensing information on genes and their protein products, the structure and organisation of genomes, and increasingly, high-throughput functional data itself, on a daily basis to an extensive and varied biological community. Reasonably enough, many of these databases are structured around individual genes or proteins (this "view" being most familiar to the majority of practitioners, but the problem of integrating this information across many genes remains problematic—for example, understanding the coherent function of even 10 of genes remains challenging using the web-based approaches. Thus the importance of developing of inherently genome-wide analysis, visualisation and interpretation remains an urgent research problem, whose solution will exploit the full power and potental of the new genome-wide technologies. To put this opinion firmly into perspective, it is worth considering the American Association for the Advancement of Science, on the occasion of their 125th anniversary in 2005, stated as one of the top 25 "Big Questions" facing scientific enquiry over the coming quarter century, "*How will coherent biological pictures emerge from a sea of biological data*".[42]

[h] www.ncbi.nih.gov

[i] www.ensembl.org

9.3.4. *Confirmation or exploration?*

Molecular and cell biology have largely been *hypothesis-driven* sciences, focused on the roles of individual genes, or or a family of related genes in, say, development or their role in the pathogenesis of a disease such as cancer (**Figure 9.2**). The utility of the hypothesis driven approach is validated by the explosive growth of biological knowledge under molecular research programs and the immense complexity of the function of any given gene that clearly demands a detailed reductionist approach. In contrast, the newer genome-wide technologies were inherently *exploratory* in their application, providing an unbiased survey of using a particular genome-wide assay. These two approaches are quite different in practice and there remains, at a disciplinary level, an unresolved tension between confirmatory and exploratory approaches to knowledge generation in molecular and cell biology, that is still being played out at the time of writing.

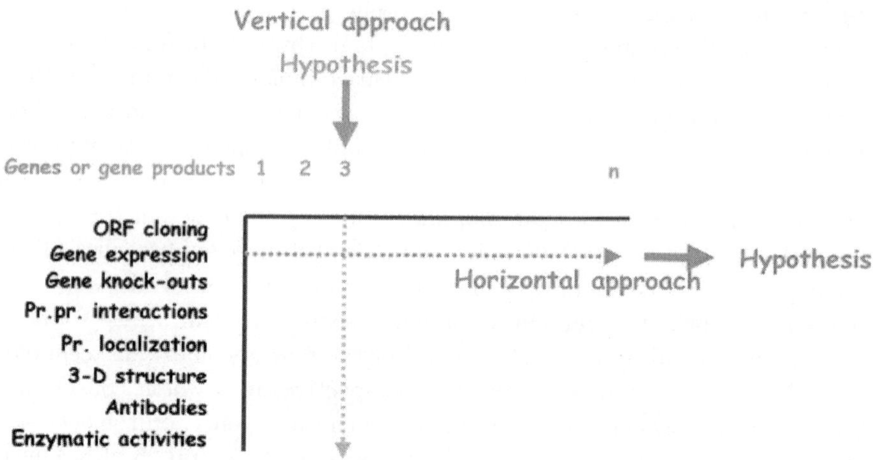

Fig. 9.2. Hypothesis-driven and exploration-driven research programs in molecular-cell biology. Contrasted are the "hypothesis" driven approach, using detailed investigation of the function of a single gene in the biological system under study, with the "exploratory" approach, of surveying the activity of all genes, with a view to generating hypotheses about the behaviour of the system. Reprinted from *Cell*, **104** , M. Vidal, An atlas of functional maps, 333-339., (2001), with permission from Elsevier.

9.4. Comprehending Complexity in Gene Regulation

9.4.1. *The eukaryotic nucleus and gene regulation*

In eukaryotic genomes, gene expression takes place in the nucleus, a specialised sub-cellular organelle of cells which contains the DNA molecule. Evolution has provided elegant, specialised "packing" solutions to the immense amount of genetic material contained within the cellular nuclei of eukaryotic organisms, whilst maintaining this material in a form in which it can be readily duplicated, expressed or repaired as the need arises. In the nucleus, the DNA molecule exists within a scaffold formed of *nucleosomes* which are periodically located along the DNA strand.[44] This DNA–nucleosome complex is referred to as *chromatin.* Chromatin is capable of highly dynamic reconfiguration depending on the requirements of the cell: for example, genes that are not required for expression in the cell can be secluded away in a highly packed, "closed" form called *heterochromatin.*[44] In contrast, genes that are routinely used tend to have a more "open" chromatin configuration, called *euchromatin.* The mechanisms by which chromatin is regulated are currently subject to intense study (see **Section 9.4.3**).

Although the amount of genetic material packed into a nucleus is breathtaking, one of the major surprises of the genome sequencing projects was the recognition of *how few genes it takes to specify a complex, multicellular organism*, like a worm or a human being. At the start of the Human Genome project, consensus projections placed the likely number of genes at 100,000–120,000, however estimates from the "first draft" were 24,000-35,000 and more recent analyses have dropped this number to as low as 20,000.[45] This number is not substantially different than many of the lower eukaryotes, for example the nematode worm (~19,000 genes[46]) or the fruit fly (~13,000 genes[29]). Interestingly, this observation was anticipated around 30 years ago in a seminal paper examining similarities between protein sequences of humans and primates: it was predicted that there would be substantial differences in protein sequence given the great phenotypic differences between human and other primates, however, the observed sequence composition was highly conserved between the two species.[53] Thus it is the *complexity of gene regulation* that appears to be subjected to substantial innovation over evolutionary time, *not only the modifications to the coding regions of the gene themselves.* As mentioned above, whilst some genes are ubiquitously expressed in all cell types and conditions, many genes need to be selectively expressed in a given tissue or cell type, or in response to a given external stimulus, and thus, very flexible regulatory architecture is required to permit this range of behaviour. In this section, we examine the current state of understanding of the regulation of gene expression, particularly highlighting the multiple levels of controls and the multiple spatial scales under which this control takes place.

We will consider three spatial levels of control, and at the risk of inventing new terminology we will describe these as *local*, *extended* and *large–scale*.[j] We use *local control* to refer to regulatory mechanisms that act on the regulatory sequence regions immediately adjacent to the protein–coding region of genes, for example, around the transcription start site, and within the untranslated region of the mRNA transcript itself. We deliberately choose to include both the gene and its transcript in this definition to permit us to encompass both transcriptional and post-transcriptional levels of control of a given gene. *Extended control* is used to describe regulation of gene expression that is conducted through changes to the chromatin "scaffold" in which the DNA molecule is maintained and finally *large–scale control* is used to describe the role of spatial organisation of genes in the nuclear space in the regulation of gene expression.

9.4.2. *Local control*

The regulation of gene expression, whether during development, processes required for the ongoing viability of cell or mediation of many responses to environment perturbation is largely mediated through the action of specialised classes of DNA–binding proteins called *transcription factors*.[10] In the human genome, there are an estimated 2000–3000 transcription factors.[10] Specialised binding domains on transcription factors interact with corresponding *motifs* in regulatory regions of genes. A motif is a generally a short characteristic sequence, for example the *SRY (sex–determining region Y)-box* family of transcription factors can preferentially bind to the sequence DNA motif `AACAAT`. Once recruited, these proteins recruit additional proteins that activate the transcription process.[15]

Further regulatory control is exerted at the level of the gene's transcript (the messenger RNA "copy" of the gene).[15] The mRNA transcript is subject to a complex series of processing steps to prepare it for translation into a protein and export it from the nucleus to the "translational machinery", called ribosomes, located in the cytoplasm of the cell (**Figure 9.1**).

[j]The reader should be warned that this terminology is non–standard.

One example of post–transcriptional control is the phenomenon of *alternative splicing*,[9] involving the removal of segments of coding sequence from a gene following the initiation of transcription. Alternative splicing expands the diversity of proteins products that can be generated from a single gene.[9] Recent genome-wide surveys for splicing events have highlighted the widespread presence of this phenomenon in eukaryotic genomes. More recently, the recognition of the widespread role of *small regulatory RNA* (*e.g.* microRNA) has added a new mechanism by which mRNA can be modified, at both post-transcriptional and translational levels.[16,17] The evident transcriptional complexity of the genome, including the seemingly widespread transcription of non-coding regions and the presence of tissue specific transcription start sites, has increased the number and diversity of regulatory elements that need to be taken into account when considering the control of gene expression.[37]

Identification of whether a genes' regulatory regions contains the motif(s) for a transcription factor or regulatory RNA is made extremely difficult by the short length of the motifs themselves.[47] In addition to experimental approaches using high–throughput DNA–binding assays,[37] *comparative genomics* has an important role to play in identifying regulatory motifs that have been conserved across eukaryotic evolution and several genome–wide maps of transcription factor binding sites have been derived.[48] Combined with targeted experimentation in specific cases, we now have an increasingly detailed and diverse view of how individual genes are controlled at both transcriptional and post–transcriptional levels (**Figure 9.3**). *How to utilise such regulatory motif data in understanding gene regulatory circuits in higher eukaryotes remains an open and challenging problem.*[49–52] Much of the recent progress in this area has been made using approaches developed in the fields of statistics and computer science (*e.g.* machine learning to predict regulatory features in sequence[54]), but increasingly modelling approaches with an explicit physico–chemical support are being employed.[55] This is increasingly important, given the recent recognition that DNA–binding proteins are capable of recognising multiple and distinct sequence motifs, and that low–affinity binding may be equally functionally important as high–affinity binding.[56]

9.4.3. *Extended control*

Nucleosomes are constructed of 4 related proteins called *histones*. These cylindrical like protein complexes were long considered as a static "scaf-

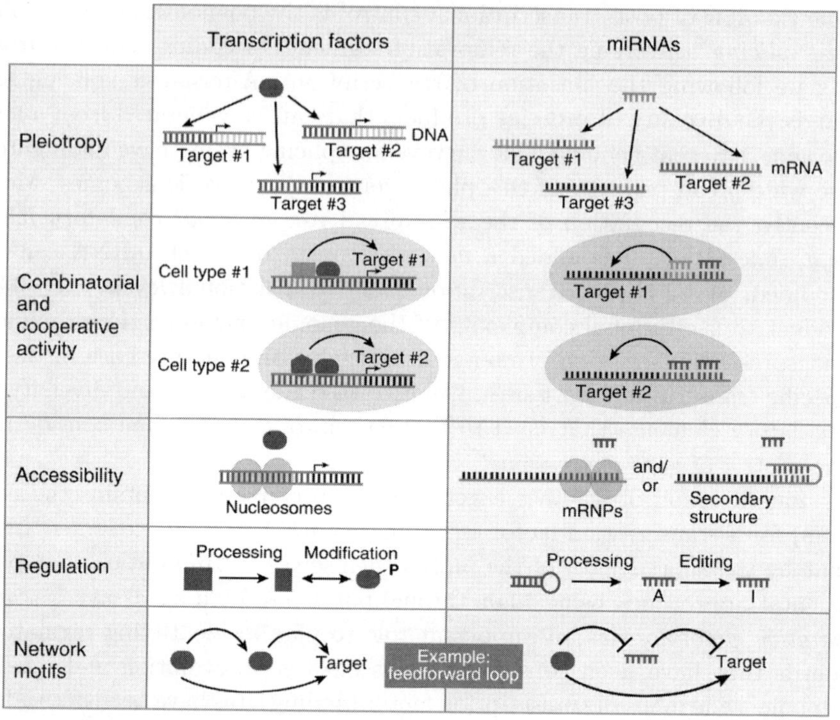

Fig. 9.3. Organisational–level phenomena associated with transcriptional and post–transcriptional control of gene expression. *Pleiotropy*: a given transcription factor or regulatory RNA (miRNA) act on multiple target genes. *Combinatorial and cooperative activity*: the regulatory regions of genes (promoter) or transcript (untranslated region) can be subject to differential regulation that is context dependent *e.g.* overlapping sets of transcription factors or miRNAs that are active in different cell types. *Accessibility*: the genomic–context (chromatin) of a gene can be subjected to regulation by changes to nucleosomes (see **Section 9.4.3**. At the transcript level, physical geometry ("secondary" structure) can influence subsequent regulation. *Regulation*: the activity of transcription factors can be modified by post–translational modification. mRNA transcripts can be modified by RNA editing mechanisms. *Network motifs*: combined action of different regulatory elements on sets of genes form large–scale regulatory networks or circuits (see **Section 9.6**). Figure taken from `Science`, Hobert, O. **319**: 1785–1786 (2008), *permission requested from the AAAS.*

fold" for the DNA molecule, however the dynamic nature of chromatin is now gaining importance, particularly through recognition of its regulatory role in gene expression and other chromosomal processes. Histones are comprised of a structural core, and a series of "tail" structures. Amino

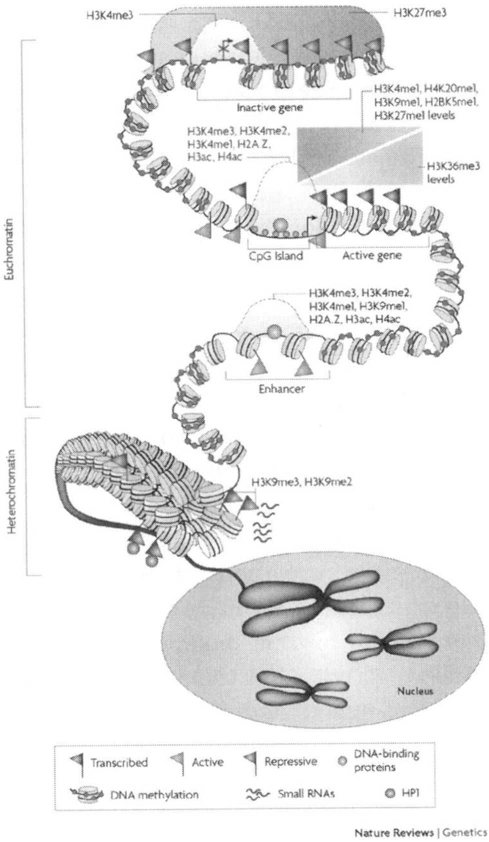

Fig. 9.4. Chromatin regulation and control of gene expression. Histone modifications are abbreviated by the histone, residue and modification type *e.g.* "H3K9ME2" refers to de–methylation of lysine 9 in histone H3. Within the context of accessible chromatin (euchromatin), different combinations of histone modifications appear to be associated with the activity levels of genes (active genes vs. inactive genes). Low levels of DNA methylation (conversion of cytosine to methyocytosine) is associated with regulatory regions such as promoters, regions with increased levels of C and G nucleotides (CpG "islands") and possibly, enhancers. Regions of "closed" chromatin are associated with a different set of histone modifications, and mediated by regulatory proteins and regulatory RNAs. Reprinted by permission from Macmillan Publishers Ltd: *Nature Reviews Genetics*, Schones and Zhao, **9**: 179–191 (2008).

acid residues in these tails are subject to post–translational modifications (reversible chemical modifications of proteins) that can confer diverse structural and functional properties.[k] An important concept is that of the *histone code* hypothesis,[58] named by loose analogy to the genetic code, that postulates that different combinations of histone modifications can be interpreted by a chromatin "reader" mechanism and modify the functional state of the surrounding chromatin: such modifications may enhance or repress the relative ability of a gene to be transcribed, via for example, making transcription factor binding sites more or less accessible.[59] At present, there is intense activity to interpret the influence of such post-translational histone modifications on gene expression. For example, an important recent study, assaying 39 histone modifications on a genome-wide basis in human CD4+ T cells (a type of immune cell) has identified a "core" histone modification pattern, comprised of a combination of 17 distinct histone modifications associated with genes that show higher than average levels of expression. These 17 "core" histone modifications are observable at the same genomic locations, down to an individual nucleosome level. While compelling, these notions need to be interpreted with care: in general, such effects are *trends*, discernable from large–scale correlations, made from a genome–wide level analyses of multiple experiments. Thus, there is a *tendency* for promoters showing this core set of histone modifications to be associated with increased transcriptional activity, but the predictive or deterministic value of these measurements in understanding the regulation of a given gene is still not clearly understood.

9.4.4. *Large–scale control*

At a higher spatial scale, the *location of genic regions within the nuclear space* appears to have functional relevance.[65] In living cells, genomes are physically organised as chromosomes and each chromosome fills a distinct and compact nuclear sub-volume, called a *chromosome territory*[66] (**Section 9.5**). The relative placement of chromosomes is highly tissue (cell-type) specific, however it is critical to recognise that the patterning of locations is *highly stochastic*. For example, in mouse liver cells, Chromosome 5 is *preferentially* located in the nuclear interior but not always observed there. The current model has these intermingling chromosome

[k]It is important to note that such post–translational effects are mediated "above" the level of gene expression itself and will not be discernible in measurements of mRNA levels (although the effects may well be indirectly apparent *e.g.* Hudson *et al.* 2009[57]). See Seets et al. (2006), Jensen (2006), Sims and Reinberg (2008) and Wilkins and Kummerfeld (2008) for reviews of the phenomenology of post-translational modifications on proteins and ideas about the hypothesised "codes" associated with these phenomena.

territories interspersed with "transcription factories" (spatially localised collections of transcription factors, RNA polymerase complexes and related "expression machinery"[67]) allowing an effective mechanism for gene expression via the sharing of transcriptional and post-transcriptional machinery and regulatory sequences[68] (**Section 9.5**). Currently, there is controversy surrounding the extent to which such phenomena are "self–organised" consequences of genome structure, or directly structured by the nuclear architecture itself,[65] but perhaps more significantly, the relative importance of this phenomenon *as an organising principle of global gene expression*, remains unclear.[69] Tantalising evidence of the functional significance of non–homologous *inter*–chromosomal interactions has emerged in recent work: particularly the co–localisation of *Ifng* and the *Th2* locus (genes) in naive T cells[70] and the demonstration that a singly expressed odorant receptor gene (of potentially thousands throughout the genome) associates with the *H* enhancer element on chr14.[71] These observations, along with previous recognition that genomically–dispersed ribosomal genes associate in the nucleous,[73] suggest that the phenomena may be widespread. Experimental approaches to the problem are either inherently limited to small numbers of genes (3D–FISH) or in the case of high–throughput extensions of the 3C technique (Chromosome Conformation Capture),[72] are extremely noisy and in a nascent state of development. Given these difficulties, it is surprising that there have been no attempts to infer the extent of inter–chromosomal organisation of gene expression from extant high throughput data, that is, *by attempting to identify groups of genes from 2 or more chromosomes that may be spatially associated and co-expressed.* In contrast, the *intra*–chromosomal level of organisation of global gene expression has been extensively studied using such data.[74]

9.5. Methods for Identifying Regulatory Control

Within this increasingly complex landscape of gene regulation, an important unifying concept is the notion of *coordinated control of gene expression.*[75] Coordinated control refers to the action of *multiple regulatory factors acting together to activate or modulate the activity of groups of genes to achieve a given biological outcome* (**Figure 9.6**). Developmental processes provide clear examples of coordinated control of gene expression at the transcriptional level. A recent dissection of the muscle development regulatory network of the (sequenced) marine animal *Ciona intestinalis*:[76] in this network, three transcription factors, *CRE*, *MyoD* and *Tbx6*, coordi-

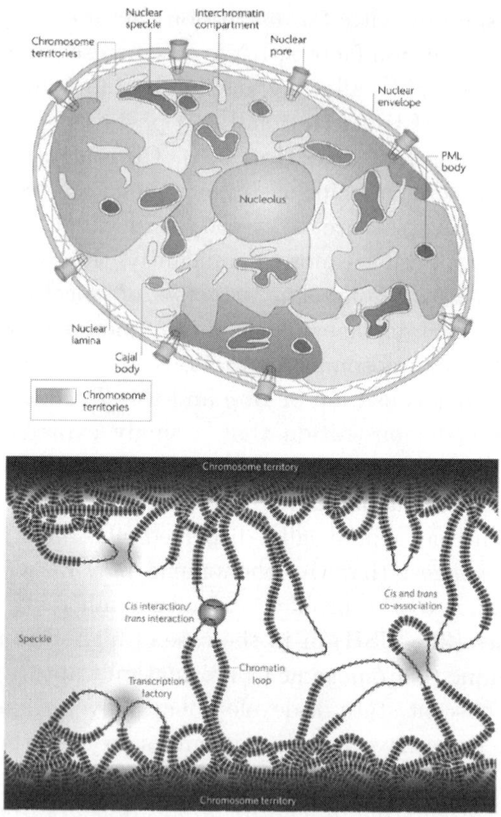

Fig. 9.5. Spatial organisation of the eukaryotic nucleus. *Top panel*: different chromosomes tend to occupy distinct regions of the nuclear space. *Bottom panel*: "Transcription factory" theory of model of gene expression: space between chromosome territories is occupied by transcriptional and post-transcriptional "machinery" (**Figure 9.1**), allowing for efficient control and regulation of gene expression. Reprinted by permission from Macmillan Publishers Ltd: *Nature Reviews Genetics*, Lanctôt *et al.*, **8**: 104–115 (2007) and *Nature*, Fraser and Bickmore, **447**: 413–417 (2007).

nately activate 19 genes required for muscle development. The 19 genes act together to produce a highly specific biological outcome, and at a molecular functional level, 17 of the 19 proteins act in the same macromolecular complex (**Figure 9.6**). In this example, the regulatory sub–network was identified using a combination of complex experimental procedures and analysis of sequence data, but in many instances, the experimental component (*e.g.* genetic manipulations to construct reporter transgenics, etc) of this work

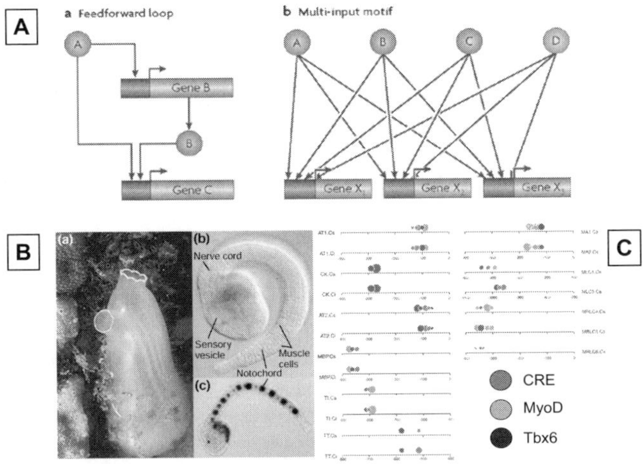

Fig. 9.6. Coordinated control of gene expression. *A*: two examples of conceptual models of coordinated control, in which regulatory factors (circles) *e.g.* transcription factors, act together regulate the expression of multiple genes (rectangles; with promoter regions in dark shading). See text for actual examples of coordinated control active during development of the marine eukaryote *Ciona intestinalis* in which three transcription factors, *CRE*, *MyoD* and *Tbx6*, co-ordinate the expression the 19 genes involved muscle cell differentiation. Reprinted by permission from Macmillan Publishers Ltd: *Nature Reviews Genetics*, Komilli and Silver, **9**: 38–48 (2008).

may not be feasible. In such circumstances, *how can we gain insight into which regulatory elements are active in a given set of genes?*

A typically scenario may be summarised by the notion of "top–down" identification of regulatory elements (**Figure 9.7**)—for simplicitly, we will discuss transcriptional level regulatory targets *i.e.* transcription factor binding sites, but there is no reason why any of the three levels of control could not be incorporated into this framework, assuming that adequate mechanistic data was available. Current approaches for identifying regulatory controls focus of the use of "*top–down*" enrichment methods: taking a set of genes considered "interesting" in a given setting (*e.g.* genes that are co–expressed or that are known to be involved in related biological processes, etc). and asking what are the regulatory motifs that are statistically enriched in this set relative to random expectation? Specifically, given a set of genes G of primary interest, and a set of genes, B, that may be considered a reference or control set, we extract the number of occurrences of a regulatory element, m (or, more generally, a DNA sequence motif) in the DNA of the regulatory regions of the genes in G and B, and denote these as n_{11} and n_{21}, respectively. This arrangement is most easily described as a 2×2 contingency table:

	Motif m Present	Motif m Absent	
Genes in G	n_{11}	n_{12}	$n_{1\bullet}$
Genes in B	n_{21}	n_{22}	$n_{2\bullet}$
	$n_{\bullet 1}$	$n_{\bullet 2}$	n

Where $n_{\bullet 1} = n_{11}+n_{21}$, $n_{\bullet 2} = n_{12}+n_{22}$, $n_{1\bullet} = n_{11}+n_{12}$, $n_{2\bullet} = n_{21}+n_{22}$ and n is the total number of genes in the sets G and B. We can then test the hypothesis that the regulatory element m is *enriched* in the gene set G by using the hypergeometric distribution:

$$P(n_{11} > x) = \sum_{k=x}^{min(n_{\bullet 1},n)} \frac{\binom{n_{\bullet 1}}{k}\binom{n-n_{\bullet 1}}{n_{1\bullet}-k}}{\binom{n}{n_{1\bullet}}} \tag{9.1}$$

where P is the probability that the motif will be observed in at least n_{11} genes in the set of interest, under the null hypothesis of no association. The general approach has been extended to multiple regulatory factors by several groups and, by searching for enrichment of sequence motifs identified without any *a priori* knowledge of their regulatory role.[50–52] While the latter approach seems more general than using known regulatory motifs, the broader problem remains of understanding which regulatory factors, if any, may be associated with the enriched motifs.

An alternative "top–down" approach, originally developed by Bussemaker *et al.*[77] is to use regression approaches (reviewed by Das[78]). As above, we could employ mRNA levels, as measured from expression microarrays, to model the relationship between the number of instaniations of a motif m of the regulatory regions of gene g and the estimated expression levels, E_g using a simple linear model:

$$E_g = \sum_m C_{mg}\beta_m + \epsilon_g \tag{9.2}$$

where the summation is over all motifs m that are thought to influence the expression of set of genes G. One issue with this approach is the choice of which motifs to include, with most papers utilising feature selection approches such as *forward–stepwise* procedures.[78,84] Another is the difficulty

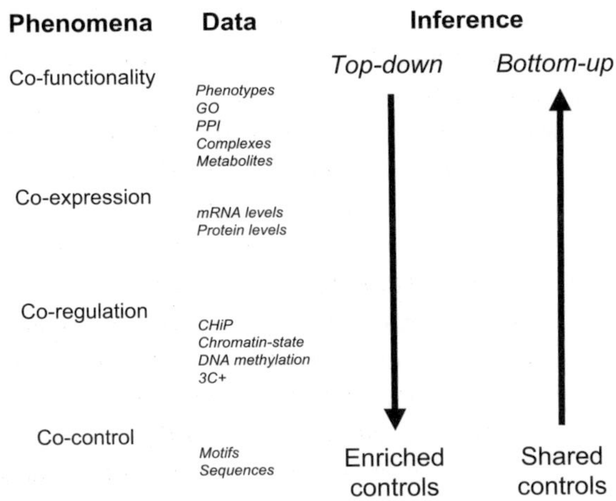

Fig. 9.7. Conceptual frameworks for identification of regulatory programs controlling gene expression. *Top–down* approaches start with sets of genes defined by shared properties, for example, involvement in the same biological processes, shared expression patterns, etc and attempt to identify the associated regulatory elements. In contrast, *bottom–up* approaches attempt to define shared regulatory elements, in order to identify groups of genes that have the potential to be co-regulated, and explore the context-specific implications of this level of organisation, using functional from higher phenotypic levels of cellular function and organisation.

is how to incorporate the effects of interactions between multiple regulatory motifs: Zhang *et al.* recently developed an extension to treating the expression data of a set of genes as a multivariate distribution.[84]

Overall, these "top–down" approaches have the advantage of being very context–specific (*e.g.* being able to identify tissue– or condition–specific regulatory factors from appropriate microarray expression data), while the principal disadvantage relates to the fact that combinatorial regulation, involving multiple regulatory elements, is difficult to detect.[78] Another alternative strategy, of which we explore the feasibility in the next section might be to try and identify genes that *share the potential to be under co-ordinated control*, and explore the consequences of this level of organisation for higher–level biological function.

9.6. A Simple Approach to Coordinated Control of Gene Expression

At the start of the last section, we employed an example[76] of a set of genes under coordinated control in development. We return this example which motivates the development of a new, simple approach for identifying modules of genes that may under coordinated control. While the sequence motifs for these 3 transcription factors were clearly clustered within promoters, there was *no additional patterning of motif order, spacing or orientation that might inform us about the organisational principles of this regulatory module.* This observation suggests that *simple quantitative models of promoter architecture may have a high level of biological plausibility in defining groups of genes that may have the potential to be under coordinated control.* While we have used an example of transcriptional level co–ordinated control, the concept could be *extended to encompass a range of regulatory mechanisms, from chromatin level, through transcriptional, post-transcriptional and structural regulatory motifs*, by using the relevant data concerning shared chromatin state or presence of regulatory motifs in sequence.

We employ a *vector space model* (similar the representations used for text mining[79]) which uses a binary matrix representation (data table of 1s and 0s) describing the presence (1) or absence (0) of a given regulatory motif in a promoter of a given gene (with genes indexed in rows and motifs indexed in columns). This model can incorporate weightings for multiple occurrences of a motif within a promoter, and global occurrence of a motif in the entire promoter set. We employ data of Xie *et al.*[48] who defined a set of 837 regulatory motifs, comprising 174 novel promoter motifs, 441 promoter motifs from TRANSFAC (not detected in[48] but included) and 222 miRNA binding sites. Additionally a set of 11653 target human genes that contain one or more of these motifs is defined. As our primary aim is to identify groups of genes that have the potential to be under coordinate control at transcriptional and/or post–transcriptional levels, we start by recasting this set into a multivariate data matrix, that is, a binary matrix, $\mathbf{M}$, i.e. a data table with genes indexed by rows $i = 1, \ldots, 11653$ and regulatory motifs indexed by columns $j = 1, \ldots, 837$ whose elements, M_{ij}, are unity if the gene i has the motif j occurring in its promoter or $3'$UTR and zero otherwise. Accordingly, we call this matrix the *motif presence matrix* (MPM), illustrated in **Equation 9.3** for a hypothetical example generated for 5 genes ($G_1, \ldots, G_5$) and 5 motifs ($R_1, \ldots, R_5$).

$$\mathbf{M} = \begin{array}{c} \begin{matrix} R_1 & R_2 & R_3 & R_4 & R_5 \end{matrix} \\ \begin{pmatrix} 0 & 1 & 1 & 1 & 0 \\ 0 & 1 & 1 & 0 & 1 \\ 1 & 0 & 1 & 0 & 1 \\ 1 & 1 & 1 & 0 & 0 \\ 1 & 0 & 0 & 0 & 1 \end{pmatrix} \end{array} \begin{matrix} G_1 \\ G_2 \\ G_3 \\ G_4 \\ G_5 \end{matrix} \tag{9.3}$$

Formulating these data in this fashion permits us to use a simple dissimilarity–metric, the Jaccard co–efficient,[80] to quantify the extent to which two genes (rows) have a similar complement of regulatory motifs in their promoter regions and/or 3′UTRs. If $M_a = (a_1, \ldots, a_M)$ and $M_b = (b_1, \ldots, b_M)$ are rows of the MPM for genes a and b, the Jaccard distance, $J(a,b) = 1 - (\alpha/(\alpha+\beta+\gamma))$ where $\alpha = \#(j : a_j = b_j = 1)$, $\beta = \#(j : a_j = 1, b_j = 0)$ and $\gamma = \#(j : a_j = 0, b_j = 1)$. We compute the Jaccard co–efficient for all pairs of genes and form a gene–by–gene dissimilarity matrix, $\mathbf{J(M)}$. For the example presented above, this becomes:

$$\mathbf{J}(\mathbf{M}) = \begin{array}{c} \begin{matrix} G_1 & G_2 & G_3 & G_4 & G_5 \end{matrix} \\ \begin{pmatrix} 0.00 & & & & \\ 0.50 & 0.00 & & & \\ 0.80 & 0.50 & 0.00 & & \\ 0.50 & 0.50 & 0.50 & 0.00 & \\ 1.00 & 0.75 & 0.33 & 0.75 & 0.00 \end{pmatrix} \end{array} \begin{matrix} G_1 \\ G_2 \\ G_3 \\ G_4 \\ G_5 \end{matrix} \tag{9.4}$$

Values of the Jaccard co–efficient closer to 1 imply that two genes have fewer shared regulatory elements, while values closer to 0 imply a greater proportion of shared regulatory motifs in the two genes. In **Equation 9.4**, G_3 and G_5 show the closest similarity in regulatory composition, sharing 2 motifs (R_1 and R_5) out of all 3 present in both genes (R_1, R_3 and R_5). Performing this procedure across all pairs of genes, we can apply a suitable clustering algorithm (*e.g.* complete–linkage hierarchical clustering) to *form clusters of genes that have the potential to be under coordinated transcriptional control.* Using a cluster-stability estimator[81] we defined 332 statistically–reliable clusters of genes that share >1 regulatory motif. Our findings indicate that *(1)*: a wide distribution of cluster sizes is observed (2–74 genes, with mean size of 6 genes); *(2)*: genomic localisation (tandem duplication) or membership of gene families is *not* an organsing principle of these clusters, being distributed across 4 chromosomes on average; *(3)* these clusters exhibit a wide spectrum of tissue specificity, ranging from being ubiquitously expressed, to having highly tissue–specific behaviour and *(4)* a number of cluster are enriched for biological functions (as assessed by Gene Ontology enrichment analysis).

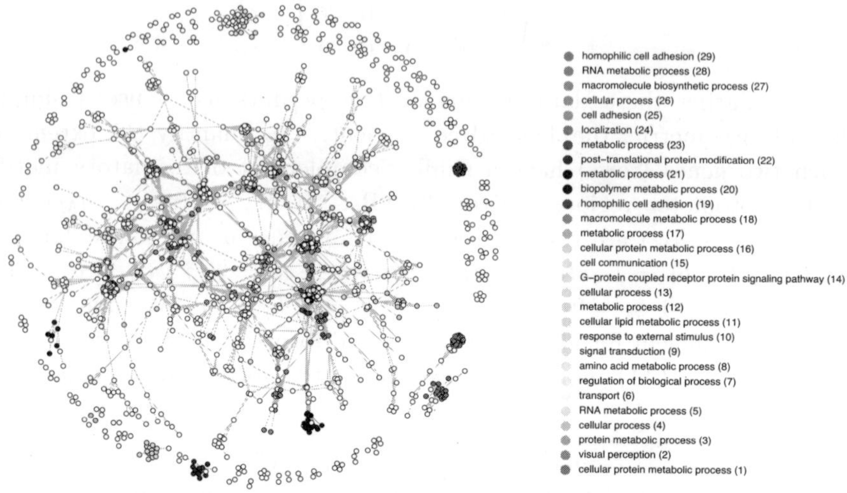

Fig. 9.8. Network views of regulatory similarity: we represent genes as nodes (circles) and connections between two genes (edges) are based on their sharing a significantly high level of motif composition. Groups of genes (clusters), numbered from 1 to 29, where identified using hierarchical clustering and the functional association of clusters was identified using Gene Ontology enrichment analysis (colours denote enriched terms for each of the 29 gene–clusters). Network visualisation was generated using a Fruchterman–Reingold layout algorthim implemented in the `igraph` library in `R`.

An example of one such cluster is a group of 15 genes which all share binding motifs for the *bicoid*–related homeodomain protein PITX2 (an enrichment of a factor of ~160 relative to the background set of 11653 genes). The cluster is significantly enriched for the GO term *visual perception* and all genes in the cluster are expressed in human retina (as determined by analysis of SAGE data from the EyeSAGE database: `neibank.nei.nih.gov/EyeSAGE/index.shtm`). The 5 genes enriching for *visual perception*, namely ABCA4, CNGB1, PRPF8, CABP4 and BBS5, all have known involvement in photoreceptor function and retinal disease. PITX2 itself is known to be involved in the eye development[82] and is a disease candidate for several diseases related to dysgenesis of the eye (*e.g.* anterior segment dysgenesis or ASD[83]), however its possible role in either photoreceptor differentiation, or later–stage maintenance and survival of these sensory neurons is less clear. Based on these findings, further experimental investigation of the role of PITX2 in this context would appear warranted. *This example clearly shows the potential of the coordinated control model to uncover context–specific, biological plausible genotype–phenotype associations.*

In **Figure 9.8** we show a network representation of the inter–relationships based on shared regulatory content: genes are represented as nodes, and edges are formed between genes if two genes have a critical proportion of common regulatory motifs in their promoters and/or 3′UTRs, defined, in this example, by the Jaccard co–efficient being less than a threshold of 0.75. Because this threshold is largely arbitrary, we shown the genes belonging to the 29 clusters defined above that were tested for functional enrichment, which have been highlighted in different colours (using a unique colour assignment per cluster). Clusters of genes that demonstrated enrichment for a given biological function appear to be clearly recapitulated in the network representation. More generally, the network appears to have a highly modular structure, showing a large number of groups of densely interconnected genes, with these individual component modules being themselves highly interconnected in the main component of the network. It is possible that using such representations will reveal significant information concerning the regulatory architecture of genes and we are currently undertaking further analyses on the properties, structure and correlates of this network.

9.7. Summary and Outlook

We have presented a survey of the major issues relating to understanding the complex nature of gene expression programs. Gene expression represents the fundamental phenomenon by which information encoded in a genome is utilised for the overall biological objectives of the organism. Gene expression is controlled by a myriad of processes, at a *variety of spatial scales* relative to the gene and its transcript. These include regulatory proteins and RNA, mediated through through control sequences in and around the gene itself, and control of the chromatin "scaffold" in which the DNA molecule is embedded, and, at a higher spatial level again, the relative position of genes within the nuclear space. Of these, DNA–embedded regulatory control sequences relating to transciptional and post-transcriptional mechanisms have been studied in great detail, but we are only beginning to appreciate the regulatory potential of chromatin and little is currently understood about the role of nuclear organisation in the regulation of gene expression.

While the genome sequencing programs have very successfully mapped the structure and content of both the coding and non–coding components of genomes, understanding the nature of the functional genome, particularly how genes act together to contribute to the generation of phenotypes at higher organisational levels, remains an open problem. Despite the relative ease with which genome–wide functional data can now be collected, mapping genotype–phenotype relationships using these new approaches has proven more challenging than originally anticipated. The fundamental importance of this issue was highlighted in 2005 by the The American Association for the Advancement of Science (AAAS) listing the question "*How Will Big Pictures Emerge From a Sea of Biological Data?*" as one of the top 25 "hard questions" that will face scientific enquiry over the next quarter century.[42] While this accumulation of post–genomic data continues at a daunting rate, the major bottleneck is now the difficulty with integration, interpretation and making mechanistic inferences concerning genome–wide phenomena from these increasingly large data depositories. Thus *new methodological approaches are required to extend the classical question of genotype–phenotype relationships using the new genome–wide approaches.*

A related aspect of the understanding the function of cellular systems in which complex systems approaches could be applicable is the need to provide more sophisticated and rich *descriptions of empirical data* collected using the new, genome–wide measurement techniques. This point is becoming particular relevant given the difficulties in defining candidate genes from recent large–scale disease mapping studies (genome-wide association studies).[85] Thus, more sophisticated attempts to describe how gene expression differences are manifested between individuals, conditions or populations, are likely to constitute an important area of future research in molecular systems biology.

References

1. Waddington C.H. *The strategy of genes: a discussion of some aspects of theoretical biology*, London, Allen and Unwin (1957).
2. Alberts B. The cell as a collection of protein machines: preparing the next generation of molecular biologists. *Cell* **92**: 291–294 (1998)
3. Weiss K.M. The phenogenetic logic of life, *Nature Reviews Genetics* **6**: 36–45 (2005).

4. Andrianantoandro E. *et al.* Synthetic biology: new engineering rules for an emerging discipline, *Mol. Syst. Biol.* **2**: 2006–0028 (2006).
5. Vilar J.M.G. Modularizing gene regulation, *Mol. Syst,. Biol.* **2**: 2006.0016 (2006).
6. Stone R. Diseases. A medical mystery in middle China. *Science* **324**:1378–1381 (2009).
7. Noble D. *The music of life: Biology beyond genes*, Oxford, Oxford University Press (2006).
8. Goodwin B.C. *How the leopard changed its spots: The evolution of complexity*, New York, C. Scribner's Sons (1994). [see also review by: Price C.S.C (1995). Structurally unsound. *Evolution* **49**: 1298–1302 (1995).]
9. Irimia M. *et al.*. Quantitative regulation of alternative splicing in evolution and development, *Bioessays* **31**: 40–50 (2009).
10. Vaquerizas J.M. *et al.* A census of human transcription factors: function, expression and evolution. *Nature Reviews Genetics* **10**: 252–263 (2009).
11. Freidman J.M. The function of leptin in nutrition, weight, and physiology. *Nutr. Rev.* **60**: S1–S14 (2002).
12. Maniatis T, Goodbourn S, Fischer JA. Regulation of inducible and tissue-specific gene expression. *Science*, **236**:1237–1245 (1987).
13. Crick F.H.C. Central dogma of molecular biology, *Nature* **227**: 561–563 (1970).
14. Hurst G.D.D., Werren J.H. The role of selfish genetic elements in eukaryotic evolution. *Nature Reviews Genetics* **2**: 597–606 (2001).
15. Orphanides G, Reinberg D. A unified theory of gene expression. *Cell*, **108**:439–451 (2002).
16. Makeyev EV, Maniatis T. Multilevel regulation of gene expression by microRNAs. *Science*, **319**: 1789–1790, 2008.
17. Hobert O. Gene regulation by transcription factors and microRNAs. *Science* **319**:1785–1786 (2008).
18. Raff R.A. *The shape of life: genes, development and the evolution of animal form*, Chicago, University of Chicago Press (1996).
19. Wray G.A. *et al.* The evolution of transcriptional regulation in eukaryotes. *Mol. Biol. Evol. 20*:1377–419 (2003).
20. Su AI *et al.* A gene atlas of the mouse and human protein-encoding transcriptomes, *Proc. Natl. Acad. Sci. U S A.* **101**:6062–6067 (2004).
21. Weaver, I.C.G. *et al.* Epigenetic programming by maternal behavior, *Nature Neuroscience* **7**: 847–854 (2004).
22. Leroi A.M. *Mutants: on the form, varieties and errors of the human body*, London, Harper Collins (2003).
23. Antonarakis SE, McKusick VA. OMIN passes the 1,000–disease-gene mark. *Nat Genet* **25**:11 (2000).
24. Frazer K.A. *et al.* A second generation human haplotype map of over 3.1 million SNPs. *Nature* **449**: 851–861 (2007).
25. Little P.F.R., Williams R.B.H. The Human Genome Project: Importance in clinical genetics, *Encylopedia of Life Sciences (online)*: a0005485 doi:10.1002/978047001590 (2007)

26. Fleischmann R.D. *et al.* Whole–genome random sequencing and assembly of Haemophilus influenzae Rd., *Science* **269**:496–512 (1995).
27. Sanger F., Coulson A.R. A rapid method for determining sequences in DNA by primed synthesis with DNA polymerase. *J. Mol. Biol.* 1975 **94**:441–448 (1975).
28. Cook-Deegan R.M. The Alta Summit, December 1984. *Genomics* **5**: 661–663 (1989).
29. Adams MD *et al.* The genome sequence of Drosophila melanogaster. *Science* **287**: 2185–2195 (2000).
30. Venter J.C. *et al.*. The sequence of the human genome, *Science.* **291**:1304–1351. Erratum in: *Science* **292**:1838 (2001).
31. Lander E.S. *et al.* Initial sequencing and analysis of the human genome, *Nature* **409**:860–921 (2001).
32. Venter JC *et al.* Environmental genome shotgun sequencing of the Sargasso Sea. *Science*, **304**: 66–74 (2004).
33. , Turnbaugh PJ *et al.* The human microbiome project. *Nature* **449**:804–810 (2004).
34. Alwine JC, Kemp DJ, Stark GR. Method for detection of specific RNAs in agarose gels by transfer to diazobenzyloxymethyl-paper and hybridization with DNA probes. *Proc. Natl. Acad. Sci. U.S.A.* **74**(12): 5350–5354 (1977).
35. Schena M *et al.* Quantitative monitoring of gene expression patterns with a complementary DNA microarray. *Science*, **270**: 467–470 (1995).
36. Schones D.E., Zhao K. Genome–wide approaches to studying chromatin modifications. *Nature Reviews Genetics* **9**: 179–191 (2008).
37. Birney E *et al.* Identification and analysis of functional elements in 1% of the human genome by the ENCODE pilot project. *Nature*, **447**:799–816 (2007).
38. Celniker S.E. *et al.* Unlocking the secrets of the genome, *Nature*, **459**:927–930 (2009).
39. Mardis E.R. Next–generation DNA sequencing methods. *Annu. Rev. Genomics Hum. Genet.* **9**: 387–402 (2008).
40. Donoho D. *High–Dimensional Data Analysis: The Curses and Blessings of Dimensionality*, lecture delivered at The American Mathematical Society Conference "*Math Challenges of the 21st Century*" Los Angeles, August 6-11, 2000 (*Aide–Memoire* available at `www-stat.stanford.edu/donoho/Lectures/AMS2000/Curses.pdf`)
41. Galperin MY, Cochrane GR. Nucleic Acids Research annual Database Issue and the NAR online Molecular Biology Database Collection in 2009. *Nucleic Acids Res.* **37**(Database issue): D1–D4 (2009).
42. Pennisi E. How will big pictures emerge from a sea of biological data? *Science* **309**: 94 (2005).
43. Vidal M.A. A biological atlas of functional maps. *Cell* **104**:333–339 (2001).
44. Tremethick D. J. Higher-order structures of chromatin: the elusive 30 nm fiber. *Cell* **128**: 651–654 (2007).
45. Clamp M. *et al.* Distinguishing protein coding and non–coding genes in the human genome, *Proceedings of the National Academy of Sciences U.S.A* **49**: 19428–19432 (2007).

46. The *C. elegans* Sequencing Consortium. Genome sequence of the nematode *C. elegans*: a platform for investigating biology. *Science* **282**: 2012–2018 (1998).
47. Tompa M *et al.* Assessing computational tools for the discovery of transcription factor binding sites. *Nat Biotechnol* **23**(1):137–144 (2005).
48. Xie XH *et al.* Systematic discovery of regulatory motifs in human promoters and 3' UTRs by comparison of several mammals, *Nature*, **434**: 338–345 (2005).
49. Walhout A.J.M. Unraveling transcription regulatory networks by protein-dna and protein-protein interaction mapping. *Genome Res* **16**:1445–1454 (2006).
50. Elemento O. *et al.* A universal framework for regulatory element discovery across all genomes and data types. *Mol Cell*, **28**: 337–350 (2007).
51. Sinha S. *et al.* Systematic functional characterization of cis–regulatory motifs in human core promoters, *Genome Res*, **18**: 477–488 (2008).
52. Warner J.B. *et al.*. Systematic identification of mammalian regulatory motifs' target genes and functions, *Nat Methods*, **5**: 347–353 (2008).
53. King M.C., Wilson A.C. Evolution at two levels in humans and chimpanzees. *Science* **188**: 107–116 (1975).
54. Davies S.R. *et al.* Computational identification and functional validation of regulatory motifs in cartilage–expressed genes. *Genome Research*, **17**: 1438–1447 (2007).
55. Segal E., Widom J. From DNA sequence top transcriptional behaviour: a quantitative approach, *Nature Reviews Genetics* **10**: 443–456 (2009).
56. Badis G. *et al.* Diversity and complexity in DNA recognition by transcription factors, *Science* **324**:1720–1723 (2009).
57. Hudson N.J. *et al.* A differential wiring analysis of expression data correctly identifies the gene containing the causal mutation, *PloS Computat. Biol.* **5**: e1000382 (2009).
58. Strahl B.D., Allis C.D. The language of covalent histone modifications. *Nature* **403**:41–45 (2000).
59. Turner B.M. Defining an epigenetic code. *Nat Cell Biol* **9**:2–6 (2007).
60. Seet B.T. *et al.* Reading protein modifications with interaction domains, *Nature Reviews Molecular Cell Biology* **7**: 473–483 (2006).
61. Jensen O.N. Interpreting the protein language using proteomics, *Nature Reviews Molecular Cell Biology* **7**: 391–403 (2006).
62. Sims R.J., Reinberg D. Is there a code embedded in proteins that is based on post–translational modifications, *Nature Reviews Molecular Cell Biology* **9**: 815–820 (2008).
63. Wilkins M.R., Kummerfeld S.K. Sticking together? Falling apart? Exploring the dynamics of the interactome. *Trends Biochem Sci.* **33**: 195–200 (2008).
64. Wang Z. *et al.* Combinatorial patterns of histone acetylations and methylations in the human genome, *Nature Genetics* **40** :897–903 (2008).
65. Schneider R, Grosschedl R. Dynamics and interplay of nuclear architecture, genome organization, and gene expression. *Genes Dev* **21**:3027–3043 (2007).
66. Parada L.A. *et al.* Tissue–specific spatial organization of genomes, *Genome Biology* **5**: R55 (2004).
67. Maciag K *et al.* Systems–level analyses identify extensive coupling among

gene expression machines. *Mol Syst Biol* **2**: 2006.0003 (2006).
68. Spudich J.A. Dynamic organization of gene loci and transcription compartments in the cell nucleus. *Biophys J* **95**:5003–5004 (2008).
69. Sutherland H., Bickmore W.A. Transcription factories: gene expression in unions? *Nature Reviews Genetics* **10**:457–466 (2009).
70. Spilianakis CG *et al.* Interchromosomal associations between alternatively expressed loci. *Nature* **435**:637–645 (2005).
71. Lomvardas S *et al.*. Interchromosomal interactions and olfactory receptor choice. *Cell* **126**: 403–413 (2006).
72. Simonis M *et al.* Nuclear organization of active and inactive chromatin domains uncovered by chromosome conformation capture-on-chip (4C). *Nat Genet* **38**:1348–1354 (2006).
73. Raska I *et al.* Structure and function of the nucleolus in the spotlight. *Curr Opin Cell Biol* **18**:325–334 (2006).
74. Hurst LD *et al.* The evolutionary dynamics of eukaryotic gene order. *Nat Rev Genet* **5**:299–310 (2004).
75. Komili S., Silver P.A. Coupling and coordination in gene expression processes: a systems biology view, *Nat Rev Genet*, **9**: 38–48 (2008).
76. Brown CD *et al.* Functional architecture and evolution of transcriptional elements that drive gene coexpression. *Science* **317**: 1557–1560 (2007).
77. Bussemaker H.J., Li H., Siggia E. D. Regulatory element detection using correlation with expression. *Nature Genetics* **27**: 167–171 (2001).
78. Das D *et al.* A primer on regression methods for decoding *cis*–regulatory logic. *PLoS Comput Biol* **5**:e1000269 (2009).
79. Sebastiani F. Machine learning in automated text categorization. *ACM Computing Surveys* **34**:1–47 (2002).
80. Gower JC and Legendre P. Metric and Euclidean properties of dissimilarity coefficients, *Journal of Classification*, **3**: 5-48, 1986
81. Hennig C. Cluster–wise assessment of cluster stability, *Computational Statistics and Data Analysis*, **52**: 258–271 (2007).
82. Gage P.J. *et al.* The bicoid–related Pitx gene family in development, *Mamm Genome*, **10**: 197–200 (1999).
83. Sowden J.C. Molecular and developmental mechanisms of anterior segment dysgenesis, *Eye*, **21**: 1310–1318 (2007).
84. Zhang N.R., Wildermuth M.C., Speed T.P. Transcription factor binding site prediction with multivariate gene expression data *Ann. Appl. Stat.* **2**: 332?365 (2008).
85. Weiss, K.M. Tilting at quixotic trait loci (QTL): an evolutionary perspective on genetic causation, *Genetics* **179**: 1741–1756 (2009).

Chapter 10

Tutorials on Agent-based Modelling with `NetLogo` and Network Analysis with `Pajek`

Matthew J. Berryman[1] and Simon D. Angus[2]

[1]*Defence and Systems Institute, SPRI Building, Mawson Lakes Campus, University of South Australia, Mawson Lakes SA 5095, Australia; Email:* `matthew.berryman@unisa.edu.au`

[2]*Department of Economics, Monash University, Clayton Vic. 3800, Australia; Email:* `simon.angus@buseco.monash.edu.au`

Complex adaptive systems typically contain multiple, heterogeneous agents, with non-trivial interactions. They tend to produce emergent (larger-scale) phenomena. Agent-based modelling allows one to readily capture the behaviour of a group of heterogeneous agents (such as people, animals, *et cetera*), with diverse behaviour and important interactions, so it is a natural fit to modelling complex systems. Many complex systems (and agent-based models thereof) can be thought of as containing networks, either explicitly or implicitly. Therefore for complex systems research it is important to have a good understanding of network analysis techniques. This chapter is aimed at beginners to complex systems modelling and network analysis, using `NetLogo` (Section 10.1) and `Pajek` (Section 10.2) respectively. It is also aimed at more advanced complex systems modellers who want an introduction to these platforms.

Contents

10.1. Agent Based Modelling

10.1.1. *Introduction*

Agent-based modelling is a type of modelling in which the focus is on representing agents (such as people or animals) and their interactions.[1] In agent-based models, as in real complex systems, a set of inductively generated local rules and behaviours of agents give rise to emergent phenomena at a group or system-wide level.[2–4] Agent-based modelling allows one to effectively capture a very rich set of complex behaviours and interactions, and is therefore highly suited to modelling complex phenomena. It has gained extensive use in the fields of economics,[5] social science,[6] ecology,[7] and biology,[8] amongst many others.

Agent-based modelling is a useful complement to more traditional model representation using a system of equations. Agent-based modelling more easily allows one to do things that are difficult to capture in traditional approaches, such as:

- Model networks of agents where the agents modify the network dynamically;
- Model agent learning and/or evolution;
- Capture a large range of different types of agents and agent behaviour; and
- Explore non-linear interactions between agents.

Agent-based modelling allows for a much richer set of behaviours than traditional variable-based modelling,[9] even if the latter is aided by the use of computers.[1,10] Agent-based modelling does not preclude the use of other styles of modelling as part of the agent-based model, or in combination with the agent-based model.

Emergent phenomena are those that arise from the low-level rules and interactions between the parts (which may be agents).[3] Sometimes the only way to observe them is to let the model run. Brock (2000) discusses the Santa Fe approach to complex systems science as stressing the identification of patterns at macro levels and trying to reproduce these using lower level rules. Agent-based models can be used to test an hypothesis in the sense that it can rule some emergent possibilities out, given a set of rules (this is discussed further in Subsection 10.1.5). Exploratory agent-based model usage, as an inductive process,[a] can generate new hypotheses, but ultimately these must be tested with other methods. As with any type of scientific modelling, it is impossible to conclusively rule hypotheses as true,[11,12] rather one tests the hypothesis by trying to prove its predictions false.[12,13] To quote Blaug (1992)[14] on this logical asymmetry, 'there is no logic of proof, but there is a logic of disproof.' Falsification is a crucial element in the demarcation of science from non-science.[12] Validation of a model's output is not sufficient, one must also verify the assumptions made.[15] In the case of agent-based modelling, both the behaviours of agents and their software implementations must be verified. The following subsection describes how to implement agent-based models in software using the `NetLogo` package.

10.1.2. *Introduction to NetLogo*

To assist in the development of agent-based models, a number of different platforms have been developed.[16] These platforms vary in how much support they provide. Some, like `Swarm` and `Mason`, offer a set of software libraries to be used in programming a model. `RePast`, in its Simphony [*sic*] version, offers a few more tools for quick construction of agent-based models. Some other agent-based modelling platforms provide fixed sets of rules that can be used with some chosen parameters, but these are often too restricted to capture the wide range of phenomena that one might want to model. `NetLogo` consists of a programming language (derived from the earlier Logo language) and a set of libraries, as well as a programming environment. Like `RePast` and `Swarm` it provides a set of programming facilities, however `NetLogo`also provides a graphical tool for quickly constructing interfaces for running agent-based models. In the following subsections, we describe the `NetLogo`interface and detail some of the basics of programming in `NetLogo`. Note that sample code can be found at the tutorial web site `http://www.complexsystems.net.au/wiki/ANUPhysicsSS2008ABM_networks`. In the following subsections, all programming commands are indicated by *italic* text.

[a] As discussed by Epstein (2007)[11] agent-based models, in and of themselves, are a form of logical deduction: given a set of rules, and initial conditions, the emergent outcomes are embedded in the rules,[4] however surprising these may be. The models remain inductive generalisations, however.

10.1.3. *NetLogo interface*

One of the benefits of using `NetLogo` is its interface. An example picture of the interface is shown in Figure 10.1. By default, the interface contains just a 2D spatial view of the model environment, which is a square lattice. Unlike `Mason` and `Repast`, `NetLogo` does not support hexagonal grids. If this is important to your model, you should investigate these other platforms. In addition to the 2D spatial view, the developer adds other elements, such as buttons to set model parameters and graphs to monitor results. Many of these elements are detailed below.

All elements of the interface may be moved around by first selecting them by right clicking and choosing select, and then clicking and dragging. Multiple items can be selected and then moved around together. To edit them in order to change the details as given below, one right clicks on an element of the interface and selects 'Edit...'. They can, except for the view, be deleted by right clicking and selecting 'Delete'.

The 2D view has several different options that prove useful in modelling. The size of the grid (the number of cells) can be changed, as can the size of the cells themselves (in pixels). The edges of the model 'world' can be changed to reflect the system the model is to represent; the choices here are to treat the edges as walls, or to treat them as wrapping around onto the cells on the opposite side of the grid. The programming implications are discussed in the Topology section of the NetLogo Programming Guide.[17] The view supports inspecting patches and turtles by right clicking on the patch and inspecting it, or selecting one of the turtles and inspecting that from the right-click menu.

To run the software, buttons are typically set up in the interface. When creating a button, the piece of code to be run is specified. By default, this is also displayed as the button name, however a different label can be specified in the button properties. The piece of code to be run can be changed by editing the button. Buttons can be set so that the code can be run repeatedly, until the button is pressed again. To specify this behaviour tick the 'forever' box. It is typical to have at least two buttons, a setup button, with a correspondingly-named procedure that clears the display and initialises the state of the model, and a go button, with the 'forever' box ticked, for running the model.

Monitors can be used to keep track of some number while the model is running. The monitor requires a reporter (procedure that returns a value) to be defined. This can either be a reporter defined in your program if

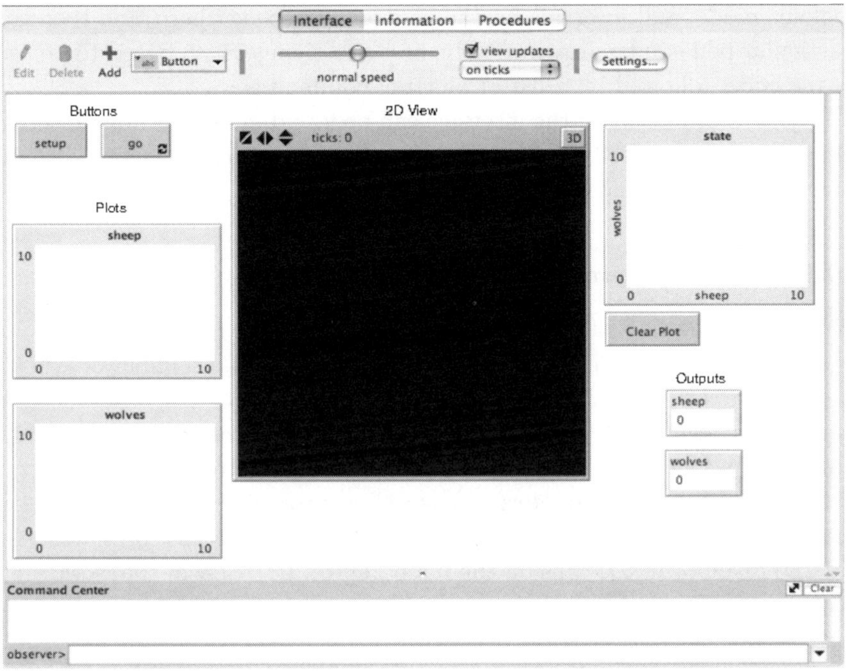

Fig. 10.1. This is the `NetLogo` interface for a sheep versus wolves predator scenario. The `NetLogo` file containing this interface, as well as the associated code, can be found at the tutorial web site.

one exists or it is something long, alternatively it could be a short piece of code defined in the monitor itself, for example *count turtles*. For more information on reporters, see Subsection 10.1.4. Monitors have an option to be labelled with something different than the piece of code, for example 'number of turtles' instead of the corresponding *count turtles* code.

Inputs, switches, choosers and sliders are all ways of specifying variables that can then be used by the code, either at initialisation, or while the code is running. The switches have two settings, on and off, and return a boolean value (true or false) when the global variable they define is used in code. The switches, choosers, and inputs all allow for some numerical input, with different constraints and interfaces.

Plots can be defined, along with their name (which can then be referenced in the code), their axes set up, and plot pens set up. The use of plot pens is for having multiple graphs on the same set of axes. The profile (colour, mode, and interval) for the plot pen can be set here, and then that particular pen can be drawn with by first specifying which pen is to be used in the code, followed by a list of plotting commands.

More information on the interface can be found in the NetLogo Interface Guide.[18] More information on linking these interface elements with the program can be found in the following subsection.

10.1.4. *NetLogo programming*

Before giving details of the commands, it is necessary to define the types of agents on which commands can operate. `NetLogo` has a number of types of agents—turtles, patches, links, and the observer agent. The observer agent is a single agent that has a view of the whole `NetLogo` "world" (turtles, links and patches), and is used for running the main parts of the program (linked to buttons on the interface) as well as providing a way of interacting command by command on the main interface. The patches represent square (in 2D) or box (in 3D) cells on the main 2D (or 3D) view of the world. The turtles are agents that can move around on the world surface, and draw. Links represent relationships between turtles.

In programming `NetLogo`, one starts with primitives—built-in `NetLogo` commands—and combines these into larger modules of code—procedures and reporters—that can be used themselves as commands, in particular as given to or used by turtles. The distinction between a procedure and a reporter is that a reporter returns some value at the end, using the command *report*, and is defined differently as follows. The block of code that forms a procedure begins with *to procedure-name* on a line by itself and on the next lines come a set of commands (primitives, or other defined procedures or reporters), and ends with the keyword *end*. It is a good idea, though not necessary in `NetLogo`, to indent code to indicate code structure. Note that the procedure name should not contain any spaces, in `NetLogo`, whitespace is used to separate out the different primitives, variables, *et cetera* that make up a line of code. The only time a space does not break up the code in this way is if it is part of a string (such as in a plot name), strings being enclosed by double quote marks. The block of code that forms a begins with *to-report reporter-name* and ends with the keyword *end*. A reporter should contain one or more lines that have the keyword *report* followed by some value, either a variable, or a call to another reporter which returns some value. Two or more *report* statements could be used depending on whether some condition is true or false—see the *if*

Table 10.1. A list of several frequently-used primitive commands available in the `NetLogo` environment. Where a shortened form of the primitive exists, this is indicated by listing the shortened form after the long form, only one (and no comma) need be used by the programmer.

Command	**Effect**
create, crt n	Create *n* turtles.
clear-all, ca	Clears all turtles, patches, plots and output, as well as resetting the tick counter and other global variables.
forward, fd and *right, rt,* and etc.	commands for moving the turtle. All of these commands take a single parameter giving the distance to be moved or angle (in degrees, clockwise) to be rotated by.

or *ifelse* statements described in Table 10.2. Once *report* has been called, it will exit the reporter immediately, so it should come as the penultimate line in the reporter, unless it is run only under some condition.

A table of frequently-used basic primitives can be found in Table 10.1. Many primitives have shortened forms, to save on typing. Where these exist they are listed separated by commas; only one form of the primitive needs to be used in each instance. Some of the primitives act globally, other ones must be used in a particular context, for example by asking a turtle or patch to run them. This can be done using the *ask* command on an agent or agentset (set of agents); more on this later.

`NetLogo` features a number of commands for working with links to represent relationships between turtles, which are the nodes of the network. They come in two different forms, undirected and directed, to represent different types of relationships. For example, directed links could represent a parent-child relationship, whereas undirected links could represent a sibling relationship. `NetLogo` primitives for directed links have either *-from*/*-to* or *-in*/*-out* in the command name. Those for directed links, have *-with* or nothing at all. `NetLogo` features a number of layout primitives for laying out networks in the 2D view, such as *layout-circle* and *layout-radial*. `NetLogo` has commands *layout-spring* and *__layout-magspring* that use force-based layout algorithms similar those that `Pajek` has, as described in Subsection 10.2.4.

Breeds are a way of specifying different classes of turtles and links. These different classes can have different state variables (described later), and can have different behaviours, by *ask*ing different breeds to run different procedures and reporters. Breeds are specified at the start of the code, by using the *breed* keyword, followed by the plural and singular forms of the breed name (in that order) enclosed in square brackets. For example, to define a breed for representing wolves, one could write: *breed [wolves wolf]*. After definition, a large number of primitives are made available automatically by `NetLogo`, starting with the singular or plural form as appropriate. To continue the example of wolves, by using that command one would obtain primitives like *create-wolves* and *is-a-wolf?*. For details of other primitives created, refer to the breeds section of the NetLogo Programming Guide.[17] You can change the breed of a turtle (using *set breed new_breed*) mid-way through running your program, if this makes sense for your model. You can also change the breed of a link, but you cannot change between a directed and undirected breeds—`NetLogo` cannot do this automatically for you, and you must program such behaviour in as required in a way that makes sense for your particular model.

Variables are placeholders for storing information. There are several different types of variables in `NetLogo`:

- Global variables hold values that are accessible anywhere in a `NetLogo` program. Global variables are defined at the start of a `NetLogo` program, using the *globals* command, followed by a list of whitespace-separated variables, the list being enclosed by square brackets. For example, the command *globals [dead_turtles found_money]* defines *dead_turtles* and *found_money* to be global variables. Any variables defined in the interface (such as sliders and monitors) do not need to be specified in the list of global variables as they are available by default.
- Local variables hold variables that are accessible only within a certain block of code. A block of code is something enclosed in square brackets, or within a procedure or reporter. Local variables are defined using the *let* keyword. This creates a variable, and assigns it an initial value, which must be specified. For example, *let food 3* creates a variable named *food* and assigns it the value of 3. Once defined with the *let* keyword, a variable can then be used on any subsequent line in that block of code.
- Variables owned by turtles, links, or patches. These can be defined by using the *-own* suffix on a type of agents or the plural form of a breed name, and then listing the variables. For example, *links-own [capacity latency]* defines all links to have variables named *capacity* and *latency*. A number of owned variables are predefined for turtles, patches and links. The NetLogo Dictionary[19] has a section that lists

all the built-in variables owned by turtles, patches and links. Some important ones to note are *xcor* and *ycor*, the co-ordinates of the turtle (for patches these are named *pxcor* and *pycor*), and *who*, the turtle's unique identity number. There is also a reporter named *who*, callable in a turtle context, that returns a turtle's who number.

All variables, once they exist, can be set (or assigned a value) using the *set* command. For those who are used to programming in languages that have a `C` (programming language) or `C`-derived syntax, note that `NetLogo` uses *set* for variable assignment and the = symbol for testing for equality (whereas in a `C`-style syntax the = symbol is used for assignment and so == is used for testing for equality).

There are two types of data structures available: agentsets (using the same terminology as the `NetLogo` manual) and lists. Agentsets are just that—sets of agents. They are useful because they then provide an easy way of giving commands to a group of agents that share some property, for example giving a set of commands to turtles that are on neighbouring patches to the current turtle. A set of commonly-used commands for constructing agentsets and for working with them can be found in the Agentsets section of the NetLogo Programming Guide.[17] Every time an agentset is used, its agents will be used in a random order. If you wish to use a specific order, then you should use a list. Lists are way of structuring agents or variables, unlike agentsets, which only store agents. A list can also store other lists, allowing for 2D or higher dimensional data structures. To change the order after a list is created, a new list must be created, and items copied over in the desired order into the new list. Primitives such as *replace-item* and *sort* simplify modification of lists, but note that they will return a new list, which must be assigned to something using *set*, or using *let* if the variable is defined and assigned in one step.

Most interesting algorithms have some control logic for repeating parts of the code, or selecting between alternative bits of code depending on some test. A summary of the `NetLogo` primitives for these can be found in Table 10.2. More details on these can be found in the NetLogo Programming Guide.[17]

`NetLogo` has a set of commands for writing output to files. To open a file for reading or writing, one uses the *file-open* command, followed by whitespace and then the filename in double quote marks, to denote that it is a string. For example, to open a file called 'output.txt', one uses *file-open "output.txt"*, where the quotes are part of the command. Whether it is

Table 10.2. Table of control logic commands, for changing the flow of execution of a program.

Command	Effect
if test [commands when true]	This runs a test, and then if true runs the command. The test must be some reporter that returns true or false, for example *xcor* > *0* or *who* = *3*.
ifelse test [commands when true] [commands when false]	Like the *if* command, except that this command has a section of commands (enclosed in a second pair of square brackets) to be run when the returns false.
while [reporter] [commands]	While the reporter returns true, this command runs the commands listed in the second set of square brackets. Note that one should be careful that the reporter will at some stage return false, otherwise this will loop forever (until the program is stopped manually).
foreach [list] [commands]	This applies the set of commands to every element of the list.
map [reporter] [list]	This returns a new list, formed from applying the reporter to each element of the list specified in the command.
repeat n *[commands]*	This repeats the block of commands *n* times.

opened for reading or writing is determined by whether the next command given is a command for reading from the file or a command for writing to the file. Once a file is opened for reading, it cannot be written to, and *vice versa*. To switch modes, use *file-close*, and then *file-open* again. The *file-open* command can also be used to select between multiple files that have been opened, by just using *file-open* on the file you want to access again (even if it was opened before). If a file is opened for writing and it already exists, then any write commands will write to the end of the file.

`NetLogo` features a number of commands for plotting. For these to work, at least one plot (distinct drawing area) must be defined as described in Subsection 10.1.3. If only one plot is in the interface, and it only has one pen (for one graph in the plot area) then commands like *plot*, for plotting points on a graph, and *plot-pen-down*, for ensuring that a point is drawn when the plotting point is set, can be used immediately. If, however, there are multiple plots and/or pens, then these should be selected using the function *set-current-plot "plot name"* to select a plot, and then if there are multiple pens in the plot the pen can be selected using *set-current-plot-pen "pen name"*. The plot name and pen name strings should match what has been defined in the interface. There are two primary commands for plotting points, *plot* and *plotxy*. The difference is that the first one uses an x-value from the last time *plot* was used (for that particular plot and pen) plus some interval. The x-value starts with a value of zero for the first time *plot* is used for that plot and pen. The interval can be changed in the pen settings. The *plotxy* command is used when one wants to specify both x and y values.

10.1.5. *The art of agent-based modelling*

As with any type of modelling, it is part art, part science. The following is a discussion of several points in modelling, mainly about agent-based modelling, that we feel are important to make. A more detailed discussion of these and other points can be found in Appendix B of Miller and Page.[1]

Part of the art of modelling is simplicity. This is important in modelling, because in modelling we are trying to gain a better understanding of the real system through appropriate simplifications. To quote from Shalizi (1998):[20] 'Models are only good if they're easier to handle and learn about than what they model, and if they really do accurately map the relations we're interested in.' Part of picking appropriate simplifications is in picking simplifications that give better understanding, while still capturing important parts of what you are trying to model, including interactions with other parts of the system / other systems. The benefit of agent-based modelling over traditional variable-based modelling is that simplifications do not need to be made just to make the model mathematisable or solvable using a set of mathematical tools. In agent-based modelling it is important not to oversimplify, however. The power of agent-based models is based on their ability to accurately and falsifiably explain and model the complexity of real-world interactions.[21,22]

In some agent-based models, humans take the roles of some agents. These participatory agent-based models are useful for educating people about complexity and emergence in general, or on specific complex systems. They are also useful for model verification (of the rules) and validation (of the types of emergent behaviour). Feedback can be provided on the agent rules, in order to verify them. For validation, participatory actors can provide the quantitative data to serve the purpose of validation. They can, however, also provide qualitative feedback on the emergent findings.

To engage people who are experienced in the real-world system to participate in your model and provide you with feedback[23] `NetLogo` supports participatory agent-based modelling through its HubNet facility. More information can be found in the NetLogo HubNet Guide[24] and HubNet Authoring Guide.[25]

Another technique that is valuable in verifying the software is to embed within the model certain cases that may be modelled with other techniques, like equations. For example, in an agent-based SIR (susceptible-infected-recovered) model of epidemiology,[26] one could try embedding within the agent-based model a simple case that could also be modelled using mean-field theory. In the example code provided on the tutorial web site, we give sample code for a predator-prey model, with a few changes this could be compared with results from the the Lotka-Volterra equations. This helps rule out major bugs in the software; however since ABMs capture a wider range of behaviours, this form of verification is somewhat limited.

Good coding practice dictates that code is commented. Try and imagine yourself reading your code in ten years time. Good comments also help others read and understand your code. This is important as code should be shared as part of the model verification process. Code should be shared more generally as part of good scientific practice, since results should be repeatable by others. Papers are usually too short to put down all the details required for replication.

10.2. Pajek

10.2.1. *Introduction to* `Pajek` *with reference to other network analysis tools*

`Pajek`, pronounced 'Payek' and meaning 'spider' in Slovenian, is a social network analysis and visualisation tool. It was specifically designed to manipulate and analyse very large networks having on the order of 10^3 – 10^6 nodes. It is not the only tool of its kind, though it has some pleasing features, which has led to its wide adoption amongst academic practioners.

Other comparable software tools include `UCINet` and the `statnet` set of packages available for the open-source statistical program `R`. Both of these tools, like `Pajek`, are able to produce visualisation *and* analysis of large networks.

`UCINet` is a commerical product and its particular strength is in linear algebra measures and graph manipulations. On the other hand, it is computationally greedy and although it claims to handle over 32,000 nodes, its practical limit due to computational requirements is of the order of only 10^4 nodes, limiting the software in some very large network applications.

In contrast to `UCINet`, the open-source `statnet` meta-package in `R` provides perhaps the most flexible analysis option. Whilst the start-up costs are high for learning a scripting language such as `R`, the benefits include the embedding of network analysis within other statistical routines, and the ability to batch run the same network analysis on many different networks using a script.

`Mathematica` has a number of combinatoric and other graph theoretic packages, along with various visualisation packages. There is also a `NetLogo-Mathematica` link package available,[27] which provides a real-time link between `Mathematica` and `NetLogo`. This allows one to analyse, collect, and display data from a `NetLogo` ABM in real time, using social network and other graph theoretic measures.

Additionally, for visualisation purposes only, the multi-platform, GPL licenced software `GUESS` is certainly worth investigating. Whilst it does not feature network analysis tools, it is able to take network data input from many sources (including `Pajek`) and output attractive visualisations, such as graph layouts over the top of an image layer (for example an airline's routing network overlayed on a map), in many popular image formats (including vector graphic formats such as EPS and SVG).

`Pajek` is situated in this context as a dedicated, flexible, but intuitive network visualisation and analysis software. It was designed with large scales in mind and is able to handle very large networks. Similarly, the `Pajek` graphical user interface (GUI) enables the management of multiple networks, components and analysis outputs at once, including those made within a `Pajek` session. Furthermore, it has various visualisation algorithms in-built and these outputs have been used in academic publications for several years.

To cope with a very large network dataset, it is useful to represent the graph in a variety of means (see Figure 10.2). Some of these representations require little analysis (for example cut-outs), whilst others require multiple processing steps (for example hierarchy analysis). `Pajek` is able to deal with many many different analytical approaches. The software is in constant development as new techniques that are derived in the academic literature make their way into the `Pajek` toolset.

In the following subsections, this tutorial discusses how to:

(1) Write and load `Pajek` format network files;
(2) Visualise a network in a number of ways;
(3) Generate random graph networks (random graphs) for comparison;

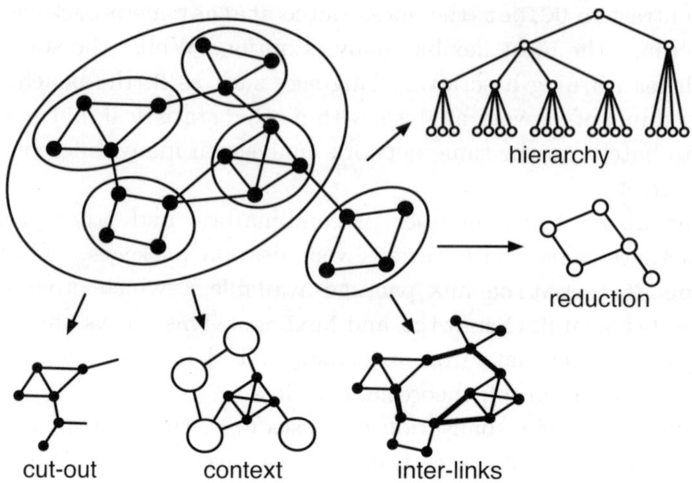

Fig. 10.2. Different ways of analysing a large graph (after the `Pajek` manual).[28]

(4) Run common graph measures over networks.

10.2.2. *Importing network data*

`Pajek` can accept a range of information about a network. There are several fundamental types of data for `Pajek`:

- Networks: Vertices and edges/arcs (are indexed by their generation number);
- Partitions: Vertex information of a particular kind—they are actually class information, such that all the vertices are part of a particular class; for example 'Male' could be partition 1, and 'Female' partition 2;
- Vectors: Information on each vertex, on a vertex by vertex basis, for example 'Male' could be represented by 1 and 'Female' by 2, but likewise, each vertex can have all kinds of unique data attached to it, such as age, height or weight. Use Partitions for data that more than one vertex naturally matches, and Vectors for information that is likely to be matched by only one vertex at a time;
- Permutations: These are produced by functions applying to the whole network that create a new network based on the original network in-

formation. A permutation is a reordering of the nodes.
- Cluster: These are produced by certain functions that group together nodes according to some measure, for example their distance in the network.
- Hierarchy: A tree structure containing the vertices of the graph, and representing relationships between the nodes; for example the subtrees (parts of the tree) could represent communities, and these subtrees when joined up could represent a larger community of communities.

The `Pajek` interface manages each of these data types separately (see Figure 10.3).

`Pajek` uses several different file formats to manage the various data types. Below we give details of the file formats `Pajek` uses to read and save data in. In specifying a network, the usual components of a network are mandatory, such as vertices (nodes) and edges (links between nodes) or arcs (directed edges). Note that `Pajek` rightly treats edges and arcs differently. See Figure 10.4 to see an example network containing both edges and arcs.

A `Pajek` network file first begins with `*vertices` followed by a space and then the number of vertices in the network. If any of the nodes are labelled, this can be specified following the `*vertices` line by specifying the number of the vertex that you want to label, then a space, and then a label. If the label has a space in it, then it needs to be surrounded by double quote marks. Edges can be specified by writing `*edges` on a new line, and on the following lines writing each edge (one per line) by specifying the two nodes, separated by a space. Arcs can be specified in one of two ways:

(1) by writing the keyword `*arcs` on a line by itself, and then on the following lines specify one arc by writing the starting node, destination node, and arc weight, all separated by whitespace; or
(2) by writing the keyword `*arcslist` on a line by itself, and then on each of the following lines specify a starting node, followed by a list of destination nodes, to be joined in separate arcs (of weight 1) from the starting node. In this case whitespace separates the nodes listed.

An example network file (`example.net`) for the graph in Figure 10.4 is given in Figure 10.5.

There are several things to note about the `Pajek` network file format

- Vertices cannot be indexed 0 (start with 1);

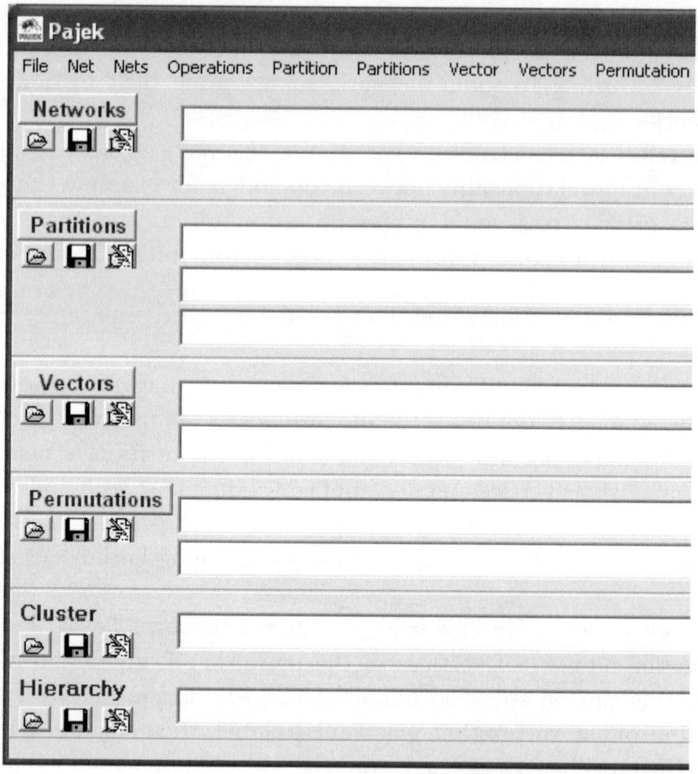

Fig. 10.3. Each of the different data types Pajek accepts has its own space in the interface where data can be opened (folder icon), saved (disk icon) and viewed or edited (edit button, or double click on the data file name). One can start with any number of files and generate the rest if needed through algorithms provided by Pajek. One could even start with no files and generate a random network as discussed in Subsection 10.2.6. Note that for image clarity reasons, not all menu options are shown here.

- Any amount of white space can be used to separate elements on a line, and whitespace consists only of space characters (not tabs);
- Note that Pajek won't draw arcs and edges that are from a node to itself, but they are still there in the data structure.
- Pajek does not support Unix text files, lines need to be ended with a carriage return and line feed (if you are working on Windows do not worry about this);
- A network file consists of just the basic information about the network (vertices and connections), other information such as each vertex's rank, cluster and/or other information for each vertex can be added as

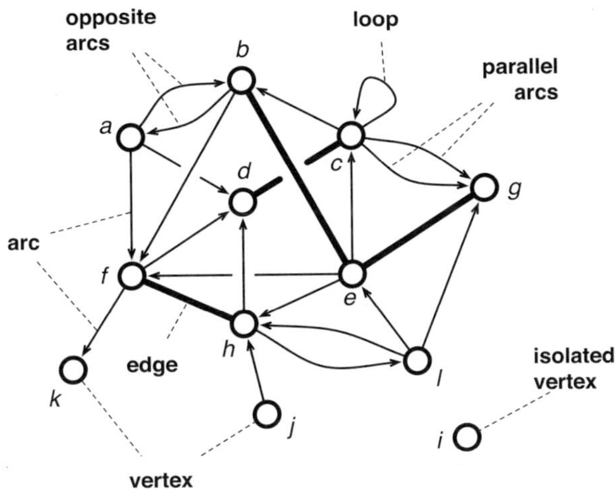

Fig. 10.4. This figure shows a set of vertices (or nodes) denoted by circles, and edges (links without arrowheads) and arcs (directed links, with arrowheads) joining them.

separate files (see Figure 10.3); and

- Each file can be read into `Pajek` and then saved as a Project: *File → Pajek Project File → Save*

In `Pajek`, one can assign vertices to clusters. The clusters can either be generated through a clustering algorithm (see Subsection 10.2.7) or loaded from a file (either one generated previously, or entered manually). To load from a file, in `Pajek` select *File → Cluster → Read.* An example clustering file (`example.cls`) for the graph given in Figure 10.4 is given in Figure 10.5. Examples of permutation and vector files for the graph in Figure 10.4 can also be found in Figure 10.5.

10.2.3. *Visualising networks*

There are several ways to visualise a network. In fact, this is an area of active current research—consider the difference between the layout of a subway network versus an electrical diagram versus a social network versus a gene expression network, *et cetera.* Each network has its own 'natural' way of laying out. Sometimes these are connected strongly to physical interpretations (for example an international flight network overlaid on the world map), but many networks do not have a physical interpretation (for

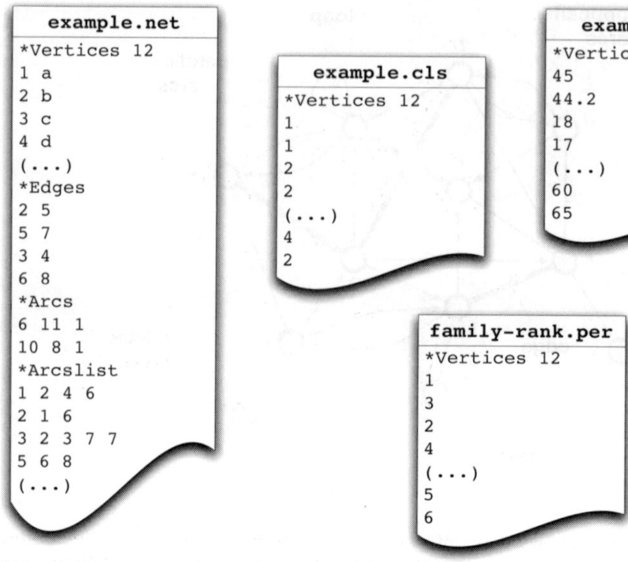

Fig. 10.5. Example files referred to in the text.

example a sexual activity network)—rather, they are just networks: objects which indicate the relationships between elements (not their location).

Some visualisation tools to try out are:

(1) With your network loaded, use *Draw* → *Draw* (or just hit Ctrl+G);
(2) You will see a simple representation, by default, the nodes are arranged on a circle—try changing this representation, or layout, by using the different options in *Layout* → ...
(3) If the network looks strange to you, try using a different starting position (*Layout* → *Energy* → *Starting Positions* → ...) (see Subsection 10.2.4);
(4) Now, try a common layout: *Layout* → *Energy* → *Fruchterman Reingold* → *3D*;
(5) Since your screen is in 2D, and the graph is now laid out in 3D, we need to alter the perspective to 'see' it. This can be done this as follows: *Spin* → *Spin Around (accept 360 Degrees)*.
(6) This likely was too quick for you .. in which case, use *Spin* → *Step in degrees* → ... and set this value to something very small (like 0.05), now use *Spin* → *Spin Around* again.

10.2.4. *Force-directed layouts*

'Force-directed' and 'spring-directed' refer to a class of 'energy' algorithms where one can conceive of the network as a collection of balls, connected by springs of a certain length (refer to Algorithm 10.1). These are a type of graph embedding algorithm.[29,30] The idea is to imagine starting with this web of balls and springs in a random configuration, and then allowing them to 'relax'. The ultimate configuration should be one where spring tension is minimised, such that connected balls are close together in the layout space. The initial position is quite important. It can be the case that this algorithm, like a human untangling many inter-connected springs, might actually not be able to untangle them all into a 'good' layout. Hence, the option is given in `Pajek` to vary the starting positions.

Algorithm 10.1 A Force-Directed Layout Algorithm (pseudo-code)

Positions $\leftarrow$ Generate-random-position-for-all-vertices (X, Y)
Energy $\leftarrow$ Calculate-total-force-in-springs
while Energy $\geq$ Energy-threshold **do**
    Displacements $\leftarrow$ Calculate-resultant-force-on-all-vertices
    Positions $\leftarrow$ Positions + Displacements
    Energy $\leftarrow$ Calculate-total-force-in-springs
end while

10.2.5. *Exporting the network layout to another program*

To get an image file of your current layout, use the *Export* $\rightarrow$ *2D* $\rightarrow$ menu item. Choose a file format that you wish, for example EPS/PS (mainly used for importing into LaTeX), or bitmap (for a web page or `MS Word` application).

10.2.6. *Generating a comparison network*

It is often useful to compare a given network to one that is like it in some ways, but different in others. Random graphs, where a random process generates a graph, with some statistics the same as the original, are useful null models.[31–34] For instance the work of Watts and Strogatz, in their influential *Nature* paper,[35] use comparisons between a given graph and its random graph equivalent (same number of nodes and average degree). We will do the same.

(1) First, we will generate a random Erdős-Rényi network: *Net → Random network → Erdos-Renyi → Undirected → General.* Specify a number of Vertices, and the Average Degree (for example 50, and 4). Obviously, for a comparison net you will want to use the same number of vertices and average degree as your network of interest.
(2) The network will appear in your 'networks' row on the home area of `Pajek`.
(3) Let us visualise it. You can do that as before.
(4) Now, since this is probably our first large network, we will do a bit more with our visualisations: Back in the home area, with the new Erdős-Rényi random network highlighted, we will create a Partition (of the vertices) by the property 'Degree' (number of incident edges): *Net → Partitions → Degree → Output.* New rows should have been added in the main area of the interface.
(5) Highlight the original Erdős-Rényi random network (networks) and use *Draw → Draw Partition* (or Ctrl+P). This should bring up a coloured visualisation.
(6) Make the vertices bigger by using *Options → Size → of Vertices* (a good size to use is 10).
(7) Now let us get a better idea of the partitions. Try *Layout → Circular → using Partition.* This will place similar vertices (by their degree/-partition) together in space.

10.2.7. *Common network measures*

In this subsection we will look at some common network measures. We will continue working with the large Erdős-Rényi random network we made earlier. A random network by itself, even visualised with one of the pretty force-directed placement algorithms, is still just a random network—it is hard to get much out of it by just visualising it. First we will generate some outputs from the network, using the sample network file for Figure 10.4. This file can be loaded by clicking on the folder icon under in the networks section of the `Pajek` interface or by clicking *File → Network → Read.* The process of generating outputs is going to get a bit confusing unless you pay close attention to the way that `Pajek` organises and operates on data. Recall from Subsection 10.2.2 the different data types for `Pajek`.

Once a network is loaded using the interface, one can calculate a number of network measures. A limitation of `Pajek` is that for some of these, it doesn't handle multiple lines and loops well, in that it doesn't take them into account when normalising. So it is a good idea to remove them from the network. It is also a good idea for some network algorithms (in general, and not just in `Pajek`) to represent edges by arcs going in both directions. These pre-processing steps can be found in the *Net* → *Transform* submenu in `Pajek`. The ones that need to be applied are *Edges −> Arcs*, *Remove* → *Multiple Lines* → *Single Line* and *Remove* → *Loops.* You can create a new network or just overwrite at each step. If you create a new network at each step, the last one is the one that will automatically be selected for the following operations.

(1) Degree distribution: What we wish to do is assign each vertex a number that describes the number of input, output or either input/output edges that are incident at the given vertex. Hence, we will be aiming to produce a vector output of (normalised) degrees. The way this is done, is actually to do a partition of the vertices into their degree numbers (we kill two birds with one operation):

  (a) Select from the menu *Net* → *Partitions* → *Degree* → *All.* Notice that this produces two new lines in Partition and Vectors;
  (b) Double-click on the text window of Vectors (see Figure 10.3), to bring up a simple text file view of the data—notice that each vertex now has a number associated with it: this is the normalised degree of each vertex;
  (c) Double-click on the text window of Partitions (see Figure 10.3), this file shows for each vertex, based on its normalised degree number, which class (or 'Partition') of vertex it falls into. Find the vertex with the highest partition number (in this case, this is the degree)—you should find it is vertex 6 ('f'). Go back to the 'Vector'; information for normalised degree and look at that vertex's normalised degree. The degrees are normalised (that is, divided) by a factor of $n - 1$.
  (d) To make a degree distribution, you will have to export the normalised degree vector into another application (for example `MATLAB`, `Excel`, `R`, `SPSS`, or another package) and run a histogram function over the data. In fact, porting to `R` and `SPSS` is integrated into `Pajek` for this purpose: *Tools* → *R* → *Send to R* → *Current Vector.*

(2) Clustering Coefficient: Now what we want to do is calculate the clustering coefficient amongst the 1-neighbourhood of each node. Each node's 1-neighbourhood is all the other nodes that it shares an edge with. For the example graph in Figure 10.4, the 1-neighbourhood of

node a is $\{b, d, f\}$. For k the total degree of the node (in this example, a has $k = 3$ neighbours), then there are at most $k(k-1)$ arcs in a directed network or $k(k-1)/2$ edges in an undirected network. The clustering coefficient is simply the number of arcs or edges between the neighbours, divided by the maximum possible for the type of network, $k(k-1)$ or $k(k-1)/2$. In the example network, there is one arc between the three neighbours (the arc from b to f, the other edges are all back to node a which isn't in its neighbourhood). So the clustering coefficient of node a is $1/(3 \times 2) = 1/6$. Note that `Pajek` doesn't compute clustering coefficients for networks with parallel arcs, these can be removed by either editing the file and selecting *Net* → *Transform* → *Remove* → *Multiple Lines* → *Single Line.* To compute the clustering coefficients in Pajek use *Net* → *Vector* → *Clustering Coefficients* → *CC1.* The *CC2* option is for computing the clustering coefficient of the 2-neighbourhood (all of a node's neighbours, and all of their neighbours excluding the node you are computing the clustering coefficient for).

(3) Shortest Paths: Now what we want to do is to find the shortest path between nodes. There are two options for this: *Net* → *Paths between 2 vertices* → *One shortest* and *Net* → *Paths between 2 vertices* → *All shortest.* The difference is that the first shows only one path if multiple exist, the latter shows both. Both options will prompt for the starting node and end node to find shortest path(s) for. Both options also ask two questions. The first is 'Forget values on lines?' If yes is selected, then any edge weights will be ignored, and a shortest path is one with the fewest edges. If no is selected, then edge weights are used, and a shortest path is one with minimum sum of edge weights. The second question asked is 'Identify vertices in source networks'. If yes is chosen, then a partition is created, partitioning the edges of the original network into belonging to the shortest path(s), indicated by a '1', or not, indicated by a '0'. The resulting network has all the nodes that are in shortest paths, with the edges that lie on shortest path(s) in the original network being the only edges in the new network.

(4) Betweeness Centrality measures the shortest paths going through each vertex.[36,37] If there are multiple (equal) shortest paths between two nodes, then instead of adding to the count by one, one adds the fraction of the shortest paths going through the vertex. The overall count can then be normalised by $(n-1)(n-2)$, the total number of paths, as is done in `Pajek`. In `Pajek` one can use *Net* → *Vector* → *Centrality* → *Betweeness* to calculate the betweenness. `Pajek` does not calculate the

newer, but similar, edge betweenness measure.[38]

(5) The diameter of a network is the longest shortest path between two vertices. So one considers every pair of vertices, finds the shortest path between each of the pairs, and then the diameter is the longest of these shortest paths. It can be computed in `Pajek` through *Net* → *Paths between 2 vertices* → *Diameter*

(6) Hierarchical Decomposition: *Net* → *Hierarchical Decomposition* → *Clustering* → *Run.* One runs a hierarchical decomposition when you wish `Pajek` to automatically group vertices based solely on the pattern of connections amongst them without reference to their type. In this way, hierarchical decomposition can be a revealing method to uncover similarities in vertex context that may otherwise not be prominent to the observer. Note that this analysis produces a 2D specific hierarchical network output which is different to `Pajek`'s normal 'Draw' output. The output is sent directly to an external, encapsulated Postscript (EPS) file that can be converted to a PDF file or other image formats using a graphics tool like `Photoshop`, `GIMP`, or `Ghostscript`.

There are many measures that you can use—the above are just a starter to get you going. Exploration in `Pajek` is rewarding!software tools!for network analysis—.

References

1. J. H. Miller and S. E. Page, *Complex Adaptive Systems: An Introduction to Computational Models of Social Life (Princeton Studies in Complexity)*. (Princeton University Press, March 2007). ISBN 0691127026.
2. M. Prokopenko, F. Boschetti, and A. J. Ryan, An information-theoretic primer on complexity, self-organisation and emergence, *Complexity.* (2008).
3. A. J. Ryan, Emergence is coupled to scope, not level, *Complexity.* **13**(2), 67–77, (2007). doi: http://dx.doi.org/10.1002/cplx.20203. URL `http://dx.doi.org/10.1002/cplx.20203`.
4. R. Abbott, Emergence explained: Abstractions: Getting epiphenomena to do real work, *Complexity.* **12**(1), 13–26, (2006). doi: http://dx.doi.org/10.1002/cplx.20146. URL `http://dx.doi.org/10.1002/cplx.20146`.
5. L. Tesfatsion, Agent-based computational economics: modeling economies as complex adaptive systems, *Information Sciences.* **149**, 263–269, (2003).
6. J. Epstein and R. Axtell, *Growing Artificial Societies. Social Science from the Bottom Up.* (MIT Press, 1996).
7. V. Grimm, E. Revilla, U. Berger, F. Jeltsch, W. M. Mooij, S. F. Railsback, H. H. Thulke, J. Weiner, T. Wiegand, and D. L. Deangelis, Pattern-oriented

modeling of agent-based complex systems: lessons from ecology, *Science.* **310**, 987–991, (2005).

8. S. L. Spencer, R. A. Gerety, K. J. Pienta, and S. Forrest, Modeling somatic evolution in tumorigenesis., *PLoS Computational Biology.* **2**(8) (August, 2006). ISSN 1553-7358. doi: 10.1371/journal.pcbi.0020108. URL `http://dx.doi.org/10.1371/journal.pcbi.0020108`.
9. E. R. Smith and F. R. Conrey, Agent-based modeling: A new approach for theory building in social psychology, *Pers Soc Psychol Rev.* **11**(1), 87–104 (February, 2007). doi: 10.1177/1088868306294789. URL `http://dx.doi.org/10.1177/1088868306294789`.
10. E. D. Beinhocker, *Origin of Wealth: Evolution, Complexity, and the Radical Remaking of Economics.* (Harvard Business School Press, September 2007). ISBN 1422121038.
11. J. M. Epstein, *Generative Social Science: Studies in Agent-Based Computational Modeling.* Princeton Studies in Complexity, (Princeton University Press, January 2007). ISBN 0691125473.
12. K. R. Popper, *The Logic of Scientific Discovery.* (Routledge, March 1972), 2nd edition. ISBN 0415278449.
13. K. R. Popper, *Conjectures and Refutations; The Growth of Scientific Knowledge.* (Routledge, August 2002). ISBN 0415285941.
14. M. Blaug, *The Methodology of Economics: Or, How Economists Explain.* Cambridge Surveys of Economic Literature, (Cambridge University Press, July 1992). ISBN 0521436788.
15. D. M. Hausman, Ed., *Testability and approximation*, In ed. D. M. Hausman, *The Philosophy of Economics: An Anthology*, chapter 8, pp. 179–181. Cambridge University Press, 3rd edition, (2008).
16. M. J. Berryman. Review of software platforms for agent based models. Technical Report DSTO-GD-0532, Defence Science and Technology Organisation, Edinburgh, Australia (April, 2008).
17. U. Wilensky. Netlogo programming guide (November, 2008). URL `http://ccl.northwestern.edu/netlogo/docs/programming.html`.
18. U. Wilensky. Netlogo interface guide (November, 2008). URL `http://ccl.northwestern.edu/netlogo/docs/interface.html`.
19. U. Wilensky. Netlogo dictionary (November, 2008). URL `http://ccl.northwestern.edu/netlogo/docs/primindex.html`.
20. C. R. Shalizi. John holland, emergence, (1998). URL `http://www.cscs.umich.edu/~crshalizi/reviews/holland-on-emergence/`.
21. K. M. Carley. Simulating society: The tension between transparency and veridicality. In *Proceedings of Agent 2002 Conference on Social Agents: Ecology, Exchange and Evolution* (October, 2002).
22. S. Odubbemi and T. Jacobson, Eds., *Governance Reform Under Real World Conditions: Citizens, Stakeholders, and Voice.* (World Bank, June 2008).
23. A. M. Ramanath and N. Gilbert, The design of participatory agent-based social simulations, *Journal of Artificial Societies and Social Simulation.* **7**(4) (October, 2004).
24. U. Wilensky. Netlogo hubnetguide (November, 2008). URL `http://ccl.`

`northwestern.edu/netlogo/docs/hubnet.html`.
25. U. Wilensky. Netlogo hubnet authoring guide (November, 2008). URL `http://ccl.northwestern.edu/netlogo/docs/hubnet.html`.
26. M. J. Berryman. *A Complex Systems Approach to Important Biological Problems*. PhD thesis, The University of Adelaide (May, 2007).
27. U. Wilensky. Netlogo-mathematica link (November, 2008). URL `http://ccl.northwestern.edu/netlogo/docs/mathematica.html`.
28. V. Batagelj and A. Mrvar. *Pajek: Program For Analysis and Visualization of Large Networks—Reference Manual*. University of Ljubljana, Slovenia, 1.24 edition (December, 2008).
29. V. Chandru and J. Hooker, *Optimization Methods for Logical Inference*. (Wiley Interscience, 1999).
30. J. L. Gross and J. Yellen, Eds., *Handbook of Graph Theory (Discrete Mathematics and Its Applications)*. (CRC, December 2003), 1 edition. ISBN 1584880902.
31. A.-L. Barabási and R. Albert, Emergence of scaling in random networks, *Science*. **286**(5439), 509–512 (October, 1999).
32. A.-L. Barabási, *Linked: How Everything Is Connected to Everything Else and What It Means for Business, Science, and Everyday Life*. (Plume Books, April 2003). ISBN 0452284392.
33. D. J. Watts, *Six Degrees: The Science of a Connected Age*. (W. W. Norton & Company, February 2004). ISBN 0393325423.
34. E. Ravasz and A.-L. Barabási, Hierarchical organization in complex networks, *Phys. Rev. E*. **67**(2), 026112 (Feb, 2003). doi: 10.1103/PhysRevE.67.026112.
35. D. J. Watts and S. H. Strogatz, Collective dynamics of 'small-world' networks., *Nature*. **393**(6684), 440–442 (June, 1998). ISSN 0028-0836.
36. L. C. Freeman, A set of measures of centrality based on betweenness, *Sociometry*. **40**, 35–41, (1977).
37. L. C. Freeman, Centrality in social networks: Conceptual clarification, *Social Networks*. **1**, 215–239, (1978–1979).
38. M. Girvan and M. E. Newman, Community structure in social and biological networks., *Proc Natl Acad Sci U S A*. **99**(12), 7821–7826 (June, 2002). ISSN 0027-8424.

Author Index

Subject Index